Lecture Notes in Computer Science   12148

More information about this series at http://www.springer.com/series/7411

Tibor Lukić · Reneta P. Barneva ·
Valentin E. Brimkov · Lidija Čomić ·
Nataša Sladoje (Eds.)

Combinatorial Image Analysis

20th International Workshop, IWCIA 2020
Novi Sad, Serbia, July 16–18, 2020
Proceedings

Springer

Editors
Tibor Lukić
Faculty of Technical Sciences
University of Novi Sad
Novi Sad, Serbia

Reneta P. Barneva
School of Business
SUNY Fredonia
Fredonia, NY, USA

Valentin E. Brimkov
SUNY Buffalo State
Buffalo, NY, USA

Institute of Mathematics and Informatics
Bulgarian Academy of Sciences
Sofia, Bulgaria

Lidija Čomić
Faculty of Technical Sciences
University of Novi Sad
Novi Sad, Serbia

Nataša Sladoje
Department of Information Technology
Uppsala University
Uppsala, Sweden

ISSN 0302-9743   ISSN 1611-3349 (electronic)
Lecture Notes in Computer Science
ISBN 978-3-030-51001-5   ISBN 978-3-030-51002-2 (eBook)
https://doi.org/10.1007/978-3-030-51002-2

LNCS Sublibrary: SL5 – Computer Communication Networks and Telecommunications

This Springer imprint is published by the registered company Springer Nature Switzerland AG
The registered company address is: Gewerbestrasse 11, 6330 Cham, Switzerland

Preface

This volume contains the Proceedings of the 20th International Workshop on Combinatorial Image Analysis (IWCIA 2020) organized in Novi Sad, Serbia, July 16–18, 2020.

Image analysis provides theoretical foundations and methods for solving real-life problems arising in various areas of human practice, such as medicine, robotics, defense, and security. Since typically the input data to be processed are discrete, the "combinatorial" approach to image analysis is a natural one and therefore its applicability is expanding. Combinatorial image analysis often provides advantages in terms of efficiency and accuracy over the more traditional approaches based on continuous models that require numerical computation.

The IWCIA workshop series provides a forum for researchers throughout the world to present cutting-edge results in combinatorial image analysis, to discuss recent advances and new challenges in this research area, and to promote interaction with researchers from other countries. IWCIA had successful prior meetings in Paris (France) 1991, Ube (Japan) 1992, Washington DC (USA) 1994, Lyon (France) 1995, Hiroshima (Japan) 1997, Madras (India) 1999, Caen (France) 2000, Philadelphia, PA (USA) 2001, Palermo (Italy) 2003, Auckland (New Zealand) 2004, Berlin (Germany) 2006, Buffalo, NY (USA) 2008, Playa del Carmen (Mexico) 2009, Madrid (Spain) 2011, Austin, TX (USA) 2012, Brno (Czech Republic) 2014, Kolkata (India) 2015, Plovdiv (Bulgaria) 2017, and Porto (Portugal) 2018. The workshop in Novi Sad retained and enriched the international spirit of these workshops. The IWCIA 2020 Program Committee members were renowned experts coming from 22 different countries from Asia, Europe, North and South America, and the authors come from 11 different countries.

Each submitted paper was sent to at least three reviewers for a double-blind review. EasyChair provided a convenient platform for smoothly carrying out the rigorous review process. The most important selection criterion for acceptance or rejection of a paper was the overall score received. Other criteria included: relevance to the workshop topics, correctness, originality, mathematical depth, clarity, and presentation quality. We believe that as a result, only papers of high quality were accepted for publication in this volume.

An excellent keynote talk was given by Prof. Gyula O. H. Katona from the Hungarian Academy of Sciences who spoke about the classification of random pictures in cryptology.

The contributed papers included in this volume are grouped into two sections. The first one consists of 12 papers devoted to theoretical foundations of combinatorial image analysis, including digital geometry and topology, array grammars, picture languages, digital tomography, and other technical tools for image analysis. The second part consists of 8 papers presenting application-driven research on topics such as image repairing, annotation of images, image reconstruction, forgery detection, and dealing with noise in images. We believe that many of these papers would be of interest to a broader audience, including researchers in scientific areas such as computer vision, shape modeling, pattern analysis and recognition, and computer graphics.

We would like to take the opportunity to honor the memory of the former IWCIA conference chairs, Program Committee members, and area leaders Rani Siromoney and Reinhard Klette who recently passed away.

Prof. Rani Siromoncy was born in India in 1929. She was on the faculty of the Department of Mathematics, Madras Christian College, Chennai, India, for 37 years and later a Professor Emeritus. She was an accomplished researcher and a devoted teacher, published widely in journals of repute. Prof. Siromoney was recognized worldwide for her research in a branch of theoretical computer science, known as *Formal Languages and Automata Theory* and was one of the leading authorities in this area. She mentored numerous researchers who are now among the most prominent scientists in the field. In 1999, she successfully organized the 6th edition of IWCIA in Chennai, India. Professor Siromoney passed away on September 28, 2019.

Prof. Dr. Reinhard Klette, a fellow of the Royal Society of New Zealand, was born in Germany in 1950. He served as the director of the Centre for Robotics & Vision at Auckland University of Technology, New Zealand. He was renowned for his work in the area of computer vision. His numerous publications include the books "Digital Geometry" (co-authored by the late Azriel Rosenfeld), "Computer Vision for Driver Assistance," "Concise Computer Vision," "Computer Vision – Three-Dimensional Data from Images," and many others. Numerous graduate students developed successfully as scientists under his supervision. He was the founding editor in chief of the *Journal of Control Engineering and Technology* (JCET) and an associate editor of *IEEE Transactions on Pattern Analysis and Machine Intelligence* (PAMI). In 2004, Prof. Klette organized the 10th edition of IWCIA in Auckland, New Zealand. He passed away on April 3, 2020.

Many individuals and organizations contributed to the success of IWCIA 2020. The Editors are indebted to IWCIA's Steering Committee for endorsing the candidacy of Novi Sad for the 20th edition of the workshop. We wish to thank everybody who submitted their work to IWCIA 2020. We are grateful to all participants and especially to the contributors of this volume. Our most sincere thanks go to the IWCIA 2020 Program Committee whose cooperation in carrying out high-quality reviews was essential in establishing a strong scientific program. We express our sincere gratitude to the keynote speaker, Gyula O. H. Katona for accepting our invitation and overall contribution to the workshop program.

The success of the workshop would not be possible without the hard work of the Local Organizing Committee. We are grateful to the host organization, the Faculty of Technical Sciences, University of Novi Sad, for their support. Finally, we wish to thank Springer Nature Computer Science Editorial, and especially Alfred Hofmann and Anna Kramer, for their efficient and kind cooperation in the timely production of this book.

May 2020

Tibor Lukić
Reneta P. Barneva
Valentin E. Brimkov
Lidija Čomić
Nataša Sladoje

Organization

The 20th International Workshop on Combinatorial Image Analysis (IWCIA 2020) was set to be held in Novi Sad, Serbia, July 16–18, 2020.

General Chair

Tibor Lukić   University of Novi Sad, Serbia

Chairs

Reneta P. Barneva   SUNY Fredonia, USA
Valentin E. Brimkov   SUNY Buffalo State, USA
Lidija Čomić   University of Novi Sad, Serbia
Nataša Sladoje   Uppsala University, Sweden

Steering Committee

Bhargab B. Bhattacharya   Indian Statistical Institute Kolkata, India
Valentin E. Brimkov   SUNY Buffalo State, USA
Gabor Herman   CUNY Graduate Center, USA
Renato M. Natal Jorge   University of Porto, Portugal
João Manuel R. S. Tavares   University of Porto, Portugal
Josef Slapal   Technical University of Brno, Czech Republic

Invited Speaker

Gyula O. H. Katona   Hungarian Academy of Sciences, Hungary

Program Committee

Eric Andres   Université de Poitiers, France
Buda Bajić   University of Novi Sad, Serbia
Péter Balaźs   University of Szeged, Hungary
Jacky Baltes   University of Manitoba, Canada
George Bebis   University of Nevada at Reno, USA
Bhargab B. Bhattacharya   Indian Statistical Institute Kharagpur, India
Partha Bhowmick   IIT Kharagpur, India
Arindam Biswas   IIEST Shibpur, India
Peter Brass   The City College of New York, USA
Boris Brimkov   Slippery Rock University, USA
Srecko Brlek   Université du Quebec à Montreal, Canada

Alfred M. Bruckstein	Technion - Israel Institute of Technology, Israel
Jean-Marc Chassery	Université de Grenoble, France
Li Chen	University of the District of Columbia, USA
Rocio Gonzalez-Diaz	University of Seville, Spain
Mousumi Dutt	St. Thomas College of Engineering and Technology, India
Fabien Feschet	Université d'Auvergne, France
Leila De Floriani	University of Genova, Italy, and University of Maryland, USA
Chiou-Shann Fuh	National Taiwan University, Taiwan
Edwin Hancock	University of York, UK
Fay Huang	National Ilan University, Taiwan
Krassimira Ivanova	Bulgarian Academy of Sciences, Bulgaria
Atsushi Imiya	IMIT, Chiba University, Japan
Maria-Jose Jimenez	University of Seville, Spain
Kamen Kanev	Shizuoka University, Japan
Tat Yung Kong	CUNY Queens College, USA
Kostadin Koroutchev	Universidad Autónoma de Madrid, Spain
Elka Korutcheva	Universidad Nacional de Educación a Distancia, Spain
Walter G. Kropatsch	Vienna University of Technology, Austria
Longin Jan Latecki	Temple University, USA
Jerome Liang	SUNY Stony Brook, USA
Joakim Lindblad	Uppsala University, Sweden
Benedek Nagy	Eastern Mediterranean University, Cyprus
Gregory M. Nielson	Arizona State University, USA
János Pach	EPFL Lausanne, Switzerland, and Renyi Institute Budapest, Hungary
Kálmán Palágyi	University of Szeged, Hungary
Petra Perner	Institute of Computer Vision and Applied Computer Sciences, Germany
Nicolai Petkov	University of Groningen, The Netherlands
Hemerson Pistori	Dom Bosco Catholic University, Brazil
Ioannis Pitas	University of Thessaloniki, Greece
Konrad Polthier	Freie Universitaet Berlin, Germany
Hong Qin	SUNY Stony Brook, USA
Paolo Remagnino	Kingston University, UK
Ralf Reulke	Humboldt University, Germany
Bodo Rosenhahn	MPI Informatik, Germany
Arun Ross	Michigan State University, USA
Nikolay Sirakov	Texas A&M University, USA
Rani Siromoney	Madras Christian College, India
Wladyslaw Skarbek	Warsaw University of Technology, Poland
Ali Shokoufandeh	Drexel University, USA
K. G. Subramanian	Madras Christian College, India
João Manuel R. S. Tavares	University of Porto, Portugal
D. G. Thomas	Madras Christian College, India

László Varga	University of Szeged, Hungary
Peter Veelaert	Ghent University, Belgium
Petra Wiederhold	CINVESTAV-IPN, Mexico
Jinhui Xu	SUNY University at Buffalo, USA

Local Organizing Committee

Buda Bajić	University of Novi Sad, Serbia
Andrija Blesić	University of Novi Sad, Serbia
Maria Kiš	University of Novi Sad, Serbia
Marina Marčeta	University of Novi Sad, Serbia

Additional Reviewers

Piyush Kanti Bhunre
Ranita Biswas
Ulderico Fugacci
Chunwei Ma
Minghua Wang

Contents

Methods and Applications

Theoretical Foundations

Euler Well-Composedness

Nicolas Boutry[1] , Rocio Gonzalez-Diaz[2] , Maria-Jose Jimenez[2(✉)] ,
and Eduardo Paluzo-Hildago[2]

[1] EPITA Research and Development Laboratory (LRDE),
Le Kremlin-Bicêtre, France
`nicolas.boutry@lrde.epita.fr`
[2] Department of Applied Math (I), Universidad de Sevilla, Sevilla, Spain
`{rogodi,majiro,epaluzo}@us.es`

Abstract. In this paper, we define a new flavour of well-composedness, called Euler well-composedness, in the general setting of regular cell complexes: A regular cell complex is Euler well-composed if the Euler characteristic of the link of each boundary vertex is 1. A cell decomposition of a picture I is a pair of regular cell complexes $(K(I), K(\bar{I}))$ such that $K(I)$ (resp. $K(\bar{I})$) is a topological and geometrical model representing I (resp. its complementary, $\bar{I}$). Then, a cell decomposition of a picture I is self-dual Euler well-composed if both $K(I)$ and $K(\bar{I})$ are Euler well-composed. We prove in this paper that, first, self-dual Euler well-composedness is equivalent to digital well-composedness in dimension 2 and 3, and second, in dimension 4, self-dual Euler well-composedness implies digital well-composedness, though the converse is not true.

Keywords: Digital topology · Discrete geometry ·
Well-composedness · Cubical complexes · Cell complexes · Manifolds ·
Euler characteristic

1 Introduction

The concept of well-composedness of a picture was first introduced in [13] for 2D pictures and extended later to 3D in [14]: a well-composed picture satisfies that the continuous analog of the given picture has a boundary surface that is a manifold. The concept is described in terms of forbidden subsets for which the picture is not well-composed. In [8], the author defines a gap in a binary object in a digital space of arbitrary dimension, an analogous concept to that of forbidden subset of Latecki et al. and similar to the notion of tunnel that had been defined in [1] for digital hyperplanes. In [3], the concept of critical configurations (i.e., forbidden subsets) was extended to nD.

3D well-composed images may have some computational advantages regarding the application of several algorithms in computer vision, computer graphics and image processing. But in general, images are not a priori well-composed. There are several "repairing" methods for turning them into well-composed

© Springer Nature Switzerland AG 2020
T. Lukić et al. (Eds.): IWCIA 2020, LNCS 12148, pp. 3–19, 2020.
https://doi.org/10.1007/978-3-030-51002-2_1

images (see, for example, [12, 15, 18, 19]). Besides, in [17], the authors extended the notion of "digital well-composedness" to nD sets.

Equivalences between different flavours of well-composedness have been studied in [4], namely: continuous well-composedness (CWCness), digital well-composedness (DWCness), well-composedness in the Alexandrov sense (AWCness), well-composedness based on the equivalence of connectivities (EWCness), and well-composedness on arbitrary grids (AGWCness). More specifically, as stated in [4], it is well-known that, in 2D, AWCness, CWCness, DWCness and EWCness are equivalent, so a 2D picture is well-composed if and only if it is XWC (X = A, C, D, E). In 3D, only AWCness, CWCness, DWCness are equivalent. Note that no link between AGWCness and the other flavours of well-composedness were known in nD, and for $n \geq 4$ the equivalences between the different flavours of well-composedness (AWCness, CWCness, DWCness and EWCness) have not been proved yet (except that AWCness implies DWCness, see [7] and that DWCness implies EWCness, see [3]).

Recently, in [6], a counterexample has been given to prove that DWCness does not imply CWCness, what is an important result since it breaks with the idea that all the flavours of well-composedness are equivalent.

In the papers [5, 9, 10], the authors developed an nD topological method for repairing digital pictures (in the cubical grid) with "pinches", turning them into weakly well-composed complexes. More specifically, such a method constructs a "simplicial decomposition" $\left(P_S(I), P_S(\bar{I})\right)$ of a given n-dimensional (nD) picture I (initially represented by a cubical complex $Q(I)$) such that: (1) $P_S(I)$ is homotopy equivalent to $Q(I)$ and $P_S(\bar{I})$ is homotopy equivalent to $Q(\bar{I})$ being $\bar{I}$ the nD picture that is the "complementary" of I; (2) $\left(P_S(I), P_S(\bar{I})\right)$ is self-dual weakly well-composed, that is, for each vertex v on the boundary of $P_S(I)$, the set of n-simplices of $P_S(I)$ incident to v are "face-connected" (defined later), as well as those of $P_S(\bar{I})$ incident to v. As we will see later, in the setting of cubical complexes canonically associated to nD pictures, self-dual weak well-composedness is equivalent to digital well-composedness.

In fact, our ultimate goal is to prove that this method provides continuously well-composed complexes, that is, the boundary of their underlying polyhedron is an $(n-1)$D topological manifold. Since this goal is not reachable yet according to us, we propose an "intermediary" flavour of well-composedness, called Euler well-composedness, that is stronger than weak well-composedness but weaker than continuous well-composedness. The aim of the present paper is then to prove that Euler well-composedness implies weak well-composedness but that the converse is not true. The plan is the following: Sect. 2 and 3 recall the background relative to self-dual weak and digital well-composedness respectively. Section 4 introduces the definition of Euler well-composedness based on what we call "χ-critical vertices" and shows that: Euler well-composedness is equivalent to self-dual weak and digital well-composedness on 2D and 3D cubical grids; and Euler well-composedness is stronger than weak and digital well-composedness on 4D cubical grids. Section 5 concludes the paper.

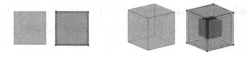

Fig. 1. Left: a 2-dimensional cube and its faces. Right: a 3-dimensional cube and its faces.

2   Background on Regular Cell Complexes

Roughly speaking, a *regular cell complex* K is a collection of cells (where k-cells are homeomorphic to k-dimensional balls) glued together by their boundaries (faces), in such a way that a non-empty intersection of any two cells of K is a cell in K. Regular cell complexes have particularly nice properties, for example, their homology is effectively computable (see [16, p. 243]). When the k-cells in K are k-dimensional cubes, we refer to K as a *cubical complex* (see Fig. 1). When they are k-dimensional simplices (points, edges, triangles, tetrahedra, etc.), we refer to K as a *simplicial complex*.

Let K be a regular cell complex. A k-cell μ is a *proper face of* an ℓ-cell $\sigma \in K$ if μ is a face of σ and $k < \ell$. A cell of K which is not a proper face of any other cell of K is said to be a *maximal* cell of K.

Let k, k' be integers such that $k < k'$. Then, the set $\{k, k+1, \ldots, k'-1, k'\}$ will be denoted by $[\![k, k']\!]$.

Definition 1 (face-connectedness). *Let μ be a cell of a regular cell complex K. Let $\mathcal{A}_K^{(\ell)}(\mu)$ be a set of ℓ-cells of K sharing μ as a face. Let σ and σ' be two ℓ-cells of $\mathcal{A}_K^{(\ell)}(\mu)$. We say that σ and σ' are face-connected in $\mathcal{A}_K^{(\ell)}(\mu)$ if there exists a path $\pi(\sigma, \sigma') = (\sigma_1 = \sigma, \sigma_2 \ldots, \sigma_{m-1}, \sigma_m = \sigma')$ of ℓ-cells of $\mathcal{A}_K^{(\ell)}(\mu)$ such that for any $i \in [\![1, m-1]\!]$, σ_i and σ_{i+1} share exactly one $(\ell-1)$-cell.*

We say that a set $\mathcal{A}_K^{(\ell)}(\mu)$ is face-connected if any two ℓ-cells σ and σ' in $\mathcal{A}_K^{(\ell)}(\mu)$ are face-connected in $\mathcal{A}_K^{(\ell)}(\mu)$.

An *external* cell of K is a proper face of exactly one maximal cell in K. A regular cell complex is *pure* if all its maximal cells have the same dimension. The *rank* of a cell complex K is the maximal dimension of its cells. The *boundary surface* of a pure regular cell complex K, denoted by ∂K, is the regular cell complex composed by the external cells of K together with all their faces. Observe that ∂K is also pure.

Definition 2 (nD cell complex). *An nD cell complex K is a pure regular cell complex of rank n embedded in $\mathbb{R}^n$. The underlying space (i.e., the union of the cells as subspaces of $\mathbb{R}^n$) will be denoted by $|K|$.*

An nD cell complex K is said to be *(continuously) well-composed* if $|\partial K|$ is an $(n-1)$-manifold, that is, each point of $|\partial K|$ has a neighborhood in $|\partial K|$ homeomorphic to $\mathbb{R}^{n-1}$.

Definition 3 (Weak well-composedness). *An nD cell complex K is weakly well-composed (wWC) if for any 0-cell (also called vertex) v in K, $\mathcal{A}_K^{(n)}(v)$ is face-connected.*

Definition 4 (Euler characteristic). *Let K be a finite regular cell complex. Let a_k denote the number of k-cells of K. The Euler characteristic of K is defined as*

$$\chi(K) = \sum_{k=0}^{\infty}(-1)^k a_k.$$

Recall that the Euler characteristic of a finite regular cell complex depends only on its homotopy type [11, p. 146].

Definition 5 (star, closed star and link). *Let v be a vertex of a given regular cell complex K.*

- *The star of v (denoted $\mathrm{St}_K(v)$) is the set of cells having v as a face. Note that the star of v is generally not a cell complex itself.*
- *The closed star of v (denoted $\mathrm{ClSt}_K(v)$) is the cell complex obtained by adding to $\mathrm{St}_K(v)$ all the faces of the cells in $\mathrm{St}_K(v)$.*
- *The link of v (denoted $\mathrm{Lk}_K(v)$) is the closed star of v minus the star of v, that is, $\mathrm{ClSt}_K(v) \setminus \mathrm{St}_K(v)$.*

3   Background on nD Pictures

Now, let us formally introduce some concepts related to digital well-composedness of nD pictures.

Definition 6 (nD picture). *Let $n \geq 2$ be an integer and $\mathbb{Z}^n$ the set of points with integer coordinates in $\mathbb{R}^n$. An nD picture is a pair $I = (\mathbb{Z}^n, F_I)$, where F_I is a subset of $\mathbb{Z}^n$. The set F_I is called the foreground of I and the set $\mathbb{Z}^n \setminus F_I$ the background of I. The picture "complement" of I is defined as $\bar{I} = (\mathbb{Z}^n, \mathbb{Z}^n \setminus F_I)$.*

Definition 7 (cubical complex $Q(I)$). *The nD cubical complex $Q(I)$ canonically associated to an nD picture $I = (\mathbb{Z}^n, F_I)$ is composed by those n-dimensional unit cubes centered at each point in F_I, whose $(n-1)$-faces are parallel to the coordinate hyperplanes, together with all their faces.*

Figure 2 shows the geometric realization of cubical complexes representing a 2D binary picture of two pixels (left) and two 3D pictures of 2 voxels each.

Roughly speaking, two topological spaces are *homotopy equivalent* if one can be continuously deformed into the other. A specific example of homotopy equivalence is a *deformation retraction* of a space X onto a subspace A which is a family of maps $f_t : X \to X$, $t \in [0, 1]$, such that: $f_0(x) = x$, $\forall x \in X$; $f_1(X) = A$; $f_t(a) = a$, $\forall a \in A$ and $t \in [0, 1]$. The family $\{f_t : X \to X\}_{t \in [0,1]}$ should be continuous in the sense that the associated map $F : X \times I \to X$, where $F(x, t) = f_t(x)$, is continuous. See [11, p. 2].

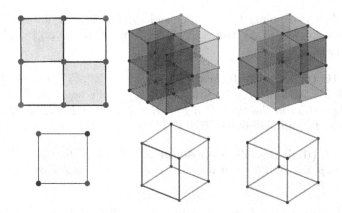

Fig. 2. Top figures: cubical complexes (in brown) of dimension 2 (left) and 3 (middle and right). Bottom figures: point representation of the cubical complexes on the top to help the intuition of Table 6 and 7. Points in red correspond to the maximal cubes, that are joined by a red edge if the corresponding cubes are face–connected. (Color figure online)

Definition 8 (cell complex over an nD picture). *A cell complex over an nD picture I is an nD cell complex, denoted by $K(I)$, such that there exists a deformation retraction from $K(I)$ onto $Q(I)$.*

In [2], the concept of blocks was introduced. For two integers $k \leq k'$, let $\mathcal{E} = \{e^1, \ldots, e^n\}$ be the canonical basis of $\mathbb{Z}^n$. Given a point $z \in \mathbb{Z}^n$ and a family of vectors $\mathcal{F} = \{f^1, \ldots, f^k\} \subseteq \mathcal{E}$, the *block of dimension k* associated to the couple $(z, \mathcal{F})$ is the set defined as:

$$B(z, \mathcal{F}) = \left\{ z + \sum_{i \in [\![1,k]\!]} \lambda_i \, f^i \ : \ \lambda_i \in \{0,1\}, \forall i \in [\![1,k]\!] \right\}.$$

This way, a 0-block is a point, a 1-block is a set of two points in $\mathbb{Z}^n$ on an unit edge, a 2-block is a set of four points on a unit square, and so on. A subset $B \subset \mathbb{Z}^n$ is called a *block* if there exists a couple $(z, \mathcal{F}) \in \mathbb{Z}^n \times \mathcal{P}(\mathcal{E})$ (where $\mathcal{P}(\mathcal{E})$ represents the set of all the subsets of $\mathcal{E}$), such that $B = B(z, \mathcal{F})$. We will denote the set of blocks of $\mathbb{Z}^n$ by $\mathcal{B}(\mathbb{Z}^n)$.

Definition 9 (antagonists). *Two points p, q belonging to a block $B \in \mathcal{B}(\mathbb{Z}^n)$ are said to be antagonists in B if their distance equals the maximum distance using the L^1-norm[1] between two points in B, that is, $\|p - q\|_1 = \max \{ \|r - s\|_1 : r, s \in B \}$.*

Remark 1. The antagonist of a point p in a block $B \in \mathcal{B}(\mathbb{Z}^n)$ containing p exists and is unique. It is denoted by $\operatorname{antag}_B(p)$.

---

[1] The L^1-norm of a vector $\alpha = (x_1, \ldots, x_n)$ is $\|\alpha\|_1 = \sum_{i \in [\![1,n]\!]} |x_i|$.

Note that when two points $(x_1, \ldots, x_n)$ and $(y_1, \ldots, y_n)$ are antagonists in a block of dimension $k \in [\![0, n]\!]$, then $|x_i - y_i| = 1$ for $i \in \{i_1, \ldots, i_k\} \subseteq [\![1, n]\!]$ and $x_i = y_i$ otherwise.

Definition 10 (critical configuration). *Let $I = (\mathbb{Z}^n, F_I)$ be an nD picture and $B \in \mathcal{B}(\mathbb{Z}^n)$ a block of dimension $k \in [\![2, n]\!]$. We say that I contains a critical configuration in the block B if $F_I \cap B = \{p, p'\}$ or $F_I \cap B = B \setminus \{p, p'\}$, with p, p' being two antagonists in B.*

Definition 11 (digital well-composedness). *An nD picture is said to be digitally well-composed (DWC) if it does not contain any critical configuration in any block $B \in \mathcal{B}(\mathbb{Z}^n)$.*

We say that a property of an nD picture I is *self-dual*, if its complement $\bar{I}$ also satisfies the property. Hence, last definition of digital well-composedness is self-dual and based on local patterns.

4   Introducing the Concept of Euler Well-Composedness

In this section, we introduce the new concept of Euler well-composedness for regular cell complexes and show that, in the cubical setting, digital well-composedness is equivalent to Euler well-composedness in 2D and 3D, but digital well-composedness is weaker than Euler well-composedness in 4D.

Definition 12 (χ-critical vertex). *Given an nD cell complex K, $n \geq 2$, a vertex $v \in K$ is χ-critical for K if:*

$$v \in \partial K \quad and \quad \chi\bigl(\mathrm{Lk}_K(v)\bigr) \neq \chi(\mathbb{B}^{n-1}) = 1,$$

where $\mathbb{B}^{n-1}$ is an $(n-1)$-dimensional ball.

In Fig. 3, different cases of χ-critical and non-χ-critical vertices are shown.

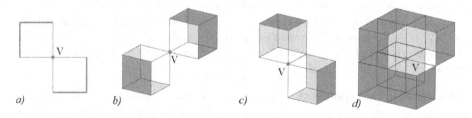

a)   *b)*   *c)*   *d)*

Fig. 3. Different cases of a vertex v on the boundary of a cubical complex $Q(I)$. $\mathrm{Lk}_Q(v)$ has been drawn in grey. a) A 2D case of a χ- critical vertex, with $\chi\bigl(\mathrm{Lk}_Q(v)\bigr) = 2$; b) a 3D case of a χ-critical vertex, with $\chi\bigl(\mathrm{Lk}_Q(v)\bigr) = 2$; c) a 3D case of a vertex on the boundary that is not a χ-critical vertex, since $\chi\bigl(\mathrm{Lk}_Q(v)\bigr) = 1$; d) complementary configuration of case (c) in which v is a χ-critical vertex, with $\chi\bigl(\mathrm{Lk}_Q(v)\bigr) = 0$.

Definition 13 (Euler well-composedness). *An nD cell complex is Euler well-composed if it has no χ-critical vertices.*

For example, in Fig. 3, only case (c) represents a cubical complex that is Euler well-composed.

Definition 14 (cell decomposition of an nD picture). *A cell decomposition of an nD picture I consists of a pair of nD cell complexes, $\bigl(K(I), K(\bar{I})\bigr)$, such that:*

- *$K(I)$ is a cell complex over I and $K(\bar{I})$ is a cell complex over $\bar{I}$.*
- *$|K(I) \cup K(\bar{I})| = \mathbb{R}^n$.*
- *$K(I) \cap K(\bar{I}) = \partial K(I) = \partial K(\bar{I})$.*

Table 1. All the possible configurations (cubical and points representations) for $U = (\mathbb{Z}^2, F_U)$ in $B(o, \mathbb{Z}^2)$ satisfying that $o \in F_U$ and $\mathrm{card}(F_U) \leq 2$, up to rotations and reflections around $v = (1/2, 1/2)$.

card(F_U) = 1,2	$Q(I)$	$Q(\bar{I})$	$Q(I)$	$Q(\bar{I})$	$Q(I)$	$Q(\bar{I})$
wWC	Yes	Yes	Yes	Yes	No	No
χWC	Yes	Yes	Yes	Yes	No	No
DWC	Yes		Yes		No	

Definition 15 (self-dual Euler well-composedness). *A cell decomposition $\bigl(K(I), K(\bar{I})\bigr)$ of an nD picture I is self-dual Euler well-composed (sχWC) if both $K(I)$ and $K(\bar{I})$ are Euler well-composed.*

The definition of self-dual weak well-composedness was introduced in [5]. We recall it here.

Definition 16 (self-dual weak well-composedness). *A cell decomposition $\bigl(K(I), K(\bar{I})\bigr)$ of an nD picture I is self-dual weakly well-composed (swWC) if both $K(I)$ and $K(\bar{I})$ are weakly well-composed.*

We recall now that self-dual weak well-composedness is equivalent to digital well-composed in the cubical setting.

Theorem 1 ([3]). *Let $I = (\mathbb{Z}^n, F_I)$ be an nD picture and $Q(I)$ the cubical complex canonically associated to I. Then, I is digitally well-composed if and only if $\bigl(Q(I), Q(\bar{I})\bigr)$ is self-dual weakly well-composed.*

We study now the possible equivalences between self-dual weak well-composedness and self-dual Euler well-composedness. We will prove that, as expected, self-dual Euler well-composedness is equivalent to digital well-composedness in 2D and 3D. Nevertheless, as we will see later, this equivalence is no longer true for 4D pictures. We will prove that self-dual Euler well-composedness implies digital well-composedness in 4D although the converse is not true. Observe that considering the definition of Euler well-composedness, we can study local patterns only. To prove such results, we should check the exhaustive lists of all the possible configurations $U = (\mathbb{Z}^n, F_U)$ in any block $B(z, \mathbb{Z}^n)$ for $z \in \mathbb{Z}^n$, for $n = 2, 3, 4$. To reduce the list and without loss of generality, we will only study configurations

Table 2. All the possible configurations for $U = (\mathbb{Z}^3, F_U)$ in $B(o, \mathbb{Z}^3)$ satisfying that $o \in F_U$ and $\mathrm{card}(F_U) \leq 3$, up to rotations and reflections around $v = (1/2, 1/2, 1/2)$.

$\mathrm{card}(F_U) = 1$						
	$Q(I)$	$Q(\bar{I})$				
wWC	Yes	Yes				
χWC	Yes	Yes				
DWC		Yes				
$\mathrm{card}(F_U) = 2$						
	$Q(I)$	$Q(\bar{I})$	$Q(I)$	$Q(\bar{I})$	$Q(I)$	$Q(\bar{I})$
wWC	No	Yes	Yes	Yes	No	Yes
χWC	Yes	No	Yes	Yes	No	No
DWC		No		Yes		No
$\mathrm{card}(F_U) = 3$						
	$Q(I)$	$Q(\bar{I})$	$Q(I)$	$Q(\bar{I})$	$Q(I)$	$Q(\bar{I})$
wWC	Yes	Yes	No	Yes	No	No
χWC	Yes	Yes	Yes	No	No	No
DWC		Yes		No		No

$U = (\mathbb{Z}^n, F_U)$ in the block $B(o, \mathbb{Z}^n)$ for o being the coordinates origin. Besides, we will also assume, again without loss of generality, that o is always in F_U. Let $\mathrm{card}(F_U)$ denote the number of points in F_U. Since $\mathrm{card}(F_{\bar{U}}) = 2^n - \mathrm{card}(F_U)$, we will only study configurations U in $B(o, \mathbb{Z}^n)$ satisfying that $\mathrm{card}(F_U) \leq 2^{n-1}$.

Table 3. All the possible configurations for $U = (\mathbb{Z}^3, F_U)$ in $B(o, \mathbb{Z}^3)$ satisfying that $o \in F_U$ and $\mathrm{card}(F_U) = 4$, up to rotations and reflections around $v = (1/2, 1/2, 1/2)$.

$\mathrm{card}(F_U) = 4$						
	$Q(I)$	$Q(\bar{I})$	$Q(I)$	$Q(\bar{I})$	$Q(I)$	$Q(\bar{I})$
wWC	Yes	Yes	Yes	Yes	No	No
χWC	Yes	Yes	Yes	Yes	No	No
DWC		Yes		Yes		No
$\mathrm{card}(F_U) = 4$						
	$Q(I)$	$Q(\bar{I})$	$Q(I)$	$Q(\bar{I})$	$Q(I)$	$Q(\bar{I})$
wWC	No	No	No	No	Yes	Yes
χWC	No	No	No	No	Yes	Yes
DWC		No		No		Yes

Fix a configuration U satisfying all the requirements listed above. Let $v \in \partial Q(U)$ be the vertex with coordinates $(1/2, \overset{n\ \mathrm{times}}{\ldots}, 1/2)$. Then, to see if such configuration is digitally well-composed, we will check if both $\mathcal{A}^{(n)}_{Q(U)}(v)$ and $\mathcal{A}^{(n)}_{Q(\bar{U})}(v)$ are face-connected or not. That is, we will check if the pair $(Q(U), Q(\bar{U}))$ is self-dual weakly well-composed. Similarly, to see if such configuration is self-dual Euler well-composed, we will check if vertex v is χ-critical in both $Q(U)$ and $Q(\bar{U})$.

Theorem 2. *Self-dual Euler well-composedness in the 2D and 3D cubical setting is equivalent to digital well-composedness.*

Table 4. Amount of configurations $U = (\mathbb{Z}^n, F_U)$ in $B(o, \mathbb{Z}^4)$ for the different cases in 4D clustered depending on the number of points in F_U and the property of being (or not) $Q(U)$ and $Q(\bar{U})$ weakly well-composed and/or Euler well-composed.

card(F_U)								$Q(U)$		$Q(\bar{U})$	
1	2	3	4	5	6	7	8	wWc	χWc	wWc	χWc
0	0	0	0	0	0	0	120	Yes	No	No	No
0	0	0	0	0	24	189	96	No	Yes	No	No
0	1	18	149	500	870	490	120	No	No	Yes	No
0	0	0	0	0	0	28	96	No	No	No	Yes
0	0	0	0	0	0	0	0	Yes	Yes	No	No
0	0	0	0	0	0	0	60	Yes	No	Yes	No
0	0	0	0	0	0	112	672	Yes	No	No	Yes
0	10	69	232	565	1074	1554	672	No	Yes	Yes	No
0	0	0	0	0	0	0	0	No	Yes	No	Yes
0	0	0	0	0	0	0	0	No	No	Yes	Yes
0	0	0	0	0	12	84	240	Yes	Yes	Yes	No
0	0	0	0	0	0	0	0	Yes	Yes	No	Yes
0	0	0	0	0	72	336	240	Yes	No	Yes	Yes
0	0	0	0	0	0	0	0	No	Yes	Yes	Yes
0	0	0	4	55	303	861	1811	No	No	No	No
1	4	18	70	245	648	1351	2308	Yes	Yes	Yes	Yes

Proof. Table 1 shows all the possible configurations for $U = (\mathbb{Z}^2, F_U)$ in $B(o, \mathbb{Z}^2)$ satisfying that $o \in F_U$ and card(F_U) ≤ 2, up to rotations and reflections around v. Looking at the table, we can check that DWCness $\Leftrightarrow$ sχWCness in 2D. Similarly, Tables 2 and 3 show that DWCness $\Leftrightarrow$ sχWCness in 3D.   □

Theorem 3. *Digital well-composedness does NOT imply self-dual Euler well-composedness in 4D.*

Proof. An exhaustive list of configurations of hypercubes in 4D incident to a vertex that are digitally well-composed but not self-dual Euler well-composed is provided in Table 6. The complete list is summed up in Table 5 and can be found in the GitHub repository: https://github.com/Cimagroup/Euler-WCness (Table 5).   □

Theorem 4. *Self-dual Euler well-composedness in the 4D cubical setting implies digital well-composedness.*

Proof. An exhaustive list of all the possible configurations $U = (\mathbb{Z}^n, F_U)$ in the block $B(o, \mathbb{Z}^4)$ satisfying that U is not digitally well-composed is given in Table 7. All such cases satisfy that they are not self-dual Euler well-composed either.   □

Table 5. Exhaustive list in 4D of configurations for U in $B(o, \mathbb{Z}^4)$ clustered in the number of points in F_U and the property of being or not DWC and/or sχWC.

card(F_U)								U	$(Q(U), Q(\bar{U}))$
1	2	3	4	5	6	7	8	DWC	sχWC
1	4	18	70	245	648	1351	2308	Yes	Yes
0	0	0	0	0	84	420	540	Yes	No
0	0	0	0	0	0	0	0	No	Yes
0	11	87	385	1120	2271	3234	3587	No	No

Table 6. Exhaustive list of configurations for U in $B(o, \mathbb{Z}^4)$ satisfying that U is DWC but $(Q(U), Q(\bar{U}))$ is not sχWC. First column indicates the number of points of F_U; second column corresponds to the amount of configurations for U in $B(o, \mathbb{Z}^4)$ with the corresponding number of points in F_U and third column is an example of such kind of configuration.

card(F_U)	Amount	Example
6	84	{(0,0,0,0),(0,0,0,1),(0,0,1,0), (0,1,0,1),(0,1,1,0),(0,1,1,1)} 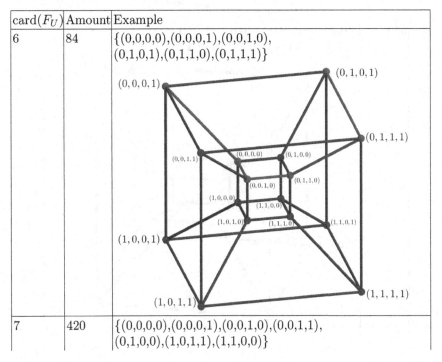
7	420	{(0,0,0,0),(0,0,0,1),(0,0,1,0),(0,0,1,1), (0,1,0,0),(1,0,1,1),(1,1,0,0)}

(continued)

Table 6. (*continued*)

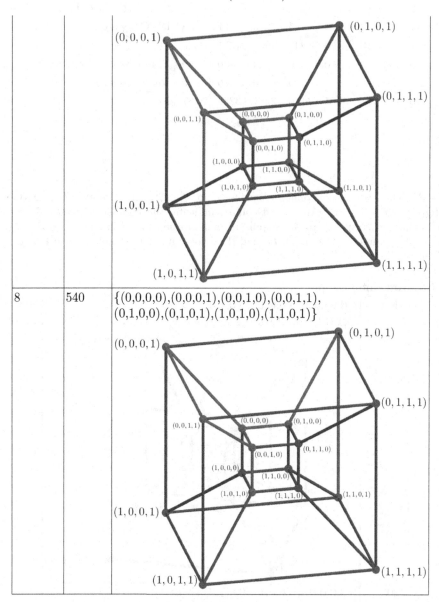

8	540	{(0,0,0,0),(0,0,0,1),(0,0,1,0),(0,0,1,1), (0,1,0,0),(0,1,0,1),(1,0,1,0),(1,1,0,1)}

Table 7. Exhaustive list of all the possible configurations $U = (\mathbb{Z}^n, F_U)$ in the block $B(o, \mathbb{Z}^4)$ satisfying that U is not digitally well-composed. All such cases satisfy that they are not self-dual Euler well-composed either. First column corresponds to the number of points in F_U, second column corresponds to the amount of configurations that there exist with such number of points in F_U. Third column shows an example of such configuration.

not DWC $\Rightarrow$ not sχWC		
card(F_U)	Amount	Example
1	0	-
2	11	$\{(0,0,0,0),(0,0,1,1)\}$

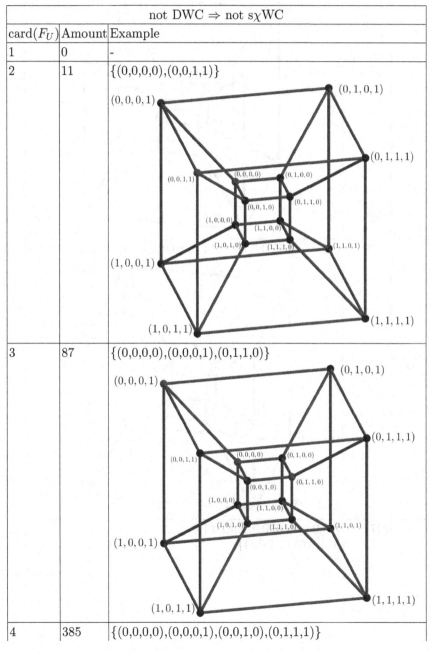

| 3 | 87 | $\{(0,0,0,0),(0,0,0,1),(0,1,1,0)\}$ |
| 4 | 385 | $\{(0,0,0,0),(0,0,0,1),(0,0,1,0),(0,1,1,1)\}$ |

(*continued*)

Table 7. (*continued*)

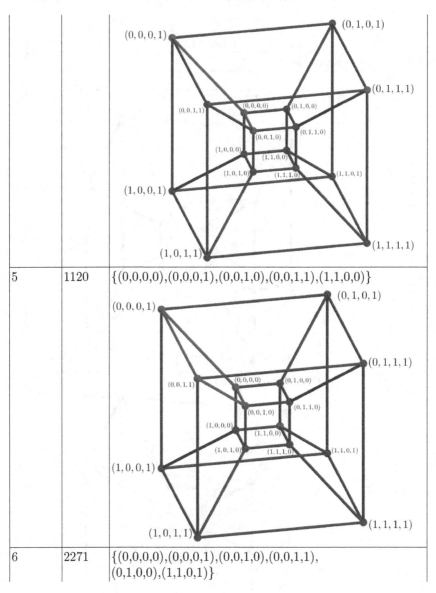

5	1120	$\{(0,0,0,0),(0,0,0,1),(0,0,1,0),(0,0,1,1),(1,1,0,0)\}$
6	2271	$\{(0,0,0,0),(0,0,0,1),(0,0,1,0),(0,0,1,1),$ $(0,1,0,0),(1,1,0,1)\}$

(*continued*)

Table 7. (*continued*)

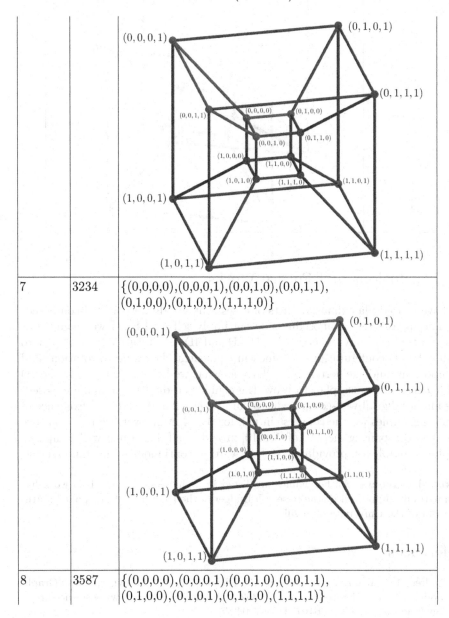

7	3234	{(0,0,0,0),(0,0,0,1),(0,0,1,0),(0,0,1,1), (0,1,0,0),(0,1,0,1),(1,1,1,0)}

8	3587	{(0,0,0,0),(0,0,0,1),(0,0,1,0),(0,0,1,1), (0,1,0,0),(0,1,0,1),(0,1,1,0),(1,1,1,1)}

(*continued*)

Table 7. (*continued*)

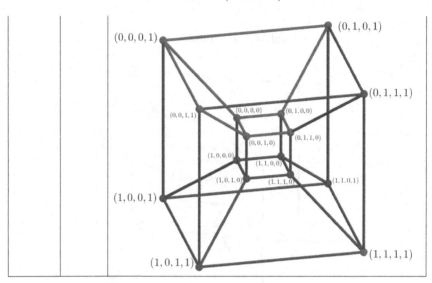

5   Conclusions and Future Works

We have proved via exhaustive lists of cases that self-dual weak well-composedness and digital well-composedness do not imply self-dual Euler well-composedness, but that the converse is true in 2D, 3D and 4D. In a future paper, we plan to cluster the 4D configurations obtained in equivalent classes up to rotations and reflections around the vertex v, similarly as what we have done to study the 2D and 3D cases. We also plan to prove the claim "self-dual Euler well-composedness implies digital well-composedness" in nD, $n \geq 2$ and study the existence of counter-examples for the converse in nD, for $n > 4$. Moreover, we plan to prove that the nD repairing method of [5,9,10] provides self-dual Euler well-composed simplicial complexes, providing a step forward to continuous well-composedness.

Acknowledgments. Author names are listed in alphabetical order. This work has been partially supported by the Research Budget of the Department of Applied Mathematics I of the University of Seville.

References

1. Andres, E., Acharya, R., Sibata, C.: Discrete analytical hyperplanes. Graph. Models Image Process. **59**(5), 302–309 (1997). http://www.sciencedirect. com/science/article/pii/S1077316997904275
2. Boutry, N., Géraud, T., Najman, L.: On making nD images well-composed by a self-dual local interpolation. In: Barcucci, E., Frosini, A., Rinaldi, S. (eds.) DGCI 2014. LNCS, vol. 8668, pp. 320–331. Springer, Cham (2014). https://doi.org/10. 1007/978-3-319-09955-2_27

3. Boutry, N., Géraud, T., Najman, L.: How to make nD functions digitally well-composed in a self-dual way. In: Benediktsson, J.A., Chanussot, J., Najman, L., Talbot, H. (eds.) ISMM 2015. LNCS, vol. 9082, pp. 561–572. Springer, Cham (2015). https://doi.org/10.1007/978-3-319-18720-4_47

4. Boutry, N., Géraud, T., Najman, L.: A tutorial on well-composedness. J. Math. Imaging Vis. **60**(3), 443–478 (2017). https://doi.org/10.1007/s10851-017-0769-6

5. Boutry, N., Gonzalez-Diaz, R., Jimenez, M.J.: Weakly well-composed cell complexes over nD pictures. Inf. Sci. **499**, 62–83 (2019)

6. Boutry, N., Gonzalez-Diaz, R., Najman, L., Géraud, T.: A 4D counter-example showing that DWCness does not imply CWCness in n-D. Research report hal-02455798 (2020). https://hal.archives-ouvertes.fr/hal-02455798

7. Boutry, N., Najman, L., Géraud, T.: Well-composedness in Alexandrov spaces implies digital well-composedness in $\mathbb{Z}^n$. In: Kropatsch, W.G., Artner, N.M., Janusch, I. (eds.) DGCI 2017. LNCS, vol. 10502, pp. 225–237. Springer, Cham (2017). https://doi.org/10.1007/978-3-319-66272-5_19

8. Brimkov, V.E.: Formulas for the number of (n-2)-gaps of binary objects in arbitrary dimension. Discrete Appl. Math. **157**(3), 452–463 (2009)

9. Gonzalez-Diaz, R., Jimenez, M.J., Medrano, B.: 3D well-composed polyhedral complexes. Discrete Appl. Math. **183**, 59–77 (2015)

10. Gonzalez-Diaz, R., Jimenez, M.-J., Medrano, B.: Efficiently storing well-composed polyhedral complexes computed over 3D binary images. J. Math. Imaging Vis. **59**(1), 106–122 (2017). https://doi.org/10.1007/s10851-017-0722-8

11. Hatcher, A.: Algebraic Topology. Cambridge University Press, Cambridge (2002)

12. Lachaud, J.O., Montanvert, A.: Continuous analogs of digital boundaries: a topological approach to iso-surfaces. Graph. Models **62**(3), 129–164 (2000)

13. Latecki, L., Eckhardt, U., Rosenfeld, A.: Well-composed sets. Comput. Vis. Image Underst. **61**(1), 70–83 (1995)

14. Latecki, L.J.: 3D well-composed pictures. Graph. Models Image Process. **59**(3), 164–172 (1997)

15. Latecki, L.J.: Discrete Representation of Spatial Objects in Computer Vision. Kluwer Academic, Dordrecht (1998)

16. Massey, W.S.: A Basic Course in Algebraic Topology. Springer, New York (1991). https://doi.org/10.1007/978-1-4939-9063-4

17. Najman, L., Géraud, T.: Discrete set-valued continuity and interpolation. In: Hendriks, C.L.L., Borgefors, G., Strand, R. (eds.) ISMM 2013. LNCS, vol. 7883, pp. 37–48. Springer, Heidelberg (2013). https://doi.org/10.1007/978-3-642-38294-9_4

18. Siqueira, M., Latecki, L., Tustison, N., Gallier, J., Gee, J.: Topological repairing of 3D digital images. J. Math. Imaging Vis. **30**, 249–274 (2008). https://doi.org/10.1007/s10851-007-0054-1

19. Stelldinger, P., Latecki, L.J.: 3D object digitization: majority interpolation and marching cubes. In: 18th International Conference on Pattern Recognition, vol. 2, pp. 1173–1176. IEEE (2006)

On Connectedness of Discretized Sets

Boris Brimkov[1] and Valentin E. Brimkov[2,3（✉）]

[1] Department of Mathematics and Statistics, Slippery Rock University
of Pennsylvania, Slippery Rock, PA 16057, USA
boris.brimkov@sru.edu
[2] Mathematics Department, SUNY Buffalo State, Buffalo, NY 14222, USA
brimkove@buffalostate.edu
[3] Institute of Mathematics and Informatics, Bulgarian Academy of Sciences,
1113 Sofia, Bulgaria

Abstract. Constructing a discretization of a given set is a major problem in various theoretical and applied disciplines. An offset discretization of a set X is obtained by taking the integer points inside a closed neighborhood of X of a certain radius. In this note we determine a minimum threshold for the offset radius, beyond which the discretization of an arbitrary (possibly disconnected) set is always connected. The results hold for a broad class of disconnected subsets of $\mathbb{R}^n$, and generalize several previous results.

Keywords: Discrete geometry · Geometrical features and analysis · Connected set · Discrete connectivity · Connectivity control · Offset discretization

1 Introduction

Constructing a discretization of a set $X \subseteq \mathbb{R}^n$ is a major problem in various theoretical and applied disciplines, such as numerical analysis, discrete geometry, computer graphics, medical imaging, and image processing. For example, in numerical analysis, one may need to transform a continuous domain of a function into its adequate discrete analogue. In raster/volume graphics, one looks for a rasterization that converts an image described in a vector graphics format into a raster image built by pixels or voxels. Such studies often elucidate interesting relations between continuous structures and their discrete counterparts.

Some of the earliest ideas and results for set discretization belong to Gauss (see, e.g., [22]); Gauss discretization is still widely used in theoretical research and applications. A number of other types of discretization have been studied by a large number of authors (see, e.g., [1,2,10,14,17–19,24,25,29] and the bibliographies therein). These works focus on special types of sets to be discretized, such as straight line segments, circles, ellipses, or some other classes of curves in the plane or on other surfaces.

An important requirement for any discretization is to preserve certain topological properties of the original object. Perhaps the most important among these

© Springer Nature Switzerland AG 2020
T. Lukić et al. (Eds.): IWCIA 2020, LNCS 12148, pp. 20–28, 2020.
https://doi.org/10.1007/978-3-030-51002-2_2

is the connectedness or disconnectedness of the discrete set obtained from a discretization process. This may be crucial for various applications ranging from medicine and bioinformatics (e.g. organ and tumor measurements in CT images, beating heart or lung simulations, protein binding simulations) to robotics and engineering (e.g. motion planning, finite element stress simulations). Most of the works cited above address issues related to the connectedness of the obtained discretizations. To be able to perform a reliable study of the topology of a digital object by means of shrinking, thinning and skeletonization algorithms (see, e.g. [4, 7, 20, 26, 28] and the bibliography therein), one needs to start from a faithful discretization of the original continuous set.

Perhaps the most natural and simple type of discretization of a set X is the one defined by the integer points within a closed neighborhood of X of radius r. This will be referred to as an r-offset discretization. Several authors have studied properties of offsets of certain curves and surfaces [3, 5, 11, 16], however without being concerned with the properties of the integer set of points enclosed within the offset. In [12, 15] results are presented on offset-like conics discretizations. Conditions for connectedness of offset discretizations of path connected or connected sets are presented in [8, 9, 27].

While all related works study conditions under which connectedness of the original set is preserved upon discretization, in the present paper we determine minimum thresholds for the offset radius, beyond which disconnectedness of a given original set is never preserved, i.e., the obtained discretization is always connected. The results hold for a broad class of disconnected subsets of $\mathbb{R}^n$. The technique we use is quantizing the set X by the minimal countable set of points containing X, which makes the use of induction feasible. To our knowledge, these are the first results concerning offset discretizations of disconnected sets. They extend a result from [9] which gives best possible bounds for an offset radius to guarantee 0- and $(n-1)$-connectedness of the offset discretization of a bounded path-connected set.

Our results can be used to model a broad range of plausible tasks related to image processing. From a theoretical perspective, they can be applied to a discretization of a set that is disconnected but whose closure is connected, e.g., a set that has been obtained by removing lower dimensional subsets from a connected set. This scenario can arise in practice when one needs to obtain a faithful digitization of an image or rendering which has been corrupted by linear or pointwise shapes. These flaws can be manifested as scratches on photographs, cracks on plasters, pottery, or bones, obstructions by hair, threads, or fibers, and point distortions such as piercings, dust, or other small particles. Our results imply that in such cases, under specified conditions on the offset radius, the offset discretization of the original object would not be affected by these corruptions.

In the next section we introduce various notions and notations to be used in the sequel. In Sect. 3 we present the main results of the paper. We conclude in Sect. 4 with final remarks.

2   Preliminaries

We recall a few basic notions of general topology and discrete geometry. For more details we refer to [13, 20, 21].

All considerations take place in $\mathbb{R}^n$ with the Euclidean norm. By $d(x, y)$ we denote the *Euclidean distance* between points $x, y \in \mathbb{R}^n$. Given two sets $A, B \subset \mathbb{R}^n$, the number $g(A, B) = \inf\{d(x, y) : x \in A, y \in B\}$ is called the *gap*[1] between the sets A and B. $B^n(x, r)$ is the closed n-ball of radius r and center x defined by the Euclidean distance (dependence on n will be omitted when it is clear from the context). Given a set $X \subseteq \mathbb{R}^n$, $|X|$ is its cardinality. The *closed r-neighborhood* of X, which we will also refer to as the *r-offset* of X, is defined by $U(X, r) = \cup_{x \in X} B(x, r)$. $Cl(X)$ is the *closure* of X, i.e., the union of X and the limit points of X. X is *connected* if it cannot be presented as a union of two nonempty subsets that are contained in two disjoint open sets. Equivalently, X is connected if and only if it cannot be presented as a union of two nonempty subsets each of which is disjoint from a closed superset of the other. A (possibly infinite) family $\mathcal{F}$ of sets is said to satisfy the *local finiteness property* if every point of a set in $\mathcal{F}$ has a neighborhood which intersects only a finite number of sets from $\mathcal{F}$.

In a discrete geometry setting, considerations take place in the *grid cell model*. In this model, the regular orthogonal grid subdivides $\mathbb{R}^n$ into n-dimensional unit hypercubes (e.g., unit squares for $n = 2$ or unit cubes for $n = 3$). These are regarded as *n-cells* and are called *hypervoxels*, or *voxels*, for short. The $(n-1)$-cells, 1-cells, and 0-cells of a voxel are referred to as *facets*, *edges*, and *vertices*, respectively.

Given a set $X \subseteq \mathbb{R}^n$, $X_{\mathbb{Z}} = X \cap \mathbb{Z}^n$ is its *Gauss discretization*, while $\Delta_r(X) = U(X, r) \cap \mathbb{Z}^n$ is its *discretization of radius r*, which we will also call the *r-offset discretization* of X.

Two integer points are *k-adjacent* for some k, $0 \leq k \leq n-1$, iff no more than $n - k$ of their coordinates differ by 1 and none by more than 1. A *k-path* (where $0 \leq k \leq n-1$) in a set $S \subset \mathbb{Z}^n$ is a sequence of integer points from S such that every two consecutive points of the path are k-adjacent. Two points of S are *k-connected* (in S) iff there is a k-path in S between them. S is *k-connected* iff there is a k-path in S connecting any two points of S. If S is not k-connected, we say that it is *k-disconnected*. A maximal (by inclusion) k-connected subset of S is called a *k-(connected) component* of S. Components of nonempty sets are nonempty and any union of distinct k-components is k-disconnected. Two voxels v, v' are *k-adjacent* if they share a k-cell. Definitions of connectedness and components of a set of voxels are analogous to those for integer points.

In the proof of our result we will use the following well-known facts (see [9]).

Fact 1. *Any closed n-ball $B \subset \mathbb{R}^n$ with a radius greater than or equal to $\sqrt{n}/2$ contains at least one integer point.*

---

[1] The function g itself, defined on the subsets of $\mathbb{R}^n$ is called a *gap functional*. See, e.g., [6] for more details.

Fact 2. *Let A and B be sets of integer points, each of which is k-connected. If there are points $p \in A$ and $q \in B$ that are k-adjacent, then $A \cup B$ is k-connected.*

Fact 3. *If A and B are sets of integer points, each of which is k-connected, and $A \cap B \neq \emptyset$, then $A \cup B$ is k-connected.*

Fact 4. *Given a closed n-ball $B \subset \mathbb{R}^n$ with $B_{\mathbb{Z}} \neq \emptyset$, $B_{\mathbb{Z}}$ is $(n-1)$-connected.*

3 Main Result

In this section we prove the following theorem.

Theorem 1. *Let $X \subset \mathbb{R}^n$, $n \geq 2$, be a set such that $Cl(X)$ is connected. Then the following hold:*

1. *$\Delta_r(X)$ is $(n-1)$-connected for all $r > \sqrt{n}/2$.*
2. *$\Delta_r(X)$ is at least 0-connected for all $r > \sqrt{n-1}/2$.*

These bounds are the best possible which always respectively guarantee $(n-1)$ and 0 connectedness of $\Delta_r(X)$.

Proof. The proof of the theorem is based on the following fact.

Claim. Let $X \subseteq \mathbb{R}^n$ be an arbitrary (possibly unbounded) set, such that $Cl(X)$ is connected. Let $W(X)$ denote the (possibly infinite) set of voxels intersected by X. Then $W(X)$ can be ordered in a sequence $\{v_1, v_2, \dots\}$ with the property that

$$Cl(X) \cap \left(v_k \cap \bigcup_{i=1}^{k-1} v_i \right) \neq \emptyset, \ \forall k \geq 2. \tag{1}$$

Proof. To simplify the notation, let $\bigcup F$ stand for the union of a family of sets F. Let $W'(X)$ be a maximal by inclusion subset of $W(X)$, such that it has a voxel ordering satisfying Property (1). In particular, $W'(X)$ can be constructed as follows. Let $W(X)$ have an arbitrary initial enumeration in which v_1 is the first element (since v_1 is intersected by X, it satisfies Property (1)). One can iteratively re-enumerate $W(X)$ (thus obtaining the desired subsequence $W'(X)$) in the following manner. At any step $k \geq 1$, if the voxels $v_1, v_2, \dots, v_k$ satisfy Property (1), then move to the next step. Otherwise swap v_k with a voxel $v \in W(X)$, if any, such that voxels $v_1, v_2, \dots, v_k = v$ satisfy Property (1). Thus v is added to the sequence $W'(X)$ as its kth term, v_k. (Respectively, voxel v_k from $W(X)$ receives the index of voxel v from $W(X)$). Then, in the case of unbounded X and infinite $W(X)$, a voxel that is not in the resulting new enumeration will not satisfy Property (1) for any k since it has been swapped infinitely many times.

Assume for contradiction that $W'(X) \neq W(X)$. By the maximality of $W'(X)$ it follows that $X_1 := Cl(X) \cap \bigcup W'(X)$ does not intersect the closed set $Y_1 := \bigcup(W(X) \setminus W'(X))$, and $X_2 := Cl(X) \cap \bigcup(W(X) \setminus W'(X))$ does not intersect the closed set $Y_2 := \bigcup W'(X)$. Note that Y_1 and Y_2 are possibly infinite unions of closed sets, and each of them satisfies the local finiteness property; it is well-known that any set (finite or infinite) with that property is closed (cf. [13, 23]). Then we have that $Cl(X)$ is the union of the nonempty sets X_1 and X_2, and each of them is disjoint from a closed superset of the other (Y_1 and Y_2, respectively), which is impossible if $Cl(X)$ is connected.   □

To establish the claimed connectedness of the offset discretization of the set X, in the proof of both parts of the theorem we use induction on k. Let $W(X) = \{v_1, v_2, \dots\}$ be defined as in Claim 3, with a voxel ordering satisfying Property (1). After handling the base case, in the induction step we show that for each next voxel v_{k+1} of the sequence $W(X)$, the current $(n-1)$-connected (resp. 0-connected) offset discretization $\Delta_r(X \cap \bigcup_{i=1}^{k} v_i)$ is locally incremented by the $(n-1)$-connected (resp. 0-connected) set $\Delta_r(X \cap v_{k+1})$, as the obtained offset discretization of $X \cap \bigcup_{i=1}^{k+1} v_i$ is $(n-1)$-connected (resp. 0-connected). Finally, in Part 3, we extend this argument to an unbounded set X, and hence an infinite sequence $W(X)$. This implies the $(n-1)$- (resp. 0-) connectedness of $\Delta_r(X)$.

Part 1. Let $k = 1$. In this case the considered set $\Delta_r(X \cap \bigcup_{i=1}^{k} v_i)$ is defined only by the part of X which is contained in the first voxel of the sequence. This also covers the eventuality when X is contained in a single voxel which is the only element of the ordered set $W(X)$.

Denote for brevity $X_1 = X \cap v_1$ and $D = \Delta_r(X_1)$. By Fact 1 we have that $D \neq \emptyset$. Assume for contradiction that D has at least two $(n-1)$-connected components. Let D_1 and D_2 be two of these components. Let $D_1 = \Delta_r(X_1^1)$ and $D_2 = \Delta_r(X_1^2)$, $X_1^1 \cup X_1^2 \subseteq X_1$. Since $Cl(X_1)$ is connected, there exists a point $x \in Cl(X_1)$, for which there are points both from X_1^1 and from X_1^2 which are arbitrarily close to x. We consider the extreme case where x is at the center of the voxel v_1, the considerations otherwise being analogous. Since $r > \sqrt{n}/2$, all vertices of v_1 are in its interior. Let p be one of these. Then we can choose points $x_1 \in X_1^1$ and $x_2 \in X_1^2$ to be at a sufficiently small distance from x to necessitate that the balls $B(x_1, r)$ and $B(x_2, r)$ centered at these points contain p. (If $x \in X_1$, then it can be used as one of the points x_1 or x_2).

Since $x_1, x_2 \in X_1$, we have that $B(x_1, r)_{\mathbb{Z}} \subseteq D_1$ and $B(x_2, r)_{\mathbb{Z}} \subseteq D_2$. Each of these two sets is $(n-1)$-connected by Fact 4 and p belongs to both. Thus the intersection of D_1 and D_2 is non-empty and Fact 3 implies that $D_1 \cup D_2$ is $(n-1)$-connected, which contradicts the assumption that D_1 and D_2 are $(n-1)$-connected components.

Now suppose that $\Delta_r(X \cap \bigcup_{i=1}^{k} v_i)$ is $(n-1)$-connected for some $k \geq 1$. We have

$$X \cap \bigcup_{i=1}^{k+1} v_i = (X \cap v_{k+1}) \cup \left(X \cap \bigcup_{i=1}^{k} v_i\right).$$

Since the closed r-neighborhood of a union of two sets equals the union of their r-neighborhoods, it follows that

$$\Delta_r(X \cap \bigcup_{i=1}^{k+1} v_i) = \Delta_r(X \cap v_{k+1}) \cup \Delta_r(X \cap \bigcup_{i=1}^{k} v_i).$$

Let us denote by f a common face of voxel $v_{k+1} \in W(X)$ and the polyhedral complex composed by the voxels $v_1, v_2, \ldots, v_k$, i.e., $f \subset v_{k+1} \cap \bigcup_{i=1}^{k} v_i$. W.l.o.g., we can consider the case where f is a facet of v_{k+1} (i.e., a cell of topological dimension $n-1$), the cases of lower dimension faces being analogous. Let H be the hyperplane in $\mathbb{R}^n$ which is the affine hull of f. Let $\mathbb{Z}_H^{n-1}$ be the subset of the set of grid-points $\mathbb{Z}^n$ contained in H.

By Claim 3, there is a point $x \in Cl(X) \cap f$. Consider the n-ball $B^n(x, r)$. Then $B^{n-1}(x, r) = B^n(x, r) \cap H$ is an $(n-1)$-ball with the same center and radius. Applying Fact 1 to $B^{n-1}(x, r)$ in the $(n-1)$-dimensional hyperplane H, we obtain that $B^{n-1}(x, r)$ contains in its interior at least one grid point $p \in \mathbb{Z}_H^{n-1}$, which is a vertex of facet f. Since $x \in Cl(X)$ is a limit point of X, there exists a point $y \in X$, such that the ball $B^{n-1}(y, r)$ contains p, too. By construction, p is common for the sets $\Delta_r(X \cap v_{k+1})$ and $\Delta_r(X \cap \bigcup_{i=1}^{k} v_i)$. The former is $(n-1)$-connected by the same argument used in the induction basis, while the latter is $(n-1)$-connected by the induction hypothesis. Then by Fact 3, their union $\Delta_r(X \cap \bigcup_{i=1}^{k+1} v_i)$ is $(n-1)$-connected, as well. This establishes Part 1.

Part 2. The proof of this part is similar to the one of Part 1. Note that in the special case where the set X is contained in a single voxel v_1 (and thus $X_1 = X \cap v_1 = X$ is contained in v_1, too), and if $\sqrt{n-1}/2 < r < \sqrt{n}/2$, then it is possible to have $\Delta_r(X) = \Delta_r(X_1) = \emptyset$ (for instance, if X consists of a single point that is the center of v_1). Then the statement follows immediately. Suppose that this is not the case. By definition, $\Delta_r(X_1) = \bigcup_{x \in X_1} B(x, r)_{\mathbb{Z}}$. For any $x \in X_1$, $B(x, r)_{\mathbb{Z}}$ is $(n-1)$-connected by Fact 4. We also have that any of the nonempty sets $B(x, r)_{\mathbb{Z}}$ contains a vertex of v_1. Since any two vertices of a grid cube are at least 0-adjacent, it follows that any subset of vertices of v_1 is at least 0-connected. Then by Fact 2, $\Delta_r(X_1)$ is at least 0-connected.

The rest of the proof parallels the one of Part 1, with the only difference that point q is common for the sets $\Delta_r(X \cap v_{k+1})$ and $\Delta_r(X \cap \bigcup_{i=1}^{k} v_i)$, each of which is 0-connected (the former by an argument used in the induction basis, and the latter by the induction hypothesis). Then Fact 3 implies that their union $\Delta_r(X \cap \bigcup_{i=1}^{k+1} v_i)$ is 0-connected, as stated.

Figure 1 illustrates that the obtained bounds for r are the best possible: if r equals $\sqrt{n}/2$ (resp. $\sqrt{n-1}/2$), then $\Delta_r(X)$ may not be $(n-1)$-connected (resp. 0-connected). This completes the proof of the theorem. $\square$

The proof of Theorem 1 implies the following corollary.

Corollary 1. *If $X \subseteq \mathbb{R}^n$ ($n \geq 2$) is connected, then $\Delta_r(X)$ is $(n-1)$-connected for all $r \geq \sqrt{n}/2$, and $\Delta_r(X)$ is at least 0-connected for all $r \geq \sqrt{n-1}/2$.*

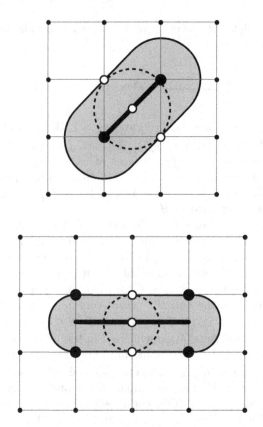

Fig. 1. *Top:* Offset radius $r = \sqrt{2}/2$, *Bottom:* offset radius $r = 1/2$. In both figures: X is the thick line segment with missing midpoint marked by a hollow dot, the shaded region is the offset, the offset discretization consists of the large thick dots, the hollow dots on the offset boundary do not belong to the offset and to the discretization.

The above in turn implies:

Corollary 2. *Let $X \subset \mathbb{R}^n$ ($n \geq 2$) be such that $U(X, r)$ is connected for some $r > 0$. Then $\Delta_{r+\sqrt{n}/2}(X)$ is $(n-1)$-connected, and $\Delta_{r+\sqrt{n-1}/2}(X)$ is at least 0-connected.*

4   Concluding Remarks

In this note we obtained theoretical conditions for connectedness of offset discretizations of sets in higher dimensions. An important future task is seen in computer implementation and testing the topological properties and visual appearance of offset discretizations of varying radius. It would also be interesting to study similar properties of other basic types of discretization.

Acknowledgments. We thank the anonymous referees for their valuable comments and suggestions.

References

1. Andres, E.: Discrete circles, rings and spheres. Comput. Graph. **18**(5), 695–706 (1994)
2. Andres, E.: Discrete linear objects in dimension n: the standard model. Graph. Models **65**(1), 92–111 (2003). Special Issue: Discrete Topology and Geometry for Image and Object Representation
3. Anton, F., Emiris, I., Mourrain, B., Teillaud, M.: The offset to an algebraic curve and an application to conics. In: Gervasi, O. (ed.) ICCSA 2005. LNCS, vol. 3480, pp. 683–696. Springer, Heidelberg (2005). https://doi.org/10.1007/11424758_71
4. Arcelli, C., di Baja, G.S.: Skeletons of planar patterns. In: Kong, T.Y., Rosenfeld, A. (eds.) Topological Algorithms for Digital Image Processing, Machine Intelligence and Pattern Recognition, vol. 19, pp. 99–143. Elsevier (1996)
5. Arrondo, E., Sendra, J., Sendra, J.: Genus formula for generalized offset curves. J. Pure Appl. Algebra **136**(3), 199–209 (1999)
6. Beer, G.: Topologies on Closed and Closed Convex Sets, Mathematics and its Applications, vol. 268. Kluwer Academic Publisher, London (1993)
7. Borgefors, G., Ramella, G., di Baja, G.S.: Shape and topology preserving multi-valued image pyramids for multi-resolution skeletonization. Pattern Recogn. Lett. **22**(6), 741–751 (2001)
8. Brimkov, V.E.: Connectedness of offset digitizations in higher dimensions. In: Barneva, R.P., Brimkov, V.E., Hauptman, H.A., Natal Jorge, R.M., Tavares, J.M.R.S. (eds.) CompIMAGE 2010. LNCS, vol. 6026, pp. 36–46. Springer, Heidelberg (2010). https://doi.org/10.1007/978-3-642-12712-0_4
9. Brimkov, V.E., Barneva, R.P., Brimkov, B.: Connected distance-based rasterization of objects in arbitrary dimension. Graph. Models **73**(6), 323–334 (2011)
10. Cohen-Or, D., Kaufman, A.: 3D line voxelization and connectivity control. IEEE Comput. Graph. Appl. **17**(6), 80–87 (1997)
11. Cox, D., Little, J., O'shea, D.: Using Algebraic Geometry, Graduate Texts in Mathematics, vol. 185. Springer, New York (1998)
12. Debled-Rennesson, I., Domenjoud, E., Jamet, D.: Arithmetic discrete parabolas. In: Bebis, G. (ed.) ISVC 2006. LNCS, vol. 4292, pp. 480–489. Springer, Heidelberg (2006). https://doi.org/10.1007/11919629_49
13. Engelking, R.: General Topology. Revised and Completed Edition. Heldermann Verlag, Berlin (1989)
14. Figueiredo, O., Reveillès, J.P.: New results about 3D digital lines. In: Vision Geometry V, vol. 2826, pp. 98–109. International Society for Optics and Photonics (1996)
15. Fiorio, C., Jamet, D., Toutant, J.L.: Discrete circles: an arithmetical approach with non-constant thickness. In: Vision Geometry XIV, vol. 6066, p. 60660C. International Society for Optics and Photonics (2006)
16. Hoffmann, C., Vermeer, P.: Eliminating extraneous solutions for the sparse resultant and the mixed volume. J. Symbolic Geom. Appl **1**(1), 47–66 (1991)
17. Jonas, A., Kiryati, N.: Digital representation schemes for 3D curves. Pattern Recogn. **30**(11), 1803–1816 (1997)
18. Kaufman, A., Cohen, D., Yagel, R.: Volume graphics. Computer **26**(7), 51–64 (1993)

19. Kim, C.E.: Three-dimensional digital line segments. IEEE Trans. Pattern Anal. Mach. Intell. **2**, 231–234 (1983)
20. Klette, R., Rosenfeld, A.: Digital Geometry: Geometric Methods for Digital Picture Analysis. Elsevier, Amsterdam (2004)
21. Kong, T.: Digital topology. In: Davis, L.S. (ed.) Foundations of Image Understanding, pp. 33–71. Kluwer Academic Publishers, Boston (2001)
22. Krätzel, E.: Zahlentheorie, vol. 19. VEB Deutscher Verlag der Wissenschaften, Berlin (1981)
23. Munkers, J.R.: Topology, 2nd edn. Prentice Hall, Upper Saddle River (2000)
24. Rosenfeld, A.: Connectivity in digital pictures. J. ACM (JACM) **17**(1), 146–160 (1970)
25. Rosenfeld, A.: Arcs and curves in digital pictures. J. ACM (JACM) **20**(1), 81–87 (1973)
26. Saha, P.K., Borgefors, G., di Baja, G.S.: A survey on skeletonization algorithms and their applications. Pattern Recogn. Lett. **76**, 3–12 (2016)
27. Stelldinger, P.: Image Digitization and its Influence on Shape Properties in Finite Dimensions, vol. 312. IOS Press, Amsterdam (2008)
28. Svensson, S., di Baja, G.S.: Simplifying curve skeletons in volume images. Comput. Vis. Image Underst. **90**(3), 242–257 (2003)
29. Tajine, M., Ronse, C.: Topological properties of hausdorff discretization, and comparison to other discretization schemes. Theor. Comput. Sci. **283**(1), 243–268 (2002)

Persistent Homology as Stopping-Criterion for Voronoi Interpolation

Luciano Melodia$^{(\boxtimes)}$ (ID) and Richard Lenz (ID)

Chair of Computer Science 6, Friedrich-Alexander University Erlangen-Nürnberg, 91058 Erlangen, Germany
{luciano.melodia,richard.lenz}@fau.de

Abstract. In this study the Voronoi interpolation is used to interpolate a set of points drawn from a topological space with higher homology groups on its filtration. The technique is based on Voronoi tessellation, which induces a natural dual map to the Delaunay triangulation. Advantage is taken from this fact calculating the persistent homology on it after each iteration to capture the changing topology of the data. The boundary points are identified as critical. The Bottleneck and Wasserstein distance serve as a measure of quality between the original point set and the interpolation. If the norm of two distances exceeds a heuristically determined threshold, the algorithm terminates. We give the theoretical basis for this approach and justify its validity with numerical experiments.

Keywords: Interpolation · Persistent homology · Voronoi triangulation

1 Introduction

Most interpolation techniques ignore global properties of the underlying topological space of a set of points. The topology of an augmented point set depends on the choice of interpolant. However, it does not depend on the topological structure of the data set. The Voronoi interpolation is a technique considering these issues [3]. The algorithm has been invented by Sibson [21]. Using Voronoi triangulation to determine the position of a new point respects the topology in terms of simple-homotopy equivalence. For this an implicit restriction to a closed subset of the embedded space is used, see Fig. 2. The closure of this subset depends on the metric, in Euclidean space it is flat. This restriction, also called *clipping*, leads to varying results for interpolation according to the choice of clip. The clip does not represent the intrinsic geometry nor the topology of the data, but that of the surrounding space. This leads to artifacts during interpolation.

Persistent homology encodes the topological properties and can be calculated in high dimensions. Thus, it is used as indicator for such artifacts [25]. In particular, this measurement of topological properties behaves stable, i.e. small

© Springer Nature Switzerland AG 2020
T. Lukić et al. (Eds.): IWCIA 2020, LNCS 12148, pp. 29–44, 2020.
https://doi.org/10.1007/978-3-030-51002-2_3

changes in the coordinate function value also cause small changes in persistent homology [9]. Efficient data structures and algorithms have been designed to compute persistent homology [24, 25] and scalable ways to compare persistence diagrams using the Wasserstein metric have been developed [8]. This paper uses persistent homology to decide whether a topological change occurs or not.

Up to this point it is an open problem to detect these errors and to terminate the algorithm in time. Our contribution to a solution is divided into three parts:

- We introduce persistent homology as a stopping-criterion for interpolation methods. The distance between two persistence diagrams is an indicator of topological changes during augmentation.
- We cover the connection of the Voronoi tessellation to the Delaunay triangulation via duality. It is shown that the Delaunay complex is simple-homotopy equivalent to the Čech complex. We further show that the Delaunay complex is sufficient to compute persistence diagrams.
- We investigate the method on a signature data set. It provides interesting and visually interpretable topological features due to the topography of letters. Higher homology groups such as H_1 and H_2 may appear on the filtration of a signature. This often represents an insurmountable hurdle for other interpolation techniques.

2   Simplicial Complexes and Filtrations

Taking into account the topology of data is beneficial for interpolation, due to the assumption that the point set lies on a topological or even smooth manifold, having a family of smooth coordinate systems to describe it. Another hypothesis says, that the mutual arrangement of every dataset forms some 'shape' [25], which characterizes the manifold. If the point set changes its shape, it is no longer identifiable with this manifold.

Embedded simplicial complexes, build out of a set of simplices, are suitable objects to detect such shapes, by computing their homology groups. Simplices, denoted by σ, are the permuted span of the set $X = \{x_0, x_1, \ldots, x_k\} \subset \mathbb{R}^d$ with $k+1$ points, which are not contained in any affine subspace of dimension smaller than k [19]. A simplex forms the convex hull

$$\sigma := \left\{ x \in X \ \middle| \ \sum_{i=0}^{k} \lambda_i x_i \text{ with } \sum_{i=0}^{k} \lambda_i = 1 \text{ and } \lambda_i \geq 0 \right\}. \tag{1}$$

Simplices are well-defined embeddings of polyhedra. 'Gluing' simplices together at their *faces*, we can construct simplicial complexes out of them. Faces are meant to be h-dimensional simplices or h-simplices. Informally, the gluing creates a series of k-simplices, which are connected by h-simplices, that satisfy $h < k$. A finite simplicial complex denoted by K and embedded into Euclidean space is a finite set of simplices with the properties, that each face of a simplex of K is again a simplex of K and the intersection of two simplices is either empty or a common face of both [19].

We want to take into account the systematic development of a simplicial complex upon a point cloud. This is called filtration and it is the decomposition of a finite simplicial complex K into a nested sequence of sub-complexes, starting with the empty set [10]:

$$\emptyset = K^0 \subset K^1 \subset \cdots \subset K^n = K, \tag{2}$$

$$K^{t+1} = K^t \cup \sigma^{t+1}, \quad \text{for } t \in \{0, \ldots, n-1\}. \tag{3}$$

In practice a parameter r is fixed to determine the step size of the nested complexes. This can be thought as a 'lens' zooming into a certain 'granularity' of the filtration. In the following, we present four different simplicial complexes and their theoretical connection.

2.1   Čech Complex

Let the radius $r \geq 0$ be a real number and $B(x,r) = \{y \in \mathbb{R}^d \mid \|x - y\| \leq r\}$ the closed ball centered around the point $x \in X \subseteq \mathbb{R}^d$. The Čech complex for a finite set of points X is defined as

$$\check{\text{C}}\text{ech}(X,r) = \{U \subseteq X \mid \bigcap_{x \in U} B(x,r) \neq \emptyset\}. \tag{4}$$

By $\|\cdot\|$ we denote consequently the L^2-norm. In terms of abstract simplicial complexes (see Sect. 7), the Čech complex is the full abstract simplex spanned over X [2]. According to the Nerve lemma it is homotopy-equivalent to the union of balls $B(X,r) = \bigcup_{x \in U} B(x,r)$ [7]. Spanning the simplicial complex for $r = \sup_{x,y \in U} \|x - y\|$, we get the full simplex for the set U. For two radii $r_1 < r_2$ we get a nested sequence $\check{\text{C}}\text{ech}(X,r_1) \subset \check{\text{C}}\text{ech}(X,r_2)$. This implies that the Čech complex forms a filtration over U and therefore a filtration over the topological space X if $U = X$ [2]. These properties make the Čech complex a very precise descriptor of the topology of a point set. The flip side of the coin is that the Čech complex is not efficiently computable for large point sets. A related complex is therefore presented next, which is slightly easier to compute.

2.2   Vietoris-Rips Complex

The Vietoris-Rips complex $\text{Rips}(X,r)$ with vertex set X and distance threshold r is defined as the set

$$\text{Rips}(X,r) = \left\{U \subseteq X \mid \|x - y\| \leq r, \text{ for all } x, y \in U\right\}. \tag{5}$$

The Vietoris-Rips complex requires only the comparison of distance measures to be obtained. It spans the same 1-skeleton as the Čech complex and fulfills for an embedding into any metric space the following relationship [5, p. 15]: $\text{Rips}(X,r) \subseteq \check{\text{C}}\text{ech}(X,r) \subseteq \text{Rips}(X,2r)$. To see this, we choose a simplex

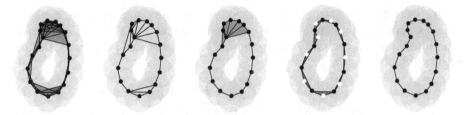

Fig. 1. Four of five geometric complexes appearing in the collapsing sequence of the
Čech-Delaunay Collapsing Theorem [2]. From *left* to *right*: a high dimensional Čech
complex projected onto the plane, the Čech-Delaunay complex, the Delaunay complex,
the Witness complex, which is an outlier in the row due to the changing shape by
different Witness sets (white bullets) and the Wrap complex.

$\sigma = \{x_0, x_1, \ldots, x_k\} \in \text{Rips}(X, r)$. The point $x_0 \in \bigcap_{i=0}^{k} B(x_i, r)$ must be within
the intersection of closed balls with radius r of all points. Now we choose a
$\sigma = \{x_0, x_1, \ldots, x_k\} \in \text{Čech}(X, r)$, then there is a point $y \in \mathbb{R}^d$ within the intersection $y \in \bigcap_{i=0}^{k} B(x_i, r)$, which is the desired condition $d(x_i - y) \leq r$ for any
$i = 0, \ldots, k$. Therefore, for all $i, j \in \{0, \ldots, k\}$ the following (in)equality applies:
$d(x_i - x_j) \leq 2r$ and $\sigma \in \text{Rips}(X, 2r)$ (Fig. 1).

The calculation time for the Vietoris-Rips complex is better than for the
Čech complex, with a bound of $\mathcal{O}(n^2)$ for n points [24]. As a third complex
we introduce the α-complex or Delaunay complex, for which the definition of
Voronoi cells and balls are prerequisite.

2.3 Delaunay Complex

If $X \subset \mathbb{R}^d$ is a finite set of points and $x \in X$, then the Voronoi cell or also
Voronoi region of a point $x \in X$ is given by

$$\text{Vor}(x) = \left\{ y \in \mathbb{R}^d \,\middle|\, ||y - x|| \leq ||y - z||, \text{ for all } z \in X \right\}. \tag{6}$$

The Voronoi ball of x with respect to X is defined as the intersection of
the Voronoi region with the closed ball of given radius around this point, i.e.
$\text{VorBall}(x, r) = B(x, r) \cap \text{Vor}(x)$ [2]. The Delaunay complex on a point set X is
defined as

$$\text{Del}(X, r) = \left\{ U \subseteq X \,\middle|\, \bigcap_{x \in U} \text{VorBall}(x, r) \neq \emptyset \right\}. \tag{7}$$

There is a fundamental connection between the union of all Voronoi balls over X
and the Delaunay complex. The idea is to find a *good cover* that does represent
the global topology. Taking the topological space X and $U = \bigcup_{i \in I} U_i$ being an
open cover, we define the Nerve of a cover as its topological structure. Therefore,
the empty set $\emptyset \in N(U)$ is part of the Nerve and if $\bigcap_{j \in J} U_j \neq \emptyset$ for a $J \subseteq I$,
then $J \in N(U)$. We consider U to be a good cover, if for each $\sigma \subset I$ the set

$\bigcap_{i\in\sigma} U_i \neq \emptyset$ is contractible, or in other words if it has the same homotopy type as a point. In this case the Nerve $N(U)$ is homotopy equivalent to $\bigcup_{i\in I} U_i$.

Most interestingly, the Delaunay complex $\mathrm{Del}_r(X)$ of a point set X is isomorphic to the Nerve of the collection of Voronoi balls. To see this, we construct Voronoi regions for two different sets. Thus, we denote the Voronoi region $\mathrm{Vor}(x, r, U)$ of a point within a set U. Be $\mathrm{Vor}(x, r, U) \subseteq \mathrm{Vor}(x, r, V)$ for each open set $U \subseteq V \subseteq X$ and all $x \in X$. We obtain the largest Voronoi ball for $U = \emptyset$ and the smallest Voronoi ball for $U = X$. In the first case each region is a ball with radius r and in the second case the Voronoi balls form a convex decomposition of the union of balls. We select a subset U and restrict the Delaunay complex to it by taking into account only the Voronoi balls around the points in U. It is called selective Delaunay complex and contains the Delaunay and Čech complex in its extremal cases:

$$\mathrm{Del}(X, r, U) = \left\{ V \subseteq X \;\middle|\; \bigcap_{x\in V} \mathrm{VorBall}(x, r, U) \neq \emptyset \right\}. \tag{8}$$

Since the union of open balls does not depend on U, the Nerve lemma implies, that for a given set of points X and a radius r all selective Delaunay complexes have the same homotopy type. This also results in $\mathrm{Del}(X, r, V) \subseteq \mathrm{Del}(X, s, U)$ for all $r \leq s$ and $U \subseteq V$. The proof has been given first by [2, §3.4].

2.4 Witness Complex

Through the restriction of the faces to randomly chosen subsets of the point cloud the filtration is carried out on a scalable complex, which is suitable for large point sets. We call these subsets Witnesses $W \subset \mathbb{R}^d$ and $L \subset \mathbb{R}^d$ landmarks. The landmarks can be part of the Witnesses $L \subseteq W$, but do not have to. Then σ is a simplex with vertices in L and some points $w \in W$. We say that w is Witnessed by σ if $||w - p|| \leq ||w - q||$, for all $p \in \sigma$ and $q \in L \setminus \sigma$. We further say it is strongly Witnessed by σ if $||w - p|| \leq ||w - q||$, for all $p \in \sigma$ and $q \in L$. The Witness complex $\mathrm{Wit}(L, W)$ consists of all simplices σ, such that any simplex $\tilde\sigma \subseteq \sigma$ has a Witness in W and the strong Witness complex analogously.

The homology groups of the Witness complex depend strongly on the landmarks. In addition to equally distributed initialization, strategies such as sequential *MaxMin* can lead to a more accurate estimate of homology groups [22]. Its time bound for construction is $\mathcal{O}\left(|W|log|W| + k|W|\right)$ [5].

3 Persistent Homology Theory

We are particularly interested in whether a topological space can be continuously transformed into another. For this purpose its k-dimensional 'holes' play a central role. Given two topological spaces M and N we say that they have the same *homotopy type*, if there exists a continuous map $h : M \times I \to N$, which deforms M over some time interval I into N. But it is very difficult to obtain homotopies. An algebraic way to compute something strongly related is homology. The connection to homotopy is established by the Hurewicz Theorem. It says, that given $\pi_k(x, X)$, the k-th homotopy group of a topological space X in a point $x \in X$, there exists a homomorphism $h : \pi_k(x, X) \rightarrow H_k(x, X)$ into the k-th homology group at x. It is an isomorphism if X is $(n-1)$-connected and $k \leq n$ when $n \geq 2$ with abelianization for $n = 1$ [17]. In this particular case, we are able to use an easier to calculate invariant to describe the topological space of the data up to homotopy. Further we need to define what a boundary and what a chain is, respectively. We want to describe the boundary of a line segment by its two endpoints, the boundary of a triangle, or 2-simplex by the union of the edges and the boundary of a tetrahedron, or 3-simplex by the union of the triangular faces. Furthermore, a boundary itself shall not have a boundary of its own. This implies the equivalence of the property to be boundaryless to the concept of a 'loop', i.e. the possibility to return from a starting point to the same point via the k-simplices, by not 'entering' a simplex twice and not 'leaving' a simplex 'unentered'.

Let σ^k be a k-simplex of a simplicial complex $K := K(X)$ over a set of points X. Further, let $k \in \mathbb{N}$. The linear combinations of k-simplices span a vector space $C_k := C_k(K) = \text{span}\left(\sigma_1^k, \ldots, \sigma_n^k\right)$. This vector space is called k-th chain group of K and contains all linear combinations of k-simplices. The coefficients of the group lie in $\mathbb{Z}$ and the group structure is established by $(C_k, +)$, with $e_{C_k} = 0$ being the neutral element and addition as group operation. A linear map $\partial : C_k \rightarrow C_{k-1}$ is induced from the k-th chain group into the $(k-1)$-th. The boundary operator $\partial_k(\sigma^k) : C_k \rightarrow C_{k-1}$ is defined by

$$\partial_k(\sigma^k) = \sum_{i=0}^{k}(-1)^i\left(v_0, \ldots, \widehat{v_i}, \ldots, v_k\right). \tag{9}$$

The vertex set of the k-simplex is $v_0, \ldots, v_k$. This group homomorphism contains an alternating sum, thus for each oriented k-simplex $(v_0, \ldots, v_k)$ one element $\widehat{v_i}$ is omitted. The boundary operator can be composed $\partial^2 := \partial \circ \partial$. We observe, that every chain, which is a boundary of higher-dimensional chains, is boundaryless. An even composition of boundary maps is zero $\partial^{2\mathbb{Z}} = 0$ [17].

The kernel of $(C_k, +)$ is the collection of elements from the k-th chain group mapped by the boundary operator to the neutral element of $(k-1)$-th: $\ker \partial_k = \partial_k^{-1}(e_{C_{k-1}}) = \{\sigma^k \in C_k \mid \partial_k(\sigma^k) = e_{C_{k-1}}\}$. A cycle should be defined by having no boundary. From this we get a group of k-cycles, denoted by Z_k, which is defined as the kernel of the k-th boundary operator $Z_k := \ker \partial_k \subseteq C_k$. Every k-simplex mapped to zero by the boundary operator is considered to be a cycle and the collection of cycles is the group of k-cycles Z_k. The k-boundaries are therefore $B_k = \text{Im}\partial_{k+1} \subset Z_k$. The k-th homology group H_k is the quotient

$$H_k := Z_k/B_k = \ker \partial_k/\text{Im}\partial_{k+1}. \tag{10}$$

We compute the k-th Betti numbers by the rank of this vector space, i.e. $\beta_k = \text{rank } H_k$. In a certain sense the Betti numbers count the amount of holes in a topological space, i.e. β_0 counts the connected components, β_1 the tunnels, β_2 voids and so forth. Using Betti numbers, the homology groups can be tracked along the filtration, representing the 'birth' and 'death' of homology classes. The filtration of a simplicial complex defines a sequence of homology groups connected by homomorphisms for each dimension. The k-th homology group over a simplicial complex K_r with parameter r is denoted by $H_k^r = H_k(K_r)$. This gives a group homomorphism $g_k^{r,r+1} : H_k^r \to H_k^{r+1}$ and the sequence [15]:

$$0 = H_k^0 \xrightarrow{g_k^{0,1}} H_k^1 \xrightarrow{g_k^{1,2}} \cdots \xrightarrow{g_k^{n,r}} H_k^r \xrightarrow{g_k^{r,r+1}} H_k^{r+1} = 0. \tag{11}$$

The image $\mathrm{Im} g_k^{r,r+1}$ consists of all k-dimensional homology classes which are born in the K_r-complex or appear before and die after K_{r+1}. The dimension k persistent homology group is the image of the homomorphisms $H_k^{n,r} = \mathrm{Im} g_k^{n,r}$, for $0 \le n \le r \le r+1$ [15]. For each dimension there is an index pair $n \le r$. Tracking the homology classes in this way yields a multi set, as elements from one homology group can appear and vanish several times for a certain parametrization. Thus, we get the following multiplicity:

$$\mu_k^{n,r} = \underbrace{(\beta_k^{n,r-1} - \beta_k^{n,r})}_{\text{Birth in } K_{r-1}, \text{ death at } K_r.} - \underbrace{(\beta_k^{n-1,r-1} - \beta_k^{n-1,r})}_{\text{Birth before } K_r, \text{ death at } K_r.} \tag{12}$$

The first difference counts the homology classes born in K_{r-1} and dying when K_r is entered. The second difference counts the homology classes born before K_{r-1} and dying by entering K_r. It follows that $\mu_k^{n,r}$ counts the k-dimensional homology classes born in K_n and dying in K_r [15].

The persistence diagram for the k-th dimension, denoted as $\mathcal{P}_K^{(\dim k)}$, is the set of points $(n,r) \in \bar{\mathbb{R}}^2$ with $\mu_k^{n,r} = 1$ where $\bar{\mathbb{R}} := \mathbb{R} \cup +\infty$. We define the general persistent diagram as the disjoint union of all k-dimensional persistence diagrams $\mathcal{P}_K = \bigsqcup_{k \in \mathbb{Z}} \mathcal{P}_K^{(\dim k)}$. In this paper we consider H_0, H_1 and H_2. We now introduce distances for comparison of persistence diagrams. In particular, it is important to resolve the distance between multiplicities in a meaningful way. Note that they are only defined for $n < r$ and that no values appear below the diagonal. This is to be interpreted such that a homology class can't disappear before it arises.

4   Bottleneck Distance

Let X be a set of points embedded in Euclidean space and K_r^1, K_r^2 two simplicial complexes forming a filtration over X. Both are finite and have in all their sub-level sets homology groups of finite rank. Note, that these groups change due to a finite set of homology-critical values. To define the bottleneck distance we use the L^∞-norm $\|x - y\|_\infty = \max \{|x_1 - y_1|, |x_2 - y_2|\}$ between two points $x = (x_1, x_2)$ and $y = (y_1, y_2)$ for $x \in \mathcal{P}_{K^1}$ and $y \in \mathcal{P}_{K^2}$. By convention, it is assumed that if $x_2 = y_2 = +\infty$, then $\|x - y\|_\infty = |x_1 - y_1|$. If $\mathcal{P}_{K^1}$ and $\mathcal{P}_{K^2}$

are two persistence diagrams and $x := (x_1, x_2) \in \mathcal{P}_{K^1}$ and $y := (y_1, y_2) \in \mathcal{P}_{K^2}$, respectively, their Bottleneck distance is defined as

$$d_B(\mathcal{P}_{K^1}, \mathcal{P}_{K^2}) = \inf_{\varphi} \sup_{x \in \mathcal{P}_{K^1}} ||x - \varphi(x)||_\infty, \qquad (13)$$

where φ is the set of all bijections from the multi set $\mathcal{P}_{K^1}$ to $\mathcal{P}_{K^2}$ [5].

4.1 Bottleneck Stability

We consider a smooth function $f : \mathbb{R} \to \mathbb{R}$ as a working example. A point $x \in \mathbb{R}$ of this function is called critical and $f(x)$ is called critical value of f if $df_x = 0$. The critical point is also said to be not degenerated if $d^2 f_x \neq 0$. Im $f(x)$ is a homology critical value, if there is a real number y for which an integer k exists, such that for a sufficiently small $\alpha > 0$ the map $H_k \left(f^{-1} \left((-\infty, y - \alpha] \right) \right) \to H_k \left(f^{-1} \left((-\infty, y + \alpha] \right) \right)$ is not an isomorphism. We call the function f tame if it has a finite number of homology critical values and the homology group $H_k \left(f^{-1} \left((-\infty, y] \right) \right)$ is finite-dimensional for all $k \in \mathbb{Z}$ and $y \in \mathbb{R}$. A persistence diagram can be generated by pairing the critical values with each other and transferring corresponding points to it.

The Bottleneck distance of the persistence diagram of two tame functions f, g is restricted to a norm between a point and its bijective projection. Therefore, not all points of a multi set can be mapped to the nearest point in another [12]. To see this, we consider f to be tame. The Hausdorff distance $d_H(X, Y)$ between two multi sets X and Y is defined by

$$\max \left\{ \sup_{x \in X} \inf_{y \in Y} ||x - y||_\infty, \sup_{y \in Y} \inf_{x \in X} ||y - x||_\infty \right\}. \qquad (14)$$

From the results of [9] it is known that the Hausdorff (in)equality $d_H(\mathcal{P}_f, \mathcal{P}_g) \leq ||f - g||_\infty = \alpha$ holds and that there must exist a point $(x_1, x_2) \in \mathcal{P}_f$ which has a maximum distance α to a second point $(y_1, y_2) \in \mathcal{P}_g$. In particular, (y_1, y_2) must be within the square $[x_1 - \alpha, x_1 + \alpha] \times [x_2 - \alpha, x_2 + \alpha]$. Let $x_1 \leq x_2 \leq x_3 \leq x_4$ be points in the extended plane $\bar{\mathbb{R}}^2$. Further, let $R = [x_1, x_2] \times [y_1, y_2]$ be a square and $R_\alpha = [x_1 + \alpha, x_2 - \alpha] \times [y_1 + \alpha, y_2 - \alpha]$ another shrinked square by some parameter α. Thus, we yield

$$\# (\mathcal{P}_f \cap R_\alpha) \leq \# (\mathcal{P}_g \cap R). \qquad (15)$$

We need the inequality to find the smallest α such that squares of side-length 2α centered at the points of one diagram cover all off-diagonal elements of the other diagram, and vice versa with the diagrams exchanged [12]. The persistence diagrams $\mathcal{P}_f$ and $\mathcal{P}_g$ satisfy for two tame functions $f, g : X \to \mathbb{R}$:

$$d_B (\mathcal{P}_f, \mathcal{P}_g) \leq ||f - g||_\infty. \qquad (16)$$

We take two points $x = (x_1, x_2), y = (y_1, y_2) \in \mathcal{P}_f$ and look at the infinite norm between them in the persistence diagram of f outside the diagonal Δ. In case that there is no such second point we consider the diagonal itself:

$$\delta_f = \min \left\{ ||x - y||_\infty \mid \mathcal{P}_f - \Delta \ni x \neq y \in \mathcal{P}_f \right\}. \tag{17}$$

We choose a second tame function g, which satisfies $||f - g||_\infty \leq \delta_f/2$. We center a square $R_\alpha(x)$ at x with radius $\alpha = ||f - g||_\infty$. Applying Eq. 15 yields

$$\mu \leq \#(\mathcal{P}_g \cap R_\alpha(x)) \leq \#(\mathcal{P}_f \cap R(x)_{2\alpha}). \tag{18}$$

Since g was chosen in such a way, that $||f - g||_\infty \leq \delta_f/2$ applies, we conclude that $2\alpha \leq \delta_f$. Thus, x is the only point of the persistence diagram $\mathcal{P}_f$ that is inside $R_{2\alpha}$ and the multiplicity μ is equal to $\#(\mathcal{P}_g \cap R(x)_\alpha)$. We can now project all points from $\mathcal{P}_g$ in $R(x)_\alpha$ onto x. As $d_H(\mathcal{P}_f, \mathcal{P}_g) \leq \alpha$ holds, the remaining points are mapped to their nearest point on the diagonal.

5   Wasserstein Distance

The Wasserstein distance is defined for separable completely metrizable topological spaces. In this particular case between the two persistence diagrams $\mathcal{P}_{K^1}$ and $\mathcal{P}_{K^2}$. The L^p-Wasserstein distance W^p is a metric arising from the examination of transport plans between two distributions and is defined for a $p \in [1, \infty)$ as

$$d_{W^p}(\mathcal{P}_{K^1}, \mathcal{P}_{K^2}) = \left(\inf_\varphi \sum_{x \in \mathcal{P}_{K^1}} ||x - \varphi(x)||_\infty^p \right)^{1/p}. \tag{19}$$

Then $\varphi : \mathcal{P}_{K^1} \to \mathcal{P}_{K^2}$ is within the set of all transportation plans from $\mathcal{P}_{K^1}$ to $\mathcal{P}_{K^2}$ over $\mathcal{P}_{K^1} \times \mathcal{P}_{K^2}$. We use the L^1-Wasserstein distance. The Wasserstein distance satisfies the axioms of a metric [23, p. 77]. The transportation problem can be stated as finding the most economical way to transfer the points from one persistence diagram into another. We assume that these two persistence diagrams are disjoint subsets of $\bar{\mathbb{R}}^2 \times \bar{\mathbb{R}}^2$. The cost of transport is given by $d : \bar{\mathbb{R}}^2 \times \bar{\mathbb{R}}^2 \to [0, \infty)$, so that $||x - \varphi(x)||$ indicates the length of a path. The transport plan is then a bijection $\varphi : \mathcal{P}_{K^1} \to \mathcal{P}_{K^2}$ from one persistence diagram to the other. The Wasserstein distance of two persistence diagrams is the optimal cost of all transport plans. Note, that the L^∞-Wasserstein distance is equivalent to the Bottleneck distance, i.e. d_B is the limit of d_{W^p} as $p \to \infty$.

5.1   Wasserstein Stability

The distance d_{W^p} is stable in a trianguliable compact metric space, which restricts it to Lipschitz continuous functions for stable results. A function $f : X \to Y$ is called Lipschitz continuous, if one distance (X, d_X) is bounded by the other (Y, d_Y) times a constant, i.e. $d_Y(f(x_1) - f(x_2)) \leq c \cdot d_X(x_1 - x_2)$, for all $x_1, x_2 \in X$. For two Lipschitz functions f, g constants b and c exist [13], which depend on X and the Lipschitz constants of f and g, such that the p-th Wasserstein distance between the two functions satisfies

$$d_{W^p}(f, g) \leq c \cdot ||f - g||_{\infty}^{1-b/p}. \tag{20}$$

For small enough perturbations of Lipschitz functions their p-th Wasserstein distance is bounded by a constant. In Fig. 3 the topological development of handwritings through interpolation is visualized. Equally colored lines represent the same user and each line represents a signature. The equally colored lines show very similar behavior and represent the small perturbations, which are caused by the slight change of letter shape when signing multiple times.

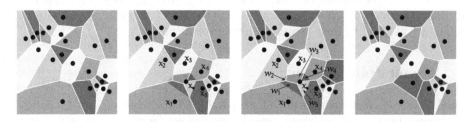

Fig. 2. From *left* to *right*: Clipped tessellation. For other manifolds the curvature should be considered; Tessellation with added point creates a new Voronoi region *stealing* area from the neighboring regions; The determined weights by the fractional amount of occupied area; Tessellation with added point.

6   The Natural Neighbor Algorithm

The algorithm re-weights the coordinates of a new point in the convex hull of a point cloud by the change of Voronoi regions relative to the Voronoi regions without the additional point, i.e. $\hat{x}_\bullet = \sum_{l=1}^{L} \lambda_l x_{\bullet l}$. For a set of points $X \subset \mathbb{R}^d$ distributed over an embedded manifold M natural neighbors behave like a local coordinate system for M with their density increasing [4].

The Voronoi tessellation is dual to the Delaunay triangulation, thus we use the latter for our computation [6, pp. 45–47]. This duality gives a bijection between the faces of one complex and the faces of the other, including incidence and reversibility of operations. Both have the same homotopy type. A Voronoi diagram $\text{dgm}_{\text{Vor}}(X)$ is defined by the union of the Voronoi regions $\text{dgm}_{\text{Vor}}(X) := \bigcup_{x \in X} \text{Vor}(x)$, for all $x \in X$ and assigns a polyhedron to each point, see Fig. 2. This interpolation method generalizes to arbitrary dimensions.

The combinatorial complexity of the Voronoi diagram of n points of $\mathbb{R}^d$ is at most the combinatorial complexity of a polyhedron defined as the intersection of n half-spaces of $\mathbb{R}^{d+1}$. Due to duality the construction of $\text{dgm}_{\text{Vor}}(X)$ takes $\mathcal{O}(n \log n + n^{d/2})$ [6].

6.1 Voronoi Tessellation

The Voronoi cells have no common interior, intersect at their boundaries and cover the entire $\mathbb{R}^d$. The resulting polygons can then be divided into Voronoi edges and vertices. The natural neighbors of a point are defined by the points of the neighboring Voronoi polygons [21].

The natural neighbor is the closest point x to two other points y and z within X. To yield the position of the added point we have to calculate the Voronoi diagram of the original signature $\mathrm{dgm}_{\mathrm{Vor}}(X)$ and one with an added point $\mathrm{dgm}_{\mathrm{Vor}}^{\bullet}(X \cup x_{\bullet}) = \bigcup_{x \in X \cup x_{\bullet}} \mathrm{Vor}(x)$. The latter consists of one Voronoi region more than the primer, see Fig. 2 (a) and (c). This polygon is part of $\mathrm{dgm}_{\mathrm{Vor}}^{\bullet}(X \cup x_{\bullet})$ and contains a certain amount of its 'area'.

The Voronoi regions sum up to one $\sum_{l=1}^{L} \lambda_l = 1$. The Voronoi interpolation is repeated until the topological stopping condition is met, measured through the W^p-distances of the persistence diagrams $\mathcal{P}_K$ and $\mathcal{P}_{K^{\bullet}}$. The weights of the coordinate representation of $x_{\bullet}$ are determined by the quotient of the 'stolen' Voronoi regions and the total 'area' of the Voronoi diagram with the additional point according to Eq. 21 [1].

$$\lambda_l = \begin{cases} \frac{\mathrm{vol}(\mathrm{Vor}(x_l) \cap \mathrm{Vor}^{\bullet}(x_{\bullet}))}{\mathrm{vol}(\mathrm{Vor}^{\bullet}(x_{\bullet}))} & \text{if } x \geq 1, \\ 0 & \text{otherwise.} \end{cases} \qquad (21)$$

But to what extent are homology groups preserved if persistent homology is computed on the Delaunay triangulation?

7 The Simplicial Collapse

The Delaunay triangulation avoids the burden of an additional simplicial structure for persistent homology. We determine how accurate the persistent homology is on this filtration. We use results from simplicial collapse [2], which show the simple-homotopy equivalence of the Čech and Delaunay complex among other related simplicial complexes. Simple-homotopy equivalence is stronger than homotopy equivalence. An elementary simplicial collapse determines a strong deformation retraction up to homotopy. Hence, simple-homotopy equivalence implies homotopy equivalence [11, §2]. Under the conditions of the Hurewicz Theorem we can draw conclusions about the homotopy groups of the data manifold.

The simplicial collapse is established using abstract simplicial complexes denoted by $\tilde{K}$. A family of simplices σ of a non-empty finite subset of a set $\tilde{K}$ is an abstract simplicial complex if for every set σ' in σ and every non-empty subset $\sigma'' \subset \sigma'$ the set σ'' also belongs to σ. We assume σ and σ' are two simplices of $\tilde{K}$, such that $\sigma \subset \sigma'$ and $\dim \sigma < \dim \sigma'$. We call the face σ' free, if it is a maximal face of $\tilde{K}$ and no other maximal face of $\tilde{K}$ contains σ. A similar notion to deformation retraction needs to be defined for the investigation of homology groups. This leads to the simplicial collapse $\searrow$ of $\tilde{K}$, which is the removal of all σ'' simplices, where $\sigma \subseteq \sigma'' \subseteq \sigma'$, with σ being a free face. Now we can define the simple-homotopy type based on the concept of simplicial collapse. Intuitively speaking, two simplicial complexes are 'combinatorial-equivalent', if it is possible to deform one complex into the other with a finite number of 'moves'. Two abstract simplicial complexes $\tilde{K}$ and $\tilde{G}$ are said to have the same simple-homotopy type, if there exists a finite sequence $\tilde{K} = \tilde{K}_0 \searrow \tilde{K}_1 \searrow \cdots \searrow \tilde{K}_n = \tilde{G}$, with each arrow representing a simplicial collapse or expansion (the inverse operation). If X is a finite set of points in general position in $\mathbb{R}^d$, then

$$\check{\mathrm{C}}\mathrm{ech}(X,r) \searrow \mathrm{Del}\check{\mathrm{C}}\mathrm{ech}(X,r) \searrow \mathrm{Del}(X,r) \searrow \mathrm{Wrap}(X,r) \qquad (22)$$

for all $r \in \mathbb{R}$. For the proof we refer to [2]. The connection in Eq. 22 establishes the simple-homotopy equivalence of the Čech- and Delaunay complex. We deduce, that if the underlying space follows the condition of a Hurewicz isomorphism, all four complexes are suitable for calculating persistent homology as a result of the simplicial collapse up to homotopy equivalence.

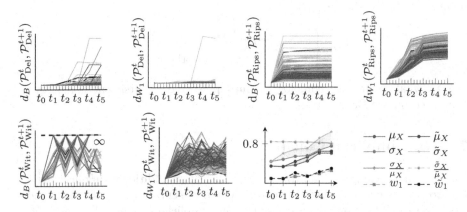

Fig. 3. Bottleneck and L^1-Wasserstein distances between the persistence diagrams in iteration t and $t+1$. The persistent homology has been computed on the Čech complex, Vietoris-Rips complex and the witness complex, respectively. A total of 250 samples from a signature collection are represented [18]. Each line corresponds to a single sample and the lines are colored corresponding to one of six selected users in • gray, • black, • blue, • yellow, • orange and • red. (Color figure online)

8   Numerical Experiments

All source code is written in Python 3.7. The GUDHI [16] library is used for the calculation of simplicial complexes, filtrations and persistent homology. We investigate 83 users, considering 45 signatures per user from the MOBISIG signature database [18], which show the same letters, but are independent writings.

For each user we have a set of 45 persistence diagrams and a set of 45 corresponding handwritings. In every iteration as many new points are added as are already in the respective example of a signature. We inserted the points uniformly within the convex hull of the initial point set, see Fig. 2.

8.1 Experimental Setting

$\text{Rips}(X)$ is expanded up to the third dimension. The *maximum edge length* is set to the average edge length between two points within the data set. We use the same r for $\text{Čech}(X,r)$ and $\text{Rips}(X,r)$, so that $\text{Rips}(X,r)$ differs topologically more from the union of closed balls around each point, but is faster to compute. Finding an optimal radius as distance threshold is considered open [25], thus we use $r = \max_{x,y \in X} ||x - y||$ as empirical heuristic.

The strong $\text{Wit}(X,r)$, embedded into $\mathbb{R}^d$, is recalculated for each sample at each interpolation step. We select uniformly 5% of the points as landmarks. We set $\alpha = 0.01, \gamma = 0.1$ and $p = 1$.

We assume that the persistence diagrams are i.i.d. A free parameter α quantifies a tolerance to topological change, thus a decision must be made on the following hypotheses about the distributions of the persistence diagrams:

(a) $H_0 : \mathcal{P}_K^t$ and $\mathcal{P}_K^{t+1}$ have different underlying distributions and
(b) $H_1 : \mathcal{P}_K^t$ and $\mathcal{P}_K^{t+1}$ have the same underlying distribution.

We use an asymptotic solution for testing by trimmed Wasserstein distance [14]:

$$\hat{\Gamma}_\gamma^p(\mathcal{P}_K^t, \mathcal{P}_K^{t+1}) = \frac{1}{(1 - 2\gamma)} \left(\sum_{j=1}^{m} ||(\mathcal{P}_K^t)_j - \mathcal{P}_K^{t+1})_j||_\infty^p \Delta\gamma \right)^{1/p} . \tag{23}$$

The trimming bound $\alpha \in [0, 1/2)$ results from the integral for the continuous case as a difference in a finite weighted sum. It is computed using the expected value of the persistence diagrams and is exact in the limit $\int_\gamma^{1-\gamma} f(x)dx = \lim_{\Delta\gamma \to 0} \sum_{x \in X} f(x)\Delta\gamma$. The critical region for our hypothesis H_0 against H_1 is

$$\left(\frac{nm}{n + m} \right)^{\frac{1}{p}} \frac{\hat{\Gamma}_\gamma^p - \alpha^p}{\hat{\sigma}_\gamma} \leq z_\gamma, \tag{24}$$

where z_γ denotes the γ-quantile of the standard normal distribution and $n = m$, with n being the number of samples. The initial problem can be rephrased as

(a) $H_0 : \Gamma_\gamma^p(\mathcal{P}_K^t, \mathcal{P}_K^{t+1}) > \alpha$ and
(b) $H_1 : \Gamma_\gamma^p(\mathcal{P}_K^t, \mathcal{P}_K^{t+1}) \leq \alpha$.

8.2   Evaluation

In Fig. 3, seventh diagram, elementary statistics are computed for the entire data set such as mean μ_X, standard deviation σ_X, variation $\frac{\sigma_X}{\mu_X}$ and $d_{W_1} = d_{W_1}(X_{\mathrm{org}}, X^t)$. The statistics are also computed for the interpolated data with topological stop, respectively, marked with $\sim$.

We achieved an improvement for each measured statistic at each iteration step using topological stop. In Fig. 3 the topological similarity between the individual users are made visual. $\mathrm{Rips}(X, r)$ and $\mathrm{Del}(X, r)$ seem suitable to estimate the homology groups, whereas $\mathrm{Wit}(L, W)$ produced far less stable results, due to the small selection of landmarks.

9   Conclusions

We have discussed the connection of Voronoi diagrams to the Delaunay complex and its connection to other complexes, which should serve as a basis to explore related algorithms to the Voronoi interpolation. We investigated into metrics to measure differences in persistent homology and could visualize the changing homology groups of the users signatures during interpolation, see Fig. 3. Our result is a stopping-criterion with a hypothesis test to determine whether the persistent homology of an interpolated signature still originates from the same distribution as the source. Our measurements show an improvement of statistics compared to vanilla Voronoi interpolation. We demonstrated, that – under mild conditions – the Delaunay complex, Čech-Delaunay complex and Wrap complex can also be used for filtration up to homotopy equivalence. Following open research questions arose during our investigations:

- The intrinsic geometry of the data points is often not the Euclidean one. On the other hand side the frequently used embedding of the Voronoi tessellation is. This causes unwanted artifacts. Is there a geometrically meaningful clipping for general metric spaces, for example using geodesics in a smooth manifold setting? In which manifold should $\mathrm{Del}(X)$ be embedded?
- To our knowledge there is no evidence known that the Voronoi tessellation obtains the homology groups. According to [20], the Voronoi tessellation is stable. However, the experiments show that for increasing iterations additional homology groups appear. Does the Voronoi tessellation preserve homology groups and homotopy groups in general metric spaces?

Acknowledgements. We would like to thank David Haller and Lekshmi Beena Gopalakrishnan Nair for proofreading. Further, we thank Anton Rechenauer, Jan Frahm and Justin Noel for suggestions and ideas on the elaboration of the final version of this work. We express our gratitude to Demian Vöhringer and Melanie Sigl for pointing out related work. This work was partially supported by Siemens Gas and Power GmbH & Co. KG.

Code and Data. The implementation of the methods, the data sets and experimental results can be found at: https://github.com/karhunenloeve/SIML.

References

1. Anton, F., Mioc, D., Fournier, A.: Reconstructing 2D images with natural neighbour interpolation. Vis. Comput. **17**(3), 134–146 (2001). https://doi.org/10.1007/PL00013404
2. Bauer, U., Edelsbrunner, H.: The Morse theory of Čech and Delaunay complexes. Trans. Am. Math. Soc. **369**(5), 3741–3762 (2017). https://doi.org/10.1090/tran/6991
3. Bobach, T.A.: Natural neighbor interpolation - critical assessment and new contributions. Doctoral thesis, TU Kaiserslautern (2009)
4. Boissonnat, J., Cazals, F.: Natural neighbor coordinates of points on a surface. Comput. Geom. **19**(2–3), 155–173 (2001). https://doi.org/10.1016/S0925-7721(01)00018-9
5. Boissonnat, J.D., Chazal, F., Yvinec, M.: Geometric and Topological Inference, vol. 57. Cambridge University Press, Cambridge (2018). https://doi.org/10.1017/9781108297806.014
6. Boissonnat, J.D., Dyer, R., Ghosh, A.: Delaunay triangulation of manifolds. Found. Comput. Math. **18**(2), 399–431 (2017). https://doi.org/10.1007/s10208-017-9344-1
7. Borsuk, K.: On the imbedding of systems of compacta in simplicial complexes. Fundamenta Mathematicae **35**, 217–234 (1948). https://doi.org/10.4064/fm-35-1-217-234
8. Carrière, M., Cuturi, M., Oudot, S.: Sliced Wasserstein kernel for persistence diagrams. In: Proceedings of the 34th International Conference on Machine Learning, pp. 664–673 (2017). https://doi.org/10.5555/3305381.3305450
9. Cerri, A., Landi, C.: Hausdorff stability of persistence spaces. Found. Comput. Math. **16**(2), 343–367 (2015). https://doi.org/10.1007/s10208-015-9244-1
10. Chazal, F., Glisse, M., Labruère, C., Michel, B.: Convergence rates for persistence diagram estimation in topological data analysis. J. Mach. Learn. Res. **16**, 3603–3635 (2015). https://doi.org/10.5555/3044805.3044825
11. Cohen, M.: A Course in Simple-Homotopy Theory, vol. 10. Springer, New York (2012). https://doi.org/10.1007/978-1-4684-9372-6
12. Cohen-Steiner, D., Edelsbrunner, H., Harer, J.: Stability of persistence diagrams. Discrete Comput. Geom. **37**(1), 103–120 (2007). https://doi.org/10.1007/s00454-006-1276-5
13. Cohen-Steiner, D., Edelsbrunner, H., Harer, J., Mileyko, Y.: Lipschitz functions have L_p-stable persistence. Found. Comput. Math. **10**(2), 127–139 (2010). https://doi.org/10.1007/s10208-010-9060-6
14. Czado, C., Munk, A.: Assessing the similarity of distributions-finite sample performance of the empirical mallows distance. J. Stat. Comput. Simul. **60**(4), 319–346 (1998). https://doi.org/10.1080/00949659808811895
15. Edelsbrunner, H., Harer, J.: Persistent homology - a survey. Contemp. Math. **453**, 257–282 (2008). https://doi.org/10.1090/conm/453/08802
16. Ghudi user and reference manual (2015). http://gudhi.gforge.inria.fr
17. Hatcher, A.: Algebraic Topology. Cambridge University Press, Cambridge (2005)
18. The MOBISIG online signature database (2019). https://ms.sapientia.ro
19. Parzanchevski, O., Rosenthal, R.: Simplicial complexes: spectrum, homology and random walks. Random Struct. Algorithms **50**(2), 225–261 (2017). https://doi.org/10.1002/rsa.20657

20. Reem, D.: The geometric stability of Voronoi diagrams with respect to small changes of the sites. In: Proceedings of the 27th Annual Symposium on Computational Geometry, pp. 254–263 (2011). https://doi.org/10.1145/1998196.1998234

21. Sibson, R.: A brief description of natural neighbor interpolation. In: Barnett, V. (ed.) Interpreting Multivariate Data, pp. 21–36. Wiley, New York (1981)

22. Silva, V.D., Carlsson, G.: Topological estimation using witness complexes. In: Symposium on Point-Based Graphics, vol. 4, pp. 157–166 (2004). https://doi.org/10.2312/SPBG/SPBG04/157-166

23. Villani, C.: Optimal Transport: Old and New, vol. 338. Springer, Heidelberg (2008). https://doi.org/10.1007/978-3-540-71050-9

24. Zomorodian, A.: Fast construction of the Vietoris-Rips complex. Comput. Graph. **34**(3), 263–271 (2010). https://doi.org/10.1016/j.cag.2010.03.007

25. Zomorodian, A., Carlsson, G.: Computing persistent homology. Discrete Comput. Geom. **33**(2), 249–274 (2004). https://doi.org/10.1007/s00454-004-1146-y

Atomic Super-Resolution Tomography

Poulami Somanya Ganguly$^{1,2(\boxtimes)}$ (iD), Felix Lucka^{1,3} (iD), Hermen Jan Hupkes^{2} (iD), and Kees Joost Batenburg^{1,2} (iD)

^{1} Centrum Wiskunde & Informatica, Science Park, Amsterdam, The Netherlands
{poulami.ganguly,felix.lucka,joost.batenburg}@cwi.nl
^{2} The Mathematical Institute, Leiden University, Leiden, The Netherlands
hhupkes@math.leidenuniv.nl
^{3} Centre for Medical Image Computing, University College London, London, UK

Abstract. We consider the problem of reconstructing a nanocrystal at atomic resolution from electron microscopy images taken at a few tilt angles. A popular reconstruction approach called discrete tomography confines the atom locations to a coarse spatial grid, which is inspired by the physical *a priori* knowledge that atoms in a crystalline solid tend to form regular lattices. Although this constraint has proven to be powerful for solving this very under-determined inverse problem in many cases, its key limitation is that, in practice, defects may occur that cause atoms to deviate from regular lattice positions. Here we propose a grid-free discrete tomography algorithm that allows for continuous deviations of the atom locations similar to super-resolution approaches for microscopy. The new formulation allows us to define atomic interaction potentials explicitly, which results in a both meaningful and powerful incorporation of the available physical *a priori* knowledge about the crystal's properties. In computational experiments, we compare the proposed grid-free method to established grid-based approaches and show that our approach can indeed recover the atom positions more accurately for common lattice defects.

Keywords: Electron tomography · Discrete tomography · Mathematical super-resolution · Molecular dynamics · Crystallographic defects

1 Introduction

Electron tomography is a powerful technique for resolving the interior of nano-materials. After preparing a microscopic sample, a series of projection images (so called tilt-series) is acquired by rotating the specimen in the electron microscope, acquiring data from a range of angles. In recent years, electron tomography has been successfully applied to reconstruct the 3D positions of the individual atoms in nanocrystalline materials [14,29,31].

Since the first demonstration of atomic resolution tomography of nanocrystals in 2010 by discrete tomography [30], a range of tomographic acquisition techniques and reconstruction algorithms have been applied to reconstruct nanocrystals of increasing complexity. In the discrete tomography approach, atoms are

© Springer Nature Switzerland AG 2020
T. Lukić et al. (Eds.): IWCIA 2020, LNCS 12148, pp. 45–61, 2020.
https://doi.org/10.1007/978-3-030-51002-2_4

assumed to lie on a regular lattice and the measured projections can be considered as atom counts along lattice lines. A key advantage of this approach is its ability to exploit the constraints induced by the discrete domain and range of the image. As a consequence, a small number of projection angles (typically less than 5) can already lead to an accurate reconstruction [7,8]. The theoretical properties of the discrete reconstruction problem have been studied extensively with results on algorithm complexity, uniqueness, and stability [4,6,19]. A key drawback of the discrete lattice assumption when considering real-world applications to nanocrystal data is that in many interesting cases the atoms do not lie on a perfect lattice due to defects in the crystal structure or interfaces between different crystal lattices. In such cases the atoms do not project perfectly into columns, forming a mismatch with the discrete tomography model.

As an alternative, it has been demonstrated that a more conventional tomographic series consisting of hundreds of projections of a nanocrystal can be acquired in certain cases. An image of the nanocrystal is then reconstructed using sparsity based reconstruction techniques on a continuous model of the tomography problem. This approach does not depend on the lattice structure and allows one to reconstruct defects and interfaces [21]. As a downside, the number of required projections is large and to accurately model the atom positions the reconstruction must be represented on a high-resolution pixel grid resulting in a large-scale computational problem. This raises the question if a reconstruction problem can be defined that fills the gap between these two extremes and can exploit the discrete nature of the lattice structure while at the same time allowing for continuous deviations of atom positions from the perfect lattice.

In this paper we propose a model for the atomic resolution tomography problem that combines these two characteristics. Inspired by the algorithm proposed in [11], our model is based on representing the crystal image as a superposition of delta functions with continuous coordinates and exploiting sparsity of the image to reduce the number of required projections. We show that by incorporating a physical model for the potential energy of the atomic configuration, the reconstruction results can be further improved.

2   Problem Setting

In this section we formulate a mathematical model of the atomic resolution tomography problem and discuss several approaches to solve it. Some of these approaches assume that the atom locations are restricted to a perfect grid, the *crystal lattice*, which corresponds to only one possible local minimum of the potential energy of the atomic configuration. To overcome certain limitations of this assumption, we propose an alternative formulation where the atom locations are allowed to vary continuously and an explicit model of the potential energy of their configuration is used to regularize the image reconstruction.

An atomic configuration is characterized by a positive measure μ on a bounded subset X of $\mathbb{R}^d$. We denote the space of such measures by $\mathcal{M}(X)$. The measure represents the *electron density*, which is the probability that an electron is present at a given location. The electron density around an atomic configuration is highest in regions where atoms are present. In electron tomography, electron density is probed by irradiating a sample with a beam of electrons. The beam undergoes absorption and scattering due to its interactions with the electrons of the atomic configuration. The transmitted or scattered signal can then be used to form an image. The Radon transform provides a simplified mathematical model of this ray-based image formation process. For $d = 2$, the Radon transform $\mathcal{R}\mu$ can be expressed as integrals taken over straight rays

$$\mathcal{R}[\mu](r, \theta) := \int_{l(r,\theta)} d\mu, \tag{1}$$

$$l(r, \theta) = \{(x_1, x_2) \in \mathbb{R}^2 \mid x_1 \cos \theta + x_2 \sin \theta = r\}, \tag{2}$$

where we parametrized the rays by the projection angle θ and the distance on the detector r. The corresponding inverse problem is to recover μ from noisy observations of $y = \mathcal{R}\mu + \varepsilon$. One way to formulate a solution to this problem is via the following optimisation over the space of measures:

$$\underset{\mu \in \mathcal{M}(X)}{\text{minimize}} \quad \|\mathcal{R}\mu - y\|_2^2, \tag{3}$$

which is an infinite dimensional non-negative linear least-squares problem. In the following, we will introduce a series of discretisations of this optimisation problem. Numerical schemes to solve them will be discussed in Sect. 3.

In situations where we only have access to data from a few projection angles, introducing a suitable discretisation of (3) is essential for obtaining a stable reconstruction. One way to achieve this is to restrict the atom locations to a spatial grid with n nodes, $\boldsymbol{x}_{i=1}^n$, and model their interaction zone with the electron beam by a Gaussian with known shape G. The atom centres are then delta peaks $\delta_{\boldsymbol{x}_i}$ on the gridded image domain. The Gaussian convolution of atom centres can be viewed as the "blurring" produced by thermal motion of atoms. In fact, it is known from lattice vibration theory that, for large configurations, the probability density function of an atom around its equilibrium position is a Gaussian, whose width depends on temperature, dimensionality and interatomic forces [24]. The discretized measure μ can then be written as

$$\mu_{\text{grid}} = \sum_{i=1}^{n} w_i (G * \delta_{\boldsymbol{x}_i}), \tag{4}$$

where n is the total number of grid points and weights $w_i \geq 0$ were introduced to indicate confidence in the presence or absence of an atom at grid location i. If we insert (4) in (3) and introduce the forward projection of a single atom as $\psi_i := \mathcal{R}(G * \delta_{\boldsymbol{x}_i})$ we get

$$\|\mathcal{R}\mu_{\text{grid}} - y\|_2^2 = \|\mathcal{R} \sum_{i=1}^{n} w_i (G * \delta_{\boldsymbol{x}_i}) - y\|_2^2$$

$$= \|\sum_{i=1}^{n} w_i \mathcal{R}(G * \delta_{\boldsymbol{x}_i}) - y\|_2^2 =: \|\sum_{i=1}^{n} \psi_i w_i - y\|_2^2 =: \|\Psi w - y\|_2^2$$

The corresponding optimisation problem is given by

$$\underset{w \in \mathbb{R}^n_+}{\text{minimize}} \quad \|\Psi w - y\|_2^2, \tag{5}$$

which is a finite dimensional linear non-negative least squares problem.

The choice of the computational grid in (4) is unfortunately not trivial. Only in certain situations, one can assume that all atoms lie on a lattice of known grid size and orientation, and directly match this lattice with the computational grid. In general, one needs to pick a computational grid of much smaller grid size. With the data y given, the grid admits multiple solutions of (5) and most efficient computational schemes tend to pick a blurred, artefact-ridden solution with many non-zero weights far from the true, underlying μ, as we will demonstrate in Sect. 4. To obtain a better reconstruction, one can choose to add *sparsity constraints* which embed our physical *a priori* knowledge that μ originates from a discrete configuration of atoms. In our model (4), this corresponds to a $w \in \mathbb{R}^n_+$ with few non-zeros entries. To obtain such a sparse solution we can add a constraint on ℓ^0 norm of the weights to the optimisation problem:

$$\underset{w \in \mathbb{R}^n_+}{\text{minimize}} \quad \|\Psi w - y\|_2^2$$
$$\text{subject to} \quad |w|_0 \leq K.$$

However, this problem is NP-hard [16]. An approximate solution can be found by replacing the ℓ^0 norm with the ℓ^1 norm and adding it to the objective function:

$$\underset{w \in \mathbb{R}^n_+}{\text{minimize}} \quad \|\Psi w - y\|_2^2 + \lambda \|w\|_1, \tag{6}$$

where λ is the relative weight of the sparsity-inducing term. This particular choice of formulation is not always best and alternative formulations of the same problem exist [16].

For atomic configurations where only one type of atom is present, the weights can be considered to be one where an atom is present and zero everywhere else. This corresponds to discretising the range of the reconstructed image. The fully discrete optimisation problem then becomes:

$$\underset{w \in \{0,1\}^n}{\text{minimize}} \quad \|\Psi w - y\|_2^2. \tag{7}$$

With image range discretisation, a constraint on the number of atoms is typically no longer needed because adding an additional atom with weight 1 after all atoms have been found leads to an increase in the objective function.

Although the optimisation problems (5), (6) and (7) allow for the recovery of atomic configurations without solving (3), all of them rely crucially on discretisation of the domain of the reconstructed image, i.e. the assumption that atoms lie on a grid. However, this assumption is not always true. In particular, atomic configurations often contain defects where atom positions deviate from the perfect lattice. Figure 1 shows examples of common lattice defects.

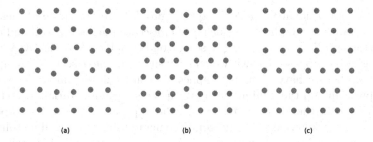

Fig. 1. Atomic configurations with (a) an interstitial point defect, (b) a vacancy and (c) an edge dislocation.

In order to resolve these defects correctly, the image domain must be discretized to higher resolutions, i.e. the grid of possible atom positions must be made finer. This introduces two main problems: First, making the grid finer for the same data makes the inverse problem more ill-posed. Second, the computational time increases significantly even for modestly sized configurations.

In order to overcome these difficulties, we revisit (4) and remove the requirement for x_i to lie on a grid. The projection of a single atom now becomes a function of its location $x \in \mathbb{R}^d$, $\psi(x) := \mathcal{R}(G * \delta_x)$. We keep the image range discretisation introduced above by requiring $w_i \in \{0, 1\}$. Now, (7) becomes

$$\underset{x \in X^n, w \in \{0,1\}^n}{\text{minimize}} \left\| \sum_{i=1}^{n} w_i \psi(x_i) - y \right\|_2^2. \tag{8}$$

The minimisation over x is a non-linear, non-convex least-squares problem which has been studied extensively in the context of mathematical super-resolution [11–13]. In these works, efficient algorithms are derived from relating it back to

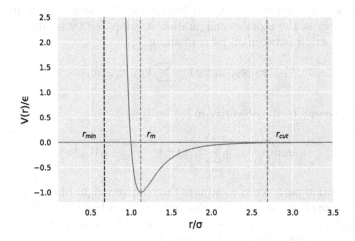

Fig. 2. The normalized Lennard-Jones pair potential as a function of normalized inter-atomic separation.

the infinite dimensional linear least-squares problem on the space of measures (3). For instance, for applications such as fluorescence microscopy [11] and ultrasound imaging [3], an alternating descent conditional gradient (ADCG) algorithm has been proposed, which we will revisit in the next section. Compared to these works, we have a more complicated non-local and under-determined inverse problem and the minimisation over w adds a combinatorial, discrete flavor to (8). To further tailor it to our specific application, we will incorporate physical *a priori* knowledge about atomic configurations of crystalline solids by adding a functional formed by the atomic interaction potentials. This will act as a regularisation of the underlying under-determined inverse problem.

2.1   Potential Energy of the Atomic Configuration

The total energy of an atomic configuration is the sum of its potential energy and kinetic energy. As we consider only static configurations, the kinetic energy of the configuration is zero and the total energy is equal to the potential energy. In order to compute the potential energy of the atomic configuration, we must prescribe the interaction between atoms. In this paper, we use the Lennard-Jones pair potential, which is a simplified model of interatomic interactions. The Lennard-Jones potential V_{LJ} as a function of interatomic separation r is given by [18]

$$V_{LJ}(r) = \begin{cases} 4\epsilon \left[\left(\dfrac{\sigma}{r} \right)^{12} - \left(\dfrac{\sigma}{r} \right)^6 \right], & r < r_{cut} \\ 0, & r \geq r_{cut} \end{cases} \tag{9}$$

where ϵ is the depth of the potential well and σ is the interatomic separation at which the potential is zero. The separation at which the potential reaches its minimum is given by $r_m = 2^{1/6}\sigma$. The parameter r_{cut} denotes a cut-off separation beyond which the potential is inactive. Figure 2 shows the form of the Lennard-Jones pair potential as a function of interatomic separation. The potential energy of the atomic configuration is computed by summing over the pairwise interaction between all pairs of atoms

$$V_{tot}(\boldsymbol{x}_1, \boldsymbol{x}_2, ..., \boldsymbol{x}_N) = \sum_{i>j} V_{LJ}(\boldsymbol{x}_i - \boldsymbol{x}_j). \tag{10}$$

Adding this energy to the objective in (8) leads to

$$\underset{\boldsymbol{x} \in \mathcal{C}, w \in \{0,1\}^n}{\text{minimize}} \quad \left\| \sum_{i=1}^n w_i \psi(\boldsymbol{x}_i) - y \right\|_2^2 + \alpha V_{tot}(\boldsymbol{x}). \tag{11}$$

The regularisation parameter, α, adjusts the relative weight of the energy term, so that by tuning it we are able to move between atomic configurations that are data-optimal and those that are energy-optimal. The constraint set $\mathcal{C} \subset X^n$ is defined by a minimum distance r_{min}, such that $|\boldsymbol{x}_i - \boldsymbol{x}_j| > r_{min}, \forall i > j$. The minimum distance, r_{min}, is chosen to be smaller than the optimal interatomic separation r_m and allows us to set α to 0 and still avoid configurations where atoms are placed exactly at the same position. For small separations, the energy is dominated by the $\left(\frac{\sigma}{r}\right)^{12}$ term and increases sharply for separations less than r_m. Thus, for non-zero α, configurations where atoms have a separation less than r_m are highly unlikely.

3 Algorithms

In this section we discuss several algorithms to solve the optimisation problems introduced in Sect. 2.

3.1 Projected Gradient Descent

The non-negative least-squares problem (5) can be solved with a simple iterative first-order optimisation scheme. At each step of the algorithm, the next iterate is computed by moving in the direction of the negative gradient of the objective function. Non-negativity of the weights is enforced by projecting negative iterates to zero. Mathematically, each iterate is given by

$$w^{k+1} = \prod_{+} \left(w^k + t\Psi^T(\Psi w^k - y) \right), \tag{12}$$

where t is the step size and the projection operator is given by

$$\prod_{+}(\cdot) = \max(\cdot, 0). \tag{13}$$

In the numerical experiments in Sect. 4, we used the SIRT algorithm [23] as implemented in the tomographic reconstruction library ASTRA [1], which is based on a minor modification of the iteration described above.

3.2 Proximal Gradient Descent

If we add the non-smooth ℓ^1 regularizer and obtain problem (6), we need to extend (12) to a proximal gradient scheme [27]

$$w^{k+1} = \text{prox}_h\left(w^k + t\Psi^T(\Psi w^k - y) \right), \tag{14}$$

where the projection operator (13) is replaced by the proximal operator of the convex function

$$h(x) := \begin{cases} \lambda\|x\|_1 & x \geq 0 \\ 0 & \text{elsewhere} \end{cases}, \tag{15}$$

which is given by the non-negative soft-thresholding operator

$$\text{prox}_h(x) = \begin{cases} x - \lambda, & x \geq \lambda \\ 0, & \text{elsewhere} \end{cases}.$$

In the numerical experiments in Sect. 4, we used the fast iterative soft-thresholding algorithm (FISTA) [9] as implemented in the Python library ODL [2], which is based on a slight modification of the iteration described above.

Algorithm 1. Discrete simulated annealing

1: **while** $\beta < \beta_{\max}$ **do**
2:     Select new atom location: $\tilde{w}^k \in \arg\min_{k \in \mathcal{C}} \Psi w^k - y$
3:     Add new atom to current configuration: $\tilde{w}^{k+1} \leftarrow \{w^k, \tilde{w}^k\}$
4:     Accept new configuration with a certain probability:
5:     **if** $\beta \|\Psi \tilde{w}^{k+1} - y\|_2^2 < \beta \|\Psi w^k - y\|_2^2$ **then**
6:         $w^{k+1} \leftarrow \tilde{w}^{k+1}$
7:     **else**
8:         Generate random number: $t \in \text{rand}[0, 1)$
9:         **if** $t < e^{-\beta \|\Psi \tilde{w}^{k+1} - y\|_2^2} / e^{-\beta \|\Psi w^k - y\|_2^2}$ **then**
10:             $w^{k+1} \leftarrow \tilde{w}^{k+1}$
11:         **end if**
12:     **end if**
13:     Move atom: $w^{k+1} \leftarrow \text{random move}(w^{k+1})$
14:     Run acceptance steps 5–13
15:     Increase β
16: **end while**

3.3   Simulated Annealing

For solving the fully discrete problem (7), we used a simulated annealing algorithm as shown in Algorithm 1, which consists of two subsequent accept-reject steps carried out with respect to the same inverse temperature parameter β. In the first one, the algorithm tries to add a new atom to the existing configuration. In the second one, the atom locations are perturbed locally. As β is increased towards $\beta_{\max}$, fewer new configurations are accepted and the algorithm converges to a minimum.

In the atom adding step at each iteration k, the algorithm tries to add an atom at one of the grid location i where the residual $\Psi w^k - y$ is minimal (this corresponds to flipping w_i^k from 0 to 1 in (7)). The allowed grid locations belong to a constraint set $\mathcal{C}$, such that no two atoms are closer than a pre-specified minimum distance $r_{\min}$. To perturb the atom positions locally, the algorithm selects an atom at random and moves it to one of its 4 nearest neighbor locations at random.

3.4   ADCG with Energy

Variants of the Frank-Wolfe algorithm (or conditional gradient method) [17,22] have been proposed for solving problems of the form (8) [3,15] without discrete constraints for w and are commonly known as alternating descent conditional gradient (ADCG) schemes (see [28] for an analysis specific to multidimensional sparse inverse problems). Here, we modify the ADCG scheme to

1. incorporate binary constraints on w
2. handle the singularities of the atomic interaction potentials
3. avoid local minima resulting from poor initialisations

The complete algorithm is shown in Algorithm 2. Essentially, the scheme also alternates between adding a new atom to the current configuration and optimising the positions of the atoms.

In the first step, the image domain is coarsely gridded and the objective function after adding an atom at each location is computed. Locations closer to existing atoms than r_{min} are excluded. In the second step, the atom coordinates are optimized by a continuous local optimisation method. Here, the Nelder-Mead method [25] implemented in SciPy [32] was used.

Algorithm 2. ADCG with energy

1: **for** $k = 1 : k_{max}$ **do**
2:     Compute next atom in grid g:
      $x_{new} \in \arg\min_{x_{new} \in g, (x^k, x_{new}) \in \mathcal{C}} \| \sum_{i=1}^{k} \psi(x_i) - y + \psi(x_{new}) \| + \alpha V_{tot}(x^k, x_{new})$
3:     Update support: $x^{k+1} \leftarrow \{x^k, x_{new}\}$
4:     Locally move atoms:
      $x^{k+1} \leftarrow \min_{x \in X} \|\Psi \mu(x^{k+1}) - y\|_2^2 + \alpha V(x^{k+1})$
5:     Break if objective function is increasing:
6:     **if** $\|\Psi \mu(x^{k+1}) - y\|_2^2 + \alpha V(x^{k+1}) > \|\Psi \mu(x^k) - y\|_2^2 + \alpha V(x^k)$ **then** break
7:     **end if**
8: **end for**

A continuation strategy is used to avoid problems resulting from poor initilisations: Algorithm 2 is run for increasing values of α, starting from $\alpha = 0$. The reconstruction obtained at the end of a run is used as initialisation for the next. In the following section, we demonstrate the effect of increasing α on the reconstructions obtained and discuss how an optimal α was selected. In the following section, we refer to Algorithm 2 as "ADCG" when used for $\alpha = 0$ and as "ADCG with energy" otherwise.

4 Numerical Experiments

We conducted numerical experiments by creating 2D atomic configurations with defects and using the algorithms discussed in Sect. 3 to resolve atom positions. In this section we describe how the ground truth configurations were generated and projected, and compare the reconstruction results of different algorithms.

4.1 Ground Truth Configurations

We generated ground truth configurations using the molecular dynamics software HOOMD-blue [5,20]. We created perfect square lattices and then induced defects by adding or removing atoms. The resulting configuration was then relaxed to an energy minimum using the FIRE energy minimizer [10] to give the configurations shown in Fig. 1. The following parameter values were used in (9) for specifying the Lennard-Jones pair potential between atoms.

Defect type	ϵ	σ	r_{cut}
Interstitial defect	0.4	0.15	0.4
Vacancy	0.4	0.14	0.4
Edge dislocation	0.4	0.13	0.17

4.2   Discretized Projection Data

We generated two 1D projections for each ground truth atomic configuration at projection angles $\theta = 0°, 90°$. As discussed in Sect. 2, the projection of a single atom centre is given by a Gaussian convolution followed by the Radon transform. The Radon transform of a Gaussian is also a Gaussian. Therefore, we interchanged the two operations in the forward transform to speed up the computations. The sum over individual projections of atom centres was used as the total (noise-free) projection. Using the Radon transform in (1), each atom centre was projected onto a 1D detector, following which it was convolved with a 1D Gaussian of the form $G(z) = e^{-(z-z_0)/\varsigma^2}$, where z_0 is the position of the atom centre on the detector and ς controls the width of the Gaussian. Finally, the continuous projection was sampled at a fixed number of points to give rise to a *discrete* projection. For our experiments, the ς of the Gaussian function was taken to be equal to the discretisation of the detector given by the detector pixel size d. Both were taken to be 0.01.

4.3   Discretisation of Reconstruction Volume

For SIRT, FISTA and simulated annealing (described in Subsects. 3.1, 3.2 and 3.3, respectively), each dimension of the reconstruction area was discretized using the detector pixel size d. Therefore, there were $1/d \times 1/d$ grid points in total.

Gridding is required for our variant of ADCG (Subsect. 3.4, Algorithm 2) at the atom adding step. We found that a coarse discretisation, with less than $1/9^{\text{th}}$ the number of grid points, was already sufficient.

4.4   Comparison Between Reconstructions

The reconstructions obtained with the different algorithms are shown in Fig. 3. For each reconstruction, data from only two projections were used. Note that two projections is far from sufficient for determining the correct atomic configuration and several different configurations have the same data discrepancy.

In the SIRT reconstructions, atom positions were blurred out and none of the defects were resolved. In all cases, the number of intensity peaks was also different from the true number of atoms. Although FISTA reconstructions, which include sparsity constraints on the weights, were less blurry, atoms still occupied more than one pixel. For both these algorithms, additional heuristic post-processing is required to output atom locations. In the edge dislocation case, both algorithms gave rise to a configuration with many more atoms than were present in the ground truth.

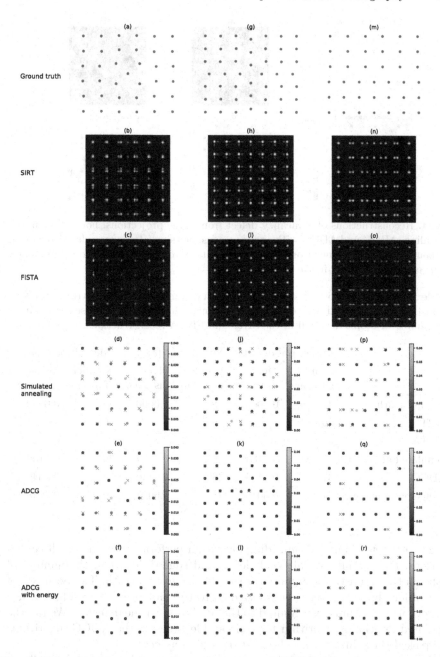

Fig. 3. Reconstructions of atomic configurations with (a)–(f) an interstitial point defect, (g)–(l) a vacancy and (m)–(r) an edge dislocation from two projections. For the simulated annealing, ADCG and ADCG with energy reconstructions, atoms are colored according to their Euclidean distance from the ground truth. The ground truth positions are marked with red crosses. In (j)–(l) an extra atom (shown in red) was present in the reconstructions but not in the ground truth. (Color figure online)

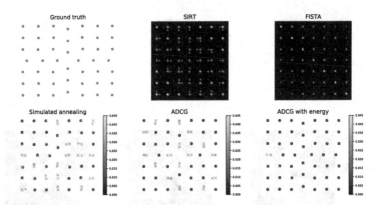

Fig. 4. Reconstructions of a vacancy defect from three projections. For the simulated annealing, ADCG and ADCG with energy reconstructions, atoms are colored according to their Euclidean distance from the ground truth. Ground truth positions are marked with red crosses. (Color figure online)

Table 1. Number of atoms and mean Euclidean distance from ground truth atoms for reconstructions obtained with different algorithms. Thresholding was used to compute the number of atoms detected in the SIRT and FISTA reconstructions.

	Interstitial defect		Vacancy (3 projs.)		Edge dislocation	
	Number of atoms	Mean distance	Number of atoms	Mean distance	Number of atoms	Mean distance
Ground truth	37	0.0000	48	0.0000	39	0.0000
SIRT	36	–	49	–	66	–
FISTA	36	–	49	–	66	–
Simulated annealing	37	0.0184	48	0.0164	39	0.0159
ADCG	37	0.0138	48	0.0130	39	0.0049
ADCG with energy	37	0.0018	48	0.0024	39	0.0048

The discrete simulated annealing algorithm performed better for all configurations. For the interstitial point defect and edge dislocation, the number of atoms in the reconstruction matched that in the ground truth. The positions of most atoms, however, were not resolved correctly. Moreover, the resolution, like in previous algorithms, was limited to the resolution on the detector. We ran the simulated annealing algorithm for comparable times as the ADCG algorithms and picked the solution with the least data discrepancy.

Already the ADCG algorithm for $\alpha = 0$ performed far better than all the previous algorithms. For the configurations with an interstitial point defect and edge dislocation, all but a few atom locations were identified correctly. For the configuration with a vacancy, all atoms were correctly placed. However, an additional atom at the centre of the configuration was placed incorrectly.

Adding the potential energy (ADCG with energy) helps to resolve atom positions that were not identified with $\alpha = 0$. For the interstitial point defect and edge dislocation, these reconstructions were the closest to the ground truth. Adding the energy to the configuration with a vacancy moved the atoms near the defect further apart but was not able to correct for the extra atom placed. For this case, we performed an additional experiment with three projections at $0°$, $45°$ and $90°$. These results are shown in Fig. 4. Taking projections at different angles (e.g. $0°$, $22.5°$ and $90°$) did not improve results. The defect was still not resolved in the SIRT and FISTA reconstructions. However, the number of atoms in the simulated annealing, ADCG and ADCG with energy reconstructions was correct. Once again, the reconstruction obtained with our algorithm was closer to the ground truth than all other reconstructions, with all but one atom placed correctly. Reconstructions with 3 projections for the interstitial point defect and edge dislocation were not significantly different from those with 2 projections. In Table 1, we report the number of atoms detected and (where applicable) the mean Euclidean distance of atoms from the ground truth. Note that for computing the mean distance, we required that the number of atoms detected in the reconstruction matched that in the ground truth. Thresholding with a pre-defined minimum distance between peaks was used to detect atoms in the SIRT and FISTA reconstructions.

4.5   Effect of Adding Energy to Optimisation

To resolve atom positions using Algorithm 2, the contribution of the potential energy was increased gradually by increasing α. In Fig. 5, we show the effect of adding energy to the optimisation problem. For $\alpha = 0$, an initial guess for the true configuration was obtained. This configuration, though data optimal, was not the ground truth. A quantitative measure of this mismatch is the Euclidean distance between the reconstructed atom locations and those in the ground truth. As α was increased, the reconstructions evolved from being data-optimal to being energy-optimal. At a certain value of α, the Euclidean distance between reconstructed and ground truth atom locations decreased to zero. Increasing α beyond this point led to a large increase in the data discrepancy term due to the addition of more atoms. For very high values of α, the configurations obtained were essentially global minima of the potential energy, such as the honeycomb configuration in Fig. 5(e) for $\alpha = 100.0$. An optimal value of the regularisation parameter was selected by increasing α to the point at which more atoms were added to the configuration and a jump in the data discrepancy was observed.

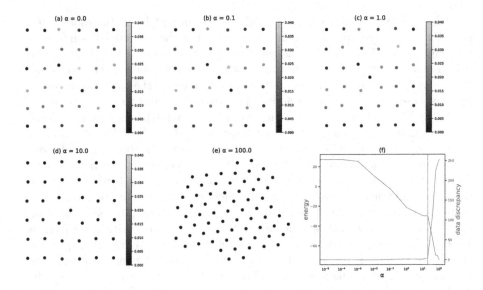

Fig. 5. (a)–(d): Increasing the weighting of the energy term from $\alpha = 0.0$ to $\alpha = 10.0$ helps to resolve the correct atomic configuration. The reconstructed atoms are colored according to their Euclidean distance from the atoms in the ground truth. (e) At high values of α, the reconstructions have a high data discrepancy and correspond to one of the global minima of the potential energy. (f) From the plots of potential energy and data discrepancy, an optimal value of α (indicated by the grey line) is selected. Increasing α beyond this optimal value leads to a large increase in the data discrepancy due to addition of more atoms.

5   Discussion

The results of our numerical experiments demonstrate that algorithms like ADCG, which do not rely on domain discretisation, are better at resolving the defects in the atomic configurations shown in Fig. 1. Moreover, the output from ADCG is a *list of coordinates* and not an image like that of SIRT or FISTA, which requires further post-processing steps to derive the atom locations. Direct access to coordinates can be particularly useful because further analysis, such as strain calculations, often require atom positions as input.

Adding the potential energy of the atomic configuration to the optimisation problem resulted in reconstructions that were closer to the ground truth. One challenge of the proposed approach (with or without adding the potential energy) is that the resulting optimisation problem is a non-convex function of the atom locations. The numerical methods we presented are not intentionally designed to escape local minima and are therefore sensitive to their initialisation. To improve this, one important extension would be to also remove atoms from the current configuration, which might make it possible to resolve the vacancy defect in Fig. 3 with two projections. More generally, one would need to include suitable features of global optimisation algorithms [26] that do not compromise

ADCG's computational efficiency (note that we could have adapted simulated annealing to solve (11) but using a cooling schedule slow enough to prevent getting trapped in local minima quickly becomes practically infeasible). A related problem is to characterize local and global minimizers of (11) to understand which configurations can be uniquely recovered by this approach and which cannot. To process experimental data, it may furthermore be important to analyze the impact of the error caused by the approximate nature of the mathematical models used for data acquisition $(\mathcal{R}, G)$ and atomic interaction (V_{LJ}).

6 Conclusions

In this paper we proposed a novel discrete tomography approach in which the locations of atoms are allowed to vary continuously and their interaction potentials are modeled explicitly. We showed in proof-of-concept numerical studies that such an approach can be better at resolving crystalline defects than image domain discretized or fully discrete algorithms. Furthermore, in situations where atom locations are desired, this approach provides access to the quantity of interest without any additional post-processing. For future work, we will extend our numerical studies on this atomic super-resolution approach to larger-sized scenarios in 3D, featuring realistic measurement noise, acquisition geometries, more suitable and accurate physical interaction potentials and different atom types. This will require additional computational effort to scale up our algorithm and will then allow us to work on real electron tomography data of nanocrystals.

Acknowledgments. This project has received funding from the European Union's Horizon 2020 research and innovation programme under the Marie Sklodowska-Curie grant agreement no. 765604.

References

1. van Aarle, W., et al.: The ASTRA Toolbox: a platform for advanced algorithm development in electron tomography. Ultramicroscopy **157**, 35–47 (2015)
2. Adler, J., Kohr, H., Oktem, O.: Operator discretization library (ODL). Software (2017). https://github.com/odlgroup/odl
3. Alberti, G.S., Ammari, H., Romero, F., Wintz, T.: Dynamic spike superresolution and applications to ultrafast ultrasound imaging. SIAM J. Imaging Sci. **12**(3), 1501–1527 (2019)
4. Alpers, A., Gritzmann, P.: On stability, error correction, and noise compensation in discrete tomography. SIAM J. Discrete Math. **20**(1), 227–239 (2006)
5. Anderson, J.A., Lorenz, C.D., Travesset, A.: General purpose molecular dynamics simulations fully implemented on graphics processing units. J. Comput. Phys. **227**(10), 5342–5359 (2008)
6. Baake, M., Huck, C., Gritzmann, P., Langfeld, B., Lord, K.: Discrete tomography of planar model sets. Acta Crystallographica A **62**(6), 419–433 (2006)
7. Batenburg, K.J.: A network flow algorithm for reconstructing binary images from discrete x-rays. J. Math. Imaging Vis. **27**(2), 175–191 (2006)

8. Batenburg, K.J., Sijbers, J.: Generic iterative subset algorithms for discrete tomography. Discrete Appl. Math. **157**(3), 438–451 (2009)

9. Beck, A., Teboulle, M.: A fast iterative shrinkage-thresholding algorithm for linear inverse problems. SIAM J. Imaging Sci. **2**(1), 183–202 (2009)

10. Bitzek, E., Koskinen, P., Gähler, F., Moseler, M., Gumbsch, P.: Structural relaxation made simple. Phys. Rev. Lett. **97**(17), 170201 (2006)

11. Boyd, N., Schiebinger, G., Recht, B.: The alternating descent conditional gradient method for sparse inverse problems. SIAM J. Optim. **27**(2), 616–639 (2017)

12. Bredies, K., Pikkarainen, H.K.: Inverse problems in spaces of measures. ESAIM Control Optim. Calc. Var. **19**(1), 190–218 (2013)

13. Candès, E.J., Fernandez-Granda, C.: Towards a mathematical theory of super-resolution. Commun. Pure Appl. Math. **67**(6), 906–956 (2014)

14. Chen, C.C., et al.: Three-dimensional imaging of dislocations in a nanoparticle at atomic resolution. Nature **496**(7443), 74–77 (2013)

15. Denoyelle, Q., Duval, V., Peyré, G., Soubies, E.: The sliding Frank-Wolfe algorithm and its application to super-resolution microscopy. Inverse Probl. **36**, 014001 (2019)

16. Foucart, S., Rauhut, H.: A Mathematical Introduction to Compressive Sensing. Applied and Numerical Harmonic Analysis. Springer, New York (2013). https://doi.org/10.1007/978-0-8176-4948-7

17. Frank, M., Wolfe, P.: An algorithm for quadratic programming. Naval Res. Logist. Q. **3**(1–2), 95–110 (1956)

18. Frenkel, D., Smit, B.: Understanding Molecular Simulation: From Algorithms to Applications, vol. 1. Elsevier, Amsterdam (2001)

19. Gardner, R.J., Gritzmann, P.: Discrete tomography: determination of finite sets by x-rays. Trans. Am, Math. Soc. **349**(6), 2271–2295 (1997)

20. Glaser, J., et al.: Strong scaling of general-purpose molecular dynamics simulations on GPUs. Comput. Phys. Commun. **192**, 97–107 (2015)

21. Goris, B., et al.: Measuring lattice strain in three dimensions through electron microscopy. Nano Lett. **15**(10), 6996–7001 (2015)

22. Jaggi, M.: Revisiting Frank-Wolfe: projection-free sparse convex optimization. In: ICML (1), pp. 427–435 (2013)

23. Kak, A.C., Slaney, M., Wang, G.: Principles of computerized tomographic imaging. Med. Phys. **29**(1), 107 (2002)

24. Montroll, E.W.: Theory of the vibration of simple cubic lattices with nearest neighbor interactions. In: Proceedings of the Third Berkeley Symposium on Mathematical Statistics and Probability, vol. 3, pp. 209–246. Univ of California Press (1956)

25. Nelder, J.A., Mead, R.: A simplex method for function minimization. Comput. J. **7**(4), 308–313 (1965)

26. Pardalos, P.M., Zhigljavsky, A., Žilinskas, J.: Advances in Stochastic and Deterministic Global Optimization. Springer, Cham (2016). https://doi.org/10.1007/978-3-319-29975-4

27. Parikh, N., Boyd, S., et al.: Proximal algorithms. Found. Trends® Optim. **1**(3), 127–239 (2014)

28. Poon, C., Peyré, G.: Multidimensional sparse super-resolution. SIAM J. Math. Anal. **51**(1), 1–44 (2019)

29. Rez, P., Treacy, M.M.: Three-dimensional imaging of dislocations. Nature **503**(7476), E1–E1 (2013)

30. Van Aert, S., Batenburg, K.J., Rossell, M.D., Erni, R., Van Tendeloo, G.: Three-dimensional atomic imaging of crystalline nanoparticles. Nature **470**(7334), 374 (2011)

31. Van Dyck, D., Jinschek, J.R., Chen, F.R.: 'Big Bang' tomography as a new route to atomic-resolution electron tomography. Nature **486**(7402), 243–246 (2012)
32. Virtanen, P., et al.: SciPy 1.0-fundamental algorithms for scientific computing in Python. arXiv preprint arXiv:1907.10121 (2019)

Characterizations of Simple Points on the Body-Centered Cubic Grid

Péter Kardos$^{(\boxtimes)}$

Department of Image Processing and Computer Graphics,
University of Szeged, Szeged, Hungary
`pkardos@inf.u-szeged.hu`

Abstract. A frequently investigated problem in various applications of binary image processing is to ensure the topology preservation of image operators, where the concept of simple points plays a key role. The literature primarily focuses on 2D and 3D images that are sampled on the conventional square and cubic grids, respectively. This paper presents some new characterizations of simple points on the body-centered cubic grid.

Keywords: BCC grid · Reduction · Topology preservation · Simple point

1 Introduction

3D digital pictures are mostly sampled on the cubic grid, whose voxel representation is based on the tessellation of the Euclidean space into unit cubes. However, an alternate structure, the *body-centered cubic grid (BCC grid)*, which tessellates the space into truncated octahedra, also attracted a remarkable scientific interest in digital image processing and digital geometry due to its advantages over the cubic model [2,9,13,14,16,17]. First, the BCC grid proves to be more efficient in the sampling of 3D images, as its packing density (i.e., the ratio between the volume of the largest ball completely enclosed in a voxel and the volume of the voxel itself) is lower than for the conventional sampling scheme [13]. Another benefit of the alternate tessellation lies in its less ambiguous connectivity structure. The 3D Jordan theorem states that a simple closed surface divides its complement into two components [10]. In the cubic grid, where neighbors may share a vertex, an edge, or a face, the discrete analog of the Jordan property can be satisfied only if we use different connectivity relations for the foreground and the background. However, in the BCC grid, there are only face-adjacencies, therefore, no connectivity paradox arises when using this type of grid [13].

Topology preservation is a key issue of digital topology. Existing 3D topological algorithms are generally assuming the cubic grid, but some algorithms working on non-standard grids have also been proposed [3,11,12,14]. The verification of topology preservation is based on the notion of simple points: an object

© Springer Nature Switzerland AG 2020
T. Lukić et al. (Eds.): IWCIA 2020, LNCS 12148, pp. 62–72, 2020.
https://doi.org/10.1007/978-3-030-51002-5_5

point is simple if its deletion does not alter the topology of the image [7]. Various characterizations of simple points were proposed for 3D images on the cubic grid [4,5,8], however, there was only one work specifically devoted to simplicity on the BCC grid [15]. The purpose of this paper is to fill the gap by presenting some visual characterizations of simple points on the mentioned non-standard grid.

The rest of this paper is organized as follows. Section 2 reviews the basic notions of 3D digital topology concerning the BCC grid. Section 3 gives some point configurations for effective verification of simplicity and shows easily visualized characterizations of simple points based on the concept of a so-called attachment set introduced by Kong in [5]. Finally, some concluding remarks are drawn.

2   Basic Notions

We apply fundamental concepts of digital topology as reviewed by Kong and Rosenfeld [7].

The BCC grid is defined as the following subset of $\mathbb{Z}^3$:

$$\mathbb{B} = \{(x, y, z) \in \mathbb{Z}^3 | \; x \equiv y \equiv z \; (\text{mod } 2)\}.$$

The elements of $\mathbb{B}$ are called *points*. We can define the following three kinds of neighborhoods of a point $p = (p_x, p_y, p_z) \in \mathbb{B}$:

$$N_6(p) = \{q = (q_x, q_y, q_z) \in \mathbb{Z}^3 \mid |p_x - q_x| + |p_y - q_y| + |p_z - q_z| = 2\},$$
$$N_8(p) = \{q = (q_x, q_y, q_z) \in \mathbb{Z}^3 \mid |p_x - q_x| + |p_y - q_y| + |p_z - q_z| = 3\},$$
$$N_{14}(p) = N_6(p) \cup N_8(p).$$

Furthermore, let $N_k^*(p) = N_k(p)\backslash\{p\}$ ($k \in \{6, 8, 14\}$). Figure 1a shows the points of $N_{14}(p)$, and an indexing scheme for these points is given in Fig. 1b, too. The volumetric elements in the tessellation of $\mathbb{B}$ representing the grid points are truncated octahedra, and we call them *voxels* (see Fig. 1c). Point p and $q \in N_{14}^*(p)$ are *14-adjacent*.

Let $X \subset \mathbb{B}$ be a non-empty set of points. The sequence of distinct points in X $\langle x_0, x_1, \ldots x_s \rangle$ is called a *14-path* of length s from point x_0 to point x_s in X if x_i is 14-adjacent to x_{i-1} ($i = 1, \ldots, s$). Note that a single point is a 14-path of length 0. Two points $p, q \in \mathbb{B}$ are said to be *14-connected* in $X \subseteq \mathbb{B}$ if there is a 14-path from p to q in X. The set X is *14-connected* in the set of points $Y \supseteq X$ if any two points in X are 14-connected in Y. The set X is a *14-component* in the set of points $Y \supseteq X$ if X is 14-connected in Y, but the set $X \cup \{y\}$ is not 14-connected in Y for any $y \in Y\backslash X$ (i.e., X is a maximal 14-connected set of points in Y). The number of elements in a set S of points is denoted by $|S|$.

Following the concept of digital pictures in [7], we define the $(14, 14)$ *binary digital picture* as a quadruple $\mathcal{P} = (\mathbb{B}, 14, 14, B)$. Each point in $B \subseteq \mathbb{B}$ is called a *black point* and has a value of 1 assigned to it. Picture $\mathcal{P}$ is finite if it contains finitely many black points. Each point in $\mathbb{B}\backslash B$ is called a *white point* and has a

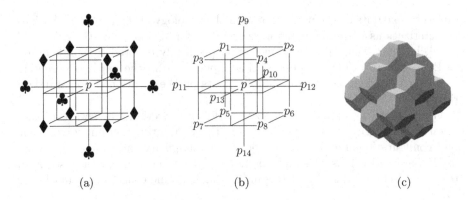

Fig. 1. Elements of $N_{14}(p)$ in $\mathbb{B}$ (a). The set N_6 of the central point $p \in \mathbb{B}$ contains p and the six points marked "♣". The set N_8 contains p and the eight points marked "♦". Indexing scheme for the points of $N_{14}^*(p)$ (b). Voxel representation of these points (c).

value of 0 assigned to it. 14-adjacency is associated with both black and white points. A *black component* or an *object* is a 14-connected set of points in B, while a *white component* is a 14-connected set of points in $\mathbb{B}\backslash B$. Here we assume that a picture contains finitely many black points. A black point is called a *border point* in a picture if it is 14-adjacent to at least one white point. A black point p is an *isolated pixel* in picture $(\mathbb{B}, k, \bar{k}, B)$ if it forms a singleton object (i.e., $N_{14}^*(p) \cap B = \emptyset$). In a finite picture, there is a unique white component that is called the *background*. A finite white component is called a *cavity*.

A *reduction* transforms a binary picture only by changing some black points to white ones (which is referred to as the deletion of 1's). There are some generally agreed conditions in [5] to verify the topological correctness of such a transform: A 3D reduction does *not* preserve topology [5] if

- any object is split or is completely deleted,
- any cavity is merged with another cavity or with the background,
- a new cavity is created, or
- a hole/tunnel (that a donut has) is eliminated or created.

The above conditions can be mathematically confirmed by investigating topological invariants such as the Betti numbers or the genus [1].

A *simple point* is a black point whose deletion is a topology preserving reduction [7].

3   Characterizations of Simple Points

For 3D pictures sampled on the cubic grid, it was earlier proved that simplicity is a local property that can be decided by investigating the $3 \times 3 \times 3$ neighborhood [5]. Later, Strand and Brunner adapted this result for the BCC grid:

Theorem 1. [15] *Let p be a border point in a* $(\mathbb{B}, 14, 14, B)$ *picture. Then p is a simple point if and only if the following conditions hold:*

1. *Point p is 14-adjacent to just one 14-component of* $N_{14}^*(p) \cap B$.
2. *Point p is 14-adjacent to just one 14-component of* $N_{14}(p) \backslash B$.

Figure 2 shows some illustrative examples for Theorem 1.

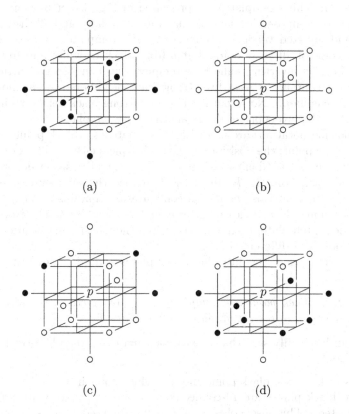

(a)   (b)

(c)   (d)

Fig. 2. Examples for simple and non-simple points. The positions marked "•" and "○" represent black and white points, respectively. Black point p is simple only in configuration (a), where both Conditions 1 and 2 of Theorem 1 are satisfied. In example (b), p is an isolated black point, while in (c) we can find two 14-components in $N_{14}^*(p) \cap B$, hence in both cases Condition 1 of Theorem 1 is violated. Configuration (d) depicts a case where there exist two 14-components in $N_{14}(p) \backslash B$, thus Condition 2 of Theorem 1 does not hold.

A straightforward consequence of Theorem 1 is the following duality theorem:

Proposition 1. *Point p is simple in picture* $(\mathbb{B}, 14, 14, B)$ *if and only if it is simple in picture* $(\mathbb{B}, 14, 14, \overline{B} \cup \{p\})$.

To make the verification of the conditions in Theorem 1 easier, this paper introduces here some configurations, so-called *matching templates*. A black point is simple if at least one template in the set of templates matches it. We can group the possible cases for the simplicity of point p by the number of black points in $N_8^*(p)$. For an easier discussion, let us refer to such black points as the *strong black neighbors* of p, and the white points in $N_8^*(p)$ as the *strong white neighbors* of p. We call a template as a *base template*, if it depicts at most four strong black neighbors.

The base matching templates are presented in Fig. 3. Let us denote $\mathcal{T}$ the set of templates composed by the base matching templates and all their rotated, reflected, and inverted versions. By inversion of a template, we mean that we change the color of all their points but p from black to white and from white to black. Thus, by inverting some base templates we can also get further templates where p contains at least five strong black neighbors (i.e., at most three white strong neighbors). Note that if p has four strong black neighbors in a base template, then the inversion results in another base template.

The following notations are used: Each "•" matches a black point; each "○" matches a white point; the position marked "." depicts either a black or a white point, which we call a "don't care" point. The central position denoted by p represents a black point in $\mathcal{P}$. In order to reduce the number of templates, additional notations are also used. If a configuration contains elements marked b and w, then b matches a black point or w matches a white point. (Note that if we invert such a template, then b and w switch their position.) Positions s and t yield two points of different colors.

Now we show that the introduced templates indeed characterize simple points.

Theorem 2. *A black point p is simple in a $(14, 14)$ picture, if and only if it matches an element of the set of matching templates $\mathcal{T}$.*

Proof. It can be readily seen that in each base matching template the following properties hold:

- There is at least one black point and one white point in it.
- Any two black points are 14-connected, and also any two white points are 14-connected. (This also holds for the points marked b, w, s, and t, if we consider choosing their color as described above.)
- Each "don't care" point is 14-adjacent to both a black and a white point, but it is not 14-adjacent to any other "don't care" point.

Therefore, by Theorem 1, each base matching template depicts a simple point, and the same can be obviously said for its rotated and reflected versions. Furthermore, note that if we invert a template, then by Proposition 1, the resulting template also matches a simple point.

It is also easy to verify that if we change the color of some points in any template, then either we get another template of $\mathcal{T}$, or the connectedness of some points of the same color will be broken. Hence, if p does not match any of the templates of $\mathcal{T}$, then p is not simple by Theorem 1.   □

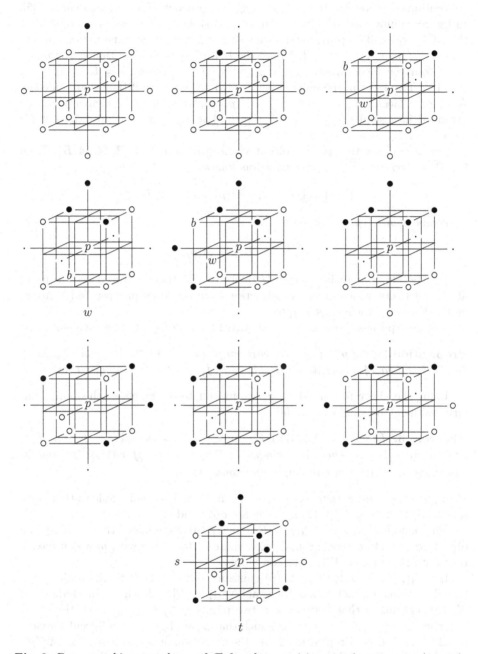

Fig. 3. Base matching templates of $\mathcal{T}$ for characterizing simple points in (14, 14) pictures. All their rotated, reflected, and inverted versions match simple points, too.

Kong proposed another easily visualized characterization of simple points on conventional orthogonal images by using the concept of an *attachment set* [5]. In the remaining part of this section, two possible adaptations of his model for the BCC grid will be provided, and for this aim, we follow some notions in [5].

Let $U^O(p) = \{f_1, \ldots, f_{14}\}$ denote the boundary of voxel p that is the union of its fourteen faces, where voxels p and p_i share face f_i ($i = 1, \ldots, 14$) (see Fig. 1b). Two faces in the universe set $U^O(p)$ are *adjacent* if they share an edge. A set of n faces ($1 \leq n \leq 14$) $F \subseteq U^O(p)$ is *connected*, if its elements can be arranged in a sequence $\langle f_{i_1}, \ldots, f_{i_n} \rangle$ such that f_{i_k} and $f_{i_{k+1}}$ are adjacent for each $k = 1, \ldots, n-1$.

Let us suppose that p is an object voxel in picture $\mathcal{P} = (\mathbb{B}, 14, 14, B)$. Then the $\mathcal{P}^O$-*attachment set* of p is defined as follows:

$$\mathcal{A}^O(p) = \{\, f_i \mid f_i \in U(p) \text{ and } p_i \in B \,\}.$$

Its *complement* $\overline{\mathcal{A}(p)}$ is defined as

$$\overline{\mathcal{A}^O(p)} = U(p) \setminus \mathcal{A}(p).$$

(In the above notation, letter O refers to the "octahedron-based" type of attachment sets, which makes a distinction from the other proposed adaptation, as we shall see later in this chapter.)

The next proposition points to a straightforward property of attachment sets:

Proposition 2. *Let p be an object voxel in picture $(\mathbb{B}, 14, 14, B)$, and let $p_i, p_j \in N_{14} \cap B$ be two 14-adjacent voxels. Then, $f_i, f_j \in \mathcal{A}^O(p)$ are adjacent faces.*

Using the above terminology, we can formulate the following theorem, similarly to the orthogonal case in [5]:

Theorem 3. *Let $\mathcal{P} = (\mathbb{B}, 14, 14, B)$, $p \in B$, and let $\mathcal{A}^O(p)$ be the $\mathcal{P}^O$-attachment set of p. Then, p is simple in $\mathcal{P}$ if and only if both $\mathcal{A}^O(p)$ and its complement, $\overline{\mathcal{A}^O(p)}$ are non-empty and connected.*

Proof. Let us suppose that p is simple. Then, from Theorem 1 follows that both sets $N_{14}^*(p) \cap B$ and $N_{14}^*(p) \cap \overline{B}$ are non-empty and connected.

The non-emptiness of $N_{14}^*(p) \cap B$ and $N_{14}^*(p) \cap \overline{B}$ implies that at least one object voxel and one background voxel shares a face each with p, which means that both $\mathcal{A}^O(p)$ and $\overline{\mathcal{A}^O(p)}$ are non-empty.

If $|N_{14}^*(p) \cap B| = 1$, then $\mathcal{A}^O(p)$ contains only one face, therefore $\mathcal{A}^O(p)$ is trivially connected. Let us assume that $|N_{14}^*(p) \cap B| \geq 2$. The connectedness of $N_{14}^*(p) \cap B$ means that between any two points $p_i, p_j \in N_{14}^*(p) \cap B$, there is a sequence $\langle p_i = p_{l_1}, p_{l_2} \ldots, p_{l_n} = p_j \rangle$ such that p_{l_k} and $p_{l_{k+1}}$ are adjacent for each $k = 1, \ldots, n-1$. By Proposition 2 and the definition of connectedness in $\mathcal{A}^O(p)$, we can also conclude that the elements of $\mathcal{A}^O(p)$ can be arranged in a sequence $\langle f_{l_1}, \ldots, f_{l_n} \rangle$ such that f_{l_k} and $f_{l_{k+1}}$ are adjacent for each $k = 1, \ldots, n-1$. Hence, $\mathcal{A}^O(p)$ is connected, which, together with Proposition 1 also concludes that $\overline{\mathcal{A}^O(p)}$ is connected.

If p is not simple, then, based on the equivalence given in Theorem 1, it can be shown similarly to the previous case that at least one of the sets $\mathcal{A}^O(p)$ or $\overline{\mathcal{A}^O(p)}$ is not connected or empty.   □

Although the voxels in the BCC grid do not form cubes, we can also give a "cube-based" adaptation of attachment sets in $\mathbb{B}$, similarly to the model proposed for the cubic grid in [5]. For this purpose, we consider a unit cube $\mathcal{C}$ in the Euclidean space whose center is a point p in $\mathbb{B}$ and whose vertices, edges, and faces are the elements of the following sets, respectively:

$V = \{v_i \mid v_i$ coincides with $p_i \in N_8^*(p)\}$,
$E = \{e_{i,j} \mid e_{i,j}$ connects v_i and v_j in $\mathcal{C}\}$,
$F = \{f_i \mid f_i$ intersects the line segment between p and $p_{i+8} \in N_6^*(p)\}$.

Let $U^C(p) = V \cup \{e_{i,j}\backslash V \mid e_{i,j} \in E\} \cup \{f_i\backslash E \mid f_i \in F\}$, i.e., this set contains altogether twenty-six elements in the boundary of the cube: eight vertices, twelve open edges (without vertices), and six open faces (without edges and vertices). We say that edge $e_{i,j}$ (with its vertices) is the *closure of open edge* $e_{i,j}\backslash V$, and face f_i (with its edges and vertices) is the *closure of open face* $f_i\backslash E$. Elements $x, y \in U^C(p)$ are *adjacent* if $y \in U^C(p)\backslash V$ (i.e. y is an open edge or an open face) and the closure of y contains x or vice versa.

A set $S \subseteq U^C(p)$ is *connected*, if its elements can be arranged in a sequence $\langle s_{i_1}, \ldots, s_{i_n} \rangle$ such that s_{i_k} and $s_{i_{k+1}}$ are adjacent for each $k = 1, \ldots, n-1$. We must introduce some further notations to refer to the special subsets of $U^C(p)$:

- $U_1 = \{v_i \mid p_i \in N_8(p) \cap B\}$,
- $U_2 = \{e_{i,j} \mid p_i, p_j \in N_8(p) \cap B\}$,
- $U_3 = \{f_i\backslash E \mid p_{i+8} \in N_6(p) \cap B\}$.

Now we proceed to define the $\mathcal{P}^C$-*attachment set* of p and it's *complement*:

$$\mathcal{A}^C(p) = U_1 \cup U_2 \cup U_3,$$
$$\overline{\mathcal{A}^C(p)} = U^C(p) \setminus \mathcal{A}^C(p).$$

(Here, the letter C refers to the "cube-based" type of attachment sets.)

By examining the possible configurations of two 14-adjacent points, we can formulate the following two observations.

Proposition 3. *Two points* $p_i, p_j \in N_8(p) \cap B$ *are 14-adjacent in* $\mathbb{B}$, *if and only if* v_i, v_j, *and the open edge* $e_{i,j}$ *are all elements of* $\mathcal{A}^C(p)$.

Proposition 4. *Two points* $p_i \in N_6(p)$ *and* $p_j \in N_8(p)$ *are 14-adjacent in* $\mathbb{B}$, *if and only if* p_i *and the open face* f_{j-8} *are adjacent elements in* $\mathcal{A}^C(p)$.

The next theorem states a similar equivalence as Theorem 3.

Theorem 4. *Let* $\mathcal{P} = (\mathbb{B}, 14, 14, B)$, $p \in B$, *and let* $\mathcal{A}^C(p)$ *be the* $\mathcal{P}^C$-*attachment set of* p. *Then,* p *is simple in* $\mathcal{P}$ *if and only if both* $\mathcal{A}^C(p)$ *and its complement,* $\overline{\mathcal{A}^C(p)}$ *are non-empty and connected.*

Proof. The proof can be done in an analogous way to that given for the previous theorem. The only critical difference lies in the verification of whether the adjacency of two points $p_i, p_j \in N_{14}^*(p) \cap B$ implies the connectedness of the corresponding subset of $\mathcal{A}^C(p)$:

- If $p_i, p_j \in N_8(p) \cap B$, then by Proposition 3, these points are connected by the open edge $e_{i,j}$ in $\mathcal{A}^C(p)$.
- If $p_i \in N_6(p) \cap B$ and $p_i \in N_8(p) \cap B$, then by Proposition 4, p_i is adjacent to the open face f_{j-8} in $\mathcal{A}^C(p)$. $\square$

Figure 4 gives some illustrative examples for Theorems 3 and 4. The drawback of the 3D representations of attachment sets is that we cannot see all faces of the voxel on them. In order to get rid of this problem, Kong proposed the use of Schlegel diagrams, which are special $n-1$ dimensional projections of n dimensional polyhedra, thereby we can get a planar representation of the boundary of voxels [6]. Figure 5 depicts the Schlegel diagrams associated with

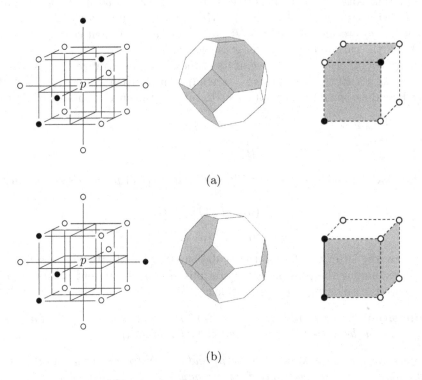

(a)

(b)

Fig. 4. Examples for $\mathcal{P}^O(p)$- and $\mathcal{P}^C(p)$-attachment sets. Point p is simple in (a), but it is not simple in (b) as $\mathcal{A}^O(p)$ and $\mathcal{A}^C(p)$ are not connected. The gray faces of the truncated octahedron represent elements of $\mathcal{A}^O(p)$, while its white faces belong to $\overline{\mathcal{A}^O(p)}$. The gray faces, thick line segment, and black vertices of the cubes at the right side represent elements of $\mathcal{A}^C(p)$, while the white faces, dashed line segments, and white vertices are members of $\overline{\mathcal{A}^C(p)}$.

the attachment sets depicted in Fig. 4. Note that in each case one face of the voxel is not shown in the interior of the diagrams: that face is considered to correspond to the outside of the diagram.

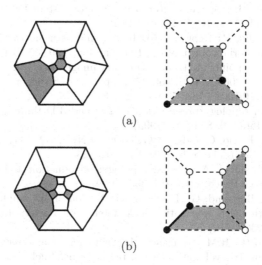

(a)

(b)

Fig. 5. The Schlegel diagrams representing the voxels in Fig. 4. The gray faces in the left diagrams show elements of $\mathcal{A}^O(p)$, while the white faces belong to $\overline{\mathcal{A}^O(p)}$. The gray faces, thick line segment, and black vertices in the right diagrams depict elements of $\mathcal{A}^C(p)$, while the white faces, dashed line segments, and white vertices are members of $\overline{\mathcal{A}^C(p)}$.

4   Conclusion and Future Work

This work discusses some new characterizations to verify the simplicity of points in $(14, 14)$ pictures. First, simple points were illustrated with some matching templates. In addition, an "octahedron-based" and a "cube-based" adaptation of Kong's attachment set were also proposed.

It should be pointed out that, although the concept of simple points helps us to validate topology preservation, simplicity in itself is generally not sufficient to ensure the topological correctness of parallel operators. Therefore, future work will focus on the establishment of some sufficient conditions for reductions to preserve topology on the BCC grid.

Acknowledgments. This research was supported by the project "Integrated program for training new generation of scientists in the fields of computer science", no. EFOP-3.6.3-VEKOP16-2017-00002. This research was supported by grant TUDFO/47138-1/2019-ITM of the Ministry for Innovation and Technology, Hungary.

References

1. Chen, L., Rong, Y.: Digital topological method for computing genus and the Betti numbers. Topol. Appl. **157**, 1931–1936 (2010)
2. Čomić, L., Nagy, B.: A combinatorial coordinate system for the body-centered cubic grid. Graph. Models **87**, 11–22 (2016)
3. Čomić, L., Magillo, P.: Repairing 3D binary images using the BCC grid with a 4-valued combinatorial coordinate system. Inf. Sci. **499**, 47–61 (2009)
4. Klette, G.: Simple points in 2D and 3D binary images. In: Petkov, N., Westenberg, M.A. (eds.) CAIP 2003. LNCS, vol. 2756, pp. 57–64. Springer, Heidelberg (2003). https://doi.org/10.1007/978-3-540-45179-2_8
5. Kong, T.Y.: On topology preservation in 2-D and 3-D thinning. Int. J. Pattern Recogn. Artif. Intell. **9**, 813–844 (1995)
6. Ahronovitz, E., Fiorio, C. (eds.): DGCI 1997. LNCS, vol. 1347. Springer, Heidelberg (1997). https://doi.org/10.1007/BFb0024825
7. Kong, T.Y., Rosenfeld, A.: Digital topology: introduction and survey. Comput. Vis. Graph. Image Process. **48**, 357–393 (1989)
8. Malandain, G., Bertrand, G.: Fast characterization of 3D simple points. In: Proceedings of 11th IEEE International Conference on Pattern Recognition, ICPR 1992, pp. 232–235 (1992)
9. Matej, S., Lewitt, R.M.: Efficient 3D grids for image reconstruction using spherically-symmetric volume elements. IEEE Trans. Nucl. Sci. **42**(4), 1361–1370 (1995)
10. Morgenthaler, D.G., Rosenfeld, A.: Surfaces in three-dimensional digital images. Inf. Control **51**(3), 227–247 (1981)
11. Stelldinger, P., Köthe, U.: Connectivity preserving digitization of blurred binary images in 2D and 3D. Comput. Graph. **30**(1), 70–76 (2006)
12. Stelldinger, P., Strand, R.: Topology preserving digitization with FCC and BCC grids. In: Reulke, R., Eckardt, U., Flach, B., Knauer, U., Polthier, K. (eds.) IWCIA 2006. LNCS, vol. 4040, pp. 226–240. Springer, Heidelberg (2006). https://doi.org/10.1007/11774938_18
13. Strand, R.: The face-centered cubic grid and the body-centered cubic grid - a literature survey. Internrapport, Centrum für Bildanalys, Uppsala University, Centre for Image Analysis (2005)
14. Strand, R.: Surface skeletons in grids with non-cubic voxels. In: Proceedings of the 17th International Conference on Pattern Recognition, 2004, ICPR 2004, vol. 1, pp. 548–551. Cambridge (2004)
15. Strand, R., Brunner, D.: Simple points on the body-centered cubic grid. Technical report 42, Centre for Image Analysis, Uppsala University, Uppsala, Sweden (2006)
16. Strand, R., Nagy, B.: Weighted neighbourhood sequences in non-standard three-dimensional grids – metricity and algorithms. In: Coeurjolly, D., Sivignon, I., Tougne, L., Dupont, F. (eds.) DGCI 2008. LNCS, vol. 4992, pp. 201–212. Springer, Heidelberg (2008). https://doi.org/10.1007/978-3-540-79126-3_19
17. Theussl, T., Moller, T., Groller, M.E.: Optimal regular volume sampling. In: Proceedings Visualization 2001, pp. 91–546 (2001)

A 4D Counter-Example Showing that DWCness Does Not Imply CWCness in nD

Nicolas Boutry[1]([✉]), Rocio Gonzalez-Diaz[2], Laurent Najman[3],
and Thierry Géraud[1]

[1] EPITA Research and Development Laboratory (LRDE), EPITA,
Le Kremlin-Bicêtre, France
[2] Universidad de Sevilla, Sevilla, Spain
[3] Université Paris-Est, LIGM,
Équipe A3SI, ESIEE, Noisy-le-Grand, France

Abstract. In this paper, we prove that the two flavours of well-composedness called Continuous Well-Composedness (shortly CWCness), stating that the boundary of the continuous analog of a discrete set is a manifold, and Digital Well-Composedness (shortly DWCness), stating that a discrete set does not contain any critical configuration, are not equivalent in dimension 4. To prove this, we exhibit the example of a configuration of 8 tesseracts (4D cubes) sharing a common corner (vertex), which is DWC but not CWC. This result is surprising since we know that CWCness and DWCness are equivalent in 2D and 3D. To reach our goal, we use local homology.

Keywords: Well-composed · Topological manifolds · Critical configurations · Digital topology · Local homology

1 Introduction

Digital well-composedness (shortly DWCness) is a nice property in digital topology, because it implies the equivalence of $2n$ and $(3^n - 1)$ connectivities [3] in a subset of $\mathbb{Z}^n$ and in its complement at the local and global points of view. A well-known application of this flavour of well-composedness is the tree of shapes [7,9]: usual natural or synthetic images generally contain many critical configurations and this way we cannot ensure that the hierarchy induced by the inclusion relationship between these shapes does not draw a graph without cycles. On the contrary, when an image is DWC, no cycle is possible and then we obtain a tree, called the *tree of shapes* [7,9,16].

On the other side, continuously well-composed (shortly CWC) images are known as "counterparts" of n-dimensional manifolds in the sense that the boundary of their continuous analog does not have singularities (called "pinches"), which is a very strong topological property. The consequence is that some geometric differential operators can be computed directly on the discrete sets, which simplifies or makes specific algorithms faster [12,13].

© Springer Nature Switzerland AG 2020
T. Lukić et al. (Eds.): IWCIA 2020, LNCS 12148, pp. 73–87, 2020.
https://doi.org/10.1007/978-3-030-51002-2_6

These two flavours of well-composednesses, known to be equivalent in 2D and in 3D, are not equivalent in 4D, and this is what we are going to prove in this paper. Section 2 recalls the material relative to discrete topology and local homology necessary to the proof detailed in Sect. 3. Section 4 concludes the paper.

Fig. 1. Examples of primary and secondary critical configurations in 2D/3D/4D blocks S. Black bullets correspond to the points of the digital set $\mathcal{X}$ and the white bullets correspond to the points of $S \setminus \mathcal{X}$.

2   Discrete Topology

As usual in discrete topology, we will only work with *digital sets*, that is, finite subsets of $\mathbb{Z}^n$ or subsets $\mathcal{X}$ of $\mathbb{Z}^n$ whose complementary set $\mathcal{X}^c = \mathbb{Z}^n \setminus \mathcal{X}$ is finite.

2.1   Digital Well-Composedness

Let $n \geq 2$ be a (finite) integer called the *dimension*. Now, let $\mathbb{B} = \{e^1, \ldots, e^n\}$ be the (orthonormal) canonical basis of $\mathbb{Z}^n$. We use the notation v_i, where i belongs to $[\![1, n]\!] := \{i \in \mathbb{Z} \; ; \; 1 \leq i \leq n\}$, to determine the ith coordinate of a vector $v \in \mathbb{Z}^n$. We recall that the L^1-*norm* of the vector $v \in \mathbb{Z}^n$ is denoted by $\| \cdot \|_1$ and is equal to $\sum_{i \in [\![1, n]\!]} |v_i|$ where $| \cdot |$ is the *absolute value*. Also, the L^∞-*norm* is denoted by $\| \cdot \|_\infty$ and is equal to $\max_{i \in [\![1, n]\!]} |v_i|$.

For a given point $p \in \mathbb{Z}^n$, the $2n$-*neighborhood* in $\mathbb{Z}^n$, denoted by $\mathcal{N}_{2n}(p)$, is equal to $\{q \in \mathbb{Z}^n \; ; \; \|p - q\|_1 \leq 1\}$. Also, the $(3^n - 1)$-*neighborhood* in $\mathbb{Z}^n$, denoted by $\mathcal{N}_{3^n-1}(p)$, is equal to $\{q \in \mathbb{Z}^n \; ; \; \|p - q\|_\infty \leq 1\}$. Let ξ be a value in $\{2n, 3^n - 1\}$. The *starred ξ-neighborhood* of $p \in \mathbb{Z}^n$ is noted $\mathcal{N}_\xi^*(p)$ and is equal to $\mathcal{N}_\xi(p) \setminus \{p\}$. An element of the starred ξ-neighborhood of $p \in \mathbb{Z}^n$ is called a ξ-*neighbor* of p in $\mathbb{Z}^n$. Two points $p, q \in \mathbb{Z}^n$ such that $p \in \mathcal{N}_\xi^*(q)$ or equivalently $q \in \mathcal{N}_\xi^*(p)$ are said to be ξ-*adjacent*. A finite sequence $(p^0, \ldots, p^k)$ of points in $\mathbb{Z}^n$ is a ξ-*path* if and only if p^0 is ξ-adjacent only to p^1, p^k is ξ-adjacent only to p^{k-1}, and if for $i \in [\![1, k-1]\!]$, p^i is ξ-adjacent only to p^{i-1} and to p^{i+1}. A digital set $\mathcal{X} \subset \mathbb{Z}^n$ is said ξ-*connected* if for any pair of points $p, q \in \mathcal{X}$, there exists a ξ-path joining them into $\mathcal{X}$. A ξ-connected subset C of $\mathcal{X}$ which is *maximal in the inclusion sense*, that is, there is no ξ-connected subset of $\mathcal{X}$ which is greater than C, is said to be a ξ-*component* of $\mathcal{X}$.

For any $p \in \mathbb{Z}^n$ and any $\mathcal{F} = (f^1, \ldots, f^k) \subseteq \mathbb{B}$, we denote by $S(p, \mathcal{F})$ the set:

$$\Big\{ p + \sum_{i \in [\![1,k]\!]} \lambda_i f^i \; ; \; \lambda_i \in \{0, 1\}, \forall i \in [\![1, k]\!] \Big\}.$$

We call this set the *block* associated with the pair $(p, \mathcal{F})$; its *dimension*, denoted by $\dim(S)$, is equal to k. More generally, a set $S \subset \mathbb{Z}^n$ is said to be a *block* if there exists a pair $(p, \mathcal{F}) \in \mathbb{Z}^n \times \mathcal{P}(\mathbb{B})$ such that $S = S(p, \mathcal{F})$. We say that two points $q, q' \in \mathbb{Z}^n$ belonging to a block S are *antagonists* in S if the distance between them equals the maximal distance using the L^1 norm between two points in S; in this case we write $q = \mathrm{antag}_S(q')$. Note that the antagonist of a point q in a block S containing q exists and is unique. Two points that are antagonists in a block of dimension $k \geq 0$ are said to be *k-antagonists*; k is then called the *order of antagonism* between these two points. We say that a digital subset $\mathcal{X}$ of $\mathbb{Z}^n$ contains a *critical configuration* in a block S of dimension $k \in [\![2, n]\!]$ if there exists two points $\{q, q'\} \in \mathbb{Z}^n$ that are antagonists in S such that $\mathcal{X} \cap S = \{q, q'\}$ (*primary case*) or such that $S \setminus \mathcal{X} = \{q, q'\}$ (*secondary case*). Figure 1 depicts examples of critical configurations.

Definition 1 (Digital Well-Composedness [3]**).** *A digital set $\mathcal{X} \subset \mathbb{Z}^n$ is said to be* digitally well-composed *(DWC) if it does not contain any critical configuration.*

This property is *self-dual*: for any digital set $\mathcal{X} \subset \mathbb{Z}^n$, $\mathcal{X}$ is digitally well-composed iff $\mathcal{X}^c$ is digitally well-composed.

2.2   Basics in Topology and Continuous Well-Composedness

Let $(X, \mathcal{U})$ be a *topological space* [1,11]. The elements of the set X are called the *points* and the elements of the topology $\mathcal{U}$ are called the *open sets*. In practice, we will abusively say that X is a topological space, assuming it is supplied with $\mathcal{U}$. An open set which contains a point of X is said to be a *neighborhood* of this point. Let X be a topological space, and let T be a subset of X. A set $T \subseteq X$ is said *closed* if it is the complement of an open set in X. A function $f : X \to Y$ between two topological spaces X and Y is *continuous* if for every open set $V \subset Y$, the inverse image $f^{-1}(V) = \{x \in X \; ; \; f(x) \in V\}$ is an open subset of X. The function f is a *homeomorphism* if it is bicontinuous and bijective. The *continuous analog* $\mathrm{CA}(p)$ of a point $p \in \mathbb{Z}^n$ is the closed unit cube centered at this point with faces parallel to the coordinate planes $\mathrm{CA}(p) = \{q \in \mathbb{R}^n \; ; \; \|p - q\|_\infty \leq 1/2\}$. The *continuous analog* $\mathrm{CA}(\mathcal{X})$ of a digital set $\mathcal{X} \subset \mathbb{Z}^n$ is the union of the continuous analogs of the points belonging to the set $\mathcal{X}$, that is, $\mathrm{CA}(\mathcal{X}) = \bigcup_{p \in \mathcal{X}} \mathrm{CA}(p)$. Then, we will denote $\mathrm{bdCA}(\mathcal{X})$ the topological boundary of $\mathrm{CA}(\mathcal{X})$, that is, $\mathrm{bdCA}(\mathcal{X}) = \mathrm{CA}(\mathcal{X}) \setminus \mathrm{Int}(\mathrm{CA}(\mathcal{X}))$, where $\mathrm{Int}(\mathrm{CA}(\mathcal{X}))$ is the union of all open subsets of $\mathrm{CA}(\mathcal{X})$.

Definition 2 (Continuous Well-Composedness [14,15]**).** *Let $\mathcal{X} \subset \mathbb{Z}^n$ be a digital set. We say that $\mathcal{X}$ is* continuously well-composed *(CWC) if the boundary* *of its continuous analog* $\mathrm{bdCA}(\mathcal{X})$ *is a topological* $(n-1)$-*manifold, that is, if for any point* $p \in \mathcal{X}$, *the (open) neighborhood of* p *in* $\mathrm{bdCA}(\mathcal{X})$ *is homeomorphic to* $\mathbb{R}^{n-1}$.

This property is *self-dual*: for any digital set $\mathcal{X} \subset \mathbb{Z}^n$, $\mathrm{bdCA}(\mathcal{X}) = \mathrm{bdCA}(\mathcal{X}^c)$ and then $\mathcal{X}$ is continuously well-composed iff $\mathcal{X}^c$ is continuously well-composed.

2.3 Homomorphisms

Recalls about *Abelian groups* and *homomorphisms* can be found in [10]. A homomorphism f is called an *isomorphism* if it is bijective. Two free Abelian groups are said *isomorphic* if there exists an isomorphism between them; for A and B two free Abelian groups, we write $A \simeq B$ when A and B are isomorphic. Let A be a free Abelian group and B a subgroup of A. For each $a \in A$, defined the *equivalence class* $[a] := \{a + b \; ; \; b \in B\}$. The *quotient group* A/B is defined as $A/B := \{[a] \; ; \; a \in A\}$.

Theorem 1 (First Isomorphism Theorem [10]). *Let* A *and* B *be two free Abelian groups and* $f : A \to B$ *a homomorphism. Then* $A/\ker f \simeq \operatorname{im} f$.

2.4 Cubical Sets

An *elementary interval* is a closed subinterval of $\mathbb{R}$ of the form $[l, l+1]$ or $\{l\}$ for some $l \in \mathbb{Z}$. Elementary intervals that consist of a single point are *degenerate*, while those of length 1 are *non-degenerate*. An *elementary cube* h is a finite product of elementary intervals, that is, $h = h_1 \times \cdots \times h_d = \times_{i \in [\![1,d]\!]} h_i \subset \mathbb{R}^d$ where each h_i is an elementary interval. The set of elementary cubes in $\mathbb{R}^d$ is denoted by $\mathcal{K}^d$. The set of all elementary cubes is $\mathcal{K} := \bigcup_{d=1}^{\infty} \mathcal{K}^d$. Let $h = \times_{i \in [\![1,d]\!]} h_i \subset \mathbb{R}^d$ be an elementary cube. The elementary interval h_i is referred to as the ith *component* of h. The *dimension* of h is defined to be the number of non-degenerate components in h and is denoted by $\dim(h)$. Also, we define $\mathcal{K}_k := \{h \in \mathcal{K} \; ; \; \dim(h) = k\}$ and $\mathcal{K}_k^d := \mathcal{K}_k \cap \mathcal{K}^d$. A set $X \subset \mathbb{R}^d$ is *cubical* if X can be written as a finite union of elementary cubes. If X is a cubical set, we adopt the following notation $\mathcal{K}(X) := \{h \in \mathcal{K} \; ; \; h \subseteq X\}$ and $\mathcal{K}_k(X) := \{h \in \mathcal{K}(X) \; ; \; \dim(h) = k\}$.

2.5 Homology

Let $X \subseteq \mathbb{R}^d$ be a cubical set. The k-chains of X, denoted by $C_k(X)$, is the free Abelian group generated by $\mathcal{K}_k(X)$. The *boundary homomorphism* $\partial_k^X :$ $C_k(X) \to C_{k-1}(X)$ is defined on the elementary cubes of $\mathcal{K}_k(X)$ and extended to $C_k(X)$ by linearity (see [10]). The chain complex $\mathcal{C}(X)$ is the graded set $\{C_k(X), \partial_k^X\}_{k \in \mathbb{Z}}$. A k-chain $z \in C_k(X)$ is called a *cycle* in X if $\partial_k^X z = 0$. The set of all k-cycles in X, which is denoted by $Z_k(X)$, is $\ker \partial_k^X$ and forms a subgroup of $C_k(X)$. A k-chain $z \in C_k(X)$ is called a *boundary* in X if there exists $c \in C_{k+1}(X)$ such that $\partial_{k+1}^X c = z$. Thus the set of boundary elements in

$C_k(X)$, which is denoted by $B_k(X)$, consists of the image of ∂^X_{k+1}. Since ∂^X_{k+1} is a homomorphism, $B_k(X)$ is a subgroup of $C_k(X)$. Since $\partial^X_k \partial^X_{k+1} = 0$, every boundary is a cycle and thus $B_k(X)$ is a subgroup of $Z_k(X)$. We say that two cycles $z_1, z_2 \in Z_k(X)$ are *homologous* and write $z_1 \sim z_2$ if $z_1 - z_2$ is a boundary in X, that is, $z_1 - z_2 \in B_k(X)$. The *equivalence classes* are then the elements of the quotient group $\mathbb{H}_k(X) = Z_k(X)/B_k(X)$ called the *k-th homology group* of X. The homology of X is the collection of all homology groups of X. The shorthand notation for this is $\mathbb{H}(X) := \{H_k(X)\}_{k\in\mathbb{Z}}$. Given $z \in Z_k(X)$, $[z]$ is the homology class of z in X. A sequence of vertices $V_0, \ldots, V_n \in \mathcal{K}_0(X)$ is an *edge path* in X if there exists edges $E_1, \ldots, E_n \in \mathcal{K}_1(X)$ such that V_{i-1}, V_i are the two faces of E_i for $i = 1, \ldots, n$. For $V, V' \in \mathcal{K}_0(X)$, we write $V \sim_X V'$ if there exists an edge path $V_0, \ldots, V_n \in \mathcal{K}_0(X)$ in X such that $V = V_0$ and $V' = V_n$. We say that X is *edge-connected* if $V \sim_X V'$ for any $V, V' \in \mathcal{K}_0(X)$. For $V \in \mathcal{K}_0(X)$ we define the *edge-connected component of* V *in* X as the union of all edge-connected cubical subsets of X that contain V. We denote it $\mathrm{ecc}_X(V)$. The following result states that in the context of cubical sets, edge-connectedness is equivalent to (topological) connectedness[1].

Theorem 2 (Theorem 2.55 of [10]). *Let X be a cubical set. Then $\mathbb{H}_0(X)$ is a free Abelian group. Furthermore, if $\{V_i \; ; \; i \in [\![1, n]\!]\}$ is a collection of vertices in X consisting of one vertex from each component of X, then*

$$\left\{ [\widehat{V}_i] \in \mathbb{H}_0(X) \; ; \; i \in [\![1, n]\!] \right\}$$

forms a basis for $\mathbb{H}_0(X)$ (where $\widehat{V}_i$ is the algebraic element associated to V_i).

This way, edge-connected components of X are (topologically) connected components of X and conversely.

2.6 Relative Homology

Now, we recall some background in matter of *relative homology*. A pair of cubical sets X and A with the property that $A \subseteq X$ is called *cubical pair* and is denoted by (X, A). Relative homology is used to compute how the two spaces A, X differ from each other. Intuitively, we want to compute the homology of X *modulo* A: we want to ignore the set A and everything connected to it. In other words, we want to work with chains belonging to $C(X)/C(A)$, which leads to the following definition.

Definition 3 (Definition 9.3 of [10]). *Let (X, A) be a cubical pair. The* relative chains *of X modula A are the elements of the quotient groups $C_k(X, A) := C_k(X)/C_k(A)$. The equivalence class of a chain $c \in \mathcal{C}(X)$ relative to $\mathcal{C}(A)$ is denoted by $[c]_A$. Note that for each k, $C_k(X, A)$ is a free Abelian group. The* relative chain complex *of X modulo A is given by $\left\{ C_k(X, A), \partial^{(X,A)}_k \right\}_{k\in\mathbb{Z}}$*

---

[1] A set X is said *connected* if it is not the union of two disjoint open non-empty sets.

where $\partial_k^{(X,A)} : C_k(X,A) \to C_{k-1}(X,A)$ is defined by $\partial_k^{(X,A)}[c]_A := [\partial_k^X c]_A$. Obviously, this map satisfies that $\partial_k^{(X,A)} \partial_{k+1}^{(X,A)} = 0$. The relative chain complex gives rise to the relative k-cycles: $Z_k(X,A) := \ker \partial_k^{(X,A)}$, the relative k-boundaries $B_k(X,A) := \operatorname{im} \partial_{k+1}^{(X,A)}$, and finally the relative homology groups: $\mathbb{H}_k(X,A) := Z_k(X,A)/B_k(X,A)$.

Proposition 4 (Proposition 9.4 of [10]). *Let X be an (edge-)connected cubical set and let A be a nonempty cubical set of X. Then, $\mathbb{H}_0(X,A) = 0$.*

2.7   Exact Sequences

A sequence of groups and homomorphisms $\cdots \to G_3 \xrightarrow{\psi_3} G_2 \xrightarrow{\psi_2} G_1 \to \ldots$ is said *exact* at G_2 if $\operatorname{im} \psi_3 = \ker \psi_2$. It is an *exact sequence* if it is exact at every group. If the sequence has a first or a last element, then it is automatically exact at that group. A *short exact sequence* is an exact sequence of the form $0 \to G_3 \xrightarrow{\psi_3} G_2 \xrightarrow{\psi_2} G_1 \to 0$. A *long exact sequence* is an exact sequence with more than three nonzero terms.

Example 5 (Example 9.21 of [10]). *The short exact sequence of the pair (X,A) is:*
$$0 \longrightarrow C_k(A) \xrightarrow{\iota_k} C_k(X) \xrightarrow{\pi_k} C_k(X,A) \longrightarrow 0$$
where ι_k is the inclusion map and π_k is the quotient map.

Lemma 6 (Exact homology sequence of a pair [10]). *Let (X,A) be a cubical pair. Then there is a long exact sequence*
$$\cdots \to \mathbb{H}_{k+1}(A) \xrightarrow{\iota_*} \mathbb{H}_{k+1}(X) \xrightarrow{\pi_*} \mathbb{H}_{k+1}(X,A) \xrightarrow{\partial_*} \mathbb{H}_k(A) \to \cdots$$
where $\iota : \mathcal{C}(A) \hookrightarrow \mathcal{C}(X)$ is the inclusion map and $\pi : \mathcal{C}(X) \to \mathcal{C}(X,A)$ is the quotient map.

2.8   Manifolds and Local Homology

A subset X of $\mathbb{R}^n$ is said to be a *(n-dimensional) homology manifold* if for any $x \in X$ the homology groups $\{\mathbb{H}_i(X, X \setminus \{x\})\}_{i \in \mathbb{Z}}$ satisfy:

$$\mathbb{H}_i(X, X \setminus \{x\}) = \begin{cases} \mathbb{Z} & \text{when } i = n, \\ 0 & \text{otherwise.} \end{cases}$$

Theorem 7 ([17]). *A topological manifold is a homology manifold.*

2.9 Homotopical Equivalence

Let X, Y be two topological spaces, and f, g be two continuous functions from X to Y. We say that f and g are *homotopic* if there exists a continuous function $H : X \times [0,1] \to Y$ such that for any $x \in X$, $H(x,0) = f(x)$ and $H(x,1) = g(x)$. Furthermore, we say that X and Y are *homotopically equivalent* if there exist $f : X \to Y$ and $g : Y \to X$ such that $g \circ f$ is homotopic to Id_X and $f \circ g$ is homotopic to Id_Y.

3 DWCness Does Not Imply CWCness

It is well-known that DWCness and CWCness are equivalent in 2D and 3D (see, for example, [4]). In this section, we prove that there exists at least one set $\mathcal{X} \subset \mathbb{Z}^4$ which is DWC but not CWC.

To this aim, we will start with the definition of the set $\mathcal{X}$ and we will check that $\mathcal{X}$ is DWC. Then, to prove that $\mathcal{X}$ is not CWC, we will prove that $X = \mathrm{bdCA}(\mathcal{X})$ (up to a translation) is not a homology manifold and conclude that it is not a topological manifold by Theorem 7. To compute the homology groups $\{\mathbb{H}_i(X, X \setminus \{x^0\})\}_{i \in \mathbb{Z}}$, where x^0 is a particular point in X (detailed hereafter), we need to compute $\{\mathbb{H}_i(X \setminus \{x^0\})\}_{i \in \mathbb{Z}}$ and $\{\mathbb{H}_i(X)\}_{i \in \mathbb{Z}}$. However, $X \setminus \{x^0\}$ is not a cubical set, then we need to find a cubical set $\widetilde{X}(x^0)$ which is homotopy equivalent to $X \setminus \{x^0\}$ to compute its homology groups using the CHomP software package [8]. After having defined $\widetilde{X}(x^0)$ and having proven that it is a cubical set, we will show that $X \setminus \{x^0\}$ and $\widetilde{X}(x^0)$ are homotopically equivalent. Then, we will compute the homology groups of $\widetilde{X}(x^0)$ and of X; this way we will deduce $\{\mathbb{H}_i(X, X \setminus \{x^0\})\}_{i \in \mathbb{Z}}$ using the long exact sequence of the pair $(X, X \setminus \{x^0\})$. At this moment, we will see that X is not a homology 3-manifold, which will make us able to conclude that $\mathcal{X}$ is not CWC since the boundary of its continuous analog is not a topological 3-manifold. This way, we will conclude that DWCness does not imply CWCness in 4D.

3.1 Choosing a Particular DWC Set $\mathcal{X} \subset \mathbb{Z}^4$

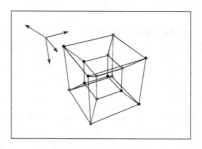

Fig. 2. A set $\mathcal{X} \subset \mathbb{Z}^4$ depicted by blue points which is DWC and not CWC. Blue lines show that the blue points are $2n$-connected ($n = 4$). (Color figure online)

We recall that it is well-known in the community of discrete topology that CWCness and DWCness are equivalent in 2D and 3D as developed in [2,4]. For this reason, we chose a digital set $\mathcal{X}$ in $\mathbb{Z}^4$ to study the relation between these two flavours of well-composedness in higher dimensions.

As depicted in Fig. 2, we can define the digital subset of $\mathbb{Z}^4$:

$$\mathcal{X} := \{\{0,0,0,0\}, \{0,0,0,1\}, \{0,0,1,1\}, \{0,1,1,1\},$$
$$\{1,1,1,1\}, \{1,1,1,0\}, \{1,1,0,0\}, \{1,0,0,0\}\}.$$

Let us check that $\mathcal{X}$ is DWC (see Fig. 2). It is easy to observe that it does not contain any 2D critical configuration. Now, to observe that there is no primary or secondary 3D critical configuration, we can simply look at the eight 3-faces (including the interior and the exterior cubes): since each one contains exactly four points of $\mathcal{X}$, they contain neither a primary critical configuration (made of two points) nor a secondary critical configuration (made of six points in the 3D case). Finally, we observe that the only 4D block that we have to consider is $\{0,1\}^4$ which contains eight points of $\mathcal{X}$, and eight points of $\mathcal{X}^c$, concluding that $\mathcal{X}$ contains neither a primary nor a secondary 4D critical configuration.

Property 8. *The digital set $\mathcal{X}$ is DWC.*

3.2 Finding a Cubical Set $\widetilde{X}(x^0)$ Homotopy Equivalent to $X \setminus \{x^0\}$

Let us start with the following proposition.

Proposition 9. *Let X be a cubical set in $\mathbb{R}^n$ and x^0 be a point of $X \cap \mathbb{Z}^n$. Then, the set:*

$$\widetilde{X}(x^0) := \{x \in X \setminus \{x^0\} \; ; \; \|x - x^0\|_\infty \geq 1\}$$

is cubical.

Proof. Our aim is to prove that $\widetilde{X}(x^0)$ is equal to $\bigcup\{h \in \mathcal{K}(X) \; ; \; x^0 \notin h\}$. This way, we will be able to conclude that $\widetilde{X}(x^0)$ is equal to $\bigcup\{h \in \mathcal{K}(X \setminus \{x^0\})\}$ and then it is a cubical set (since it is made of cubes and closed under inclusion). First, let us prove that:

$$\widetilde{X}(x^0) \subseteq \cup\{h \in \mathcal{K}(X) \; ; \; x^0 \notin h\}.$$

Let $x \in X \setminus \{x^0\}$ be a point such that $\|x - x^0\|_\infty \geq 1$. Then, there exists $i^* \in [\![1, n]\!]$ such that $\|x - x^0\|_\infty = |x_{i^*} - x_{i^*}^0| \geq 1$. Then two cases are possible:

(1) $x_{i^*} > x_{i^*}^0$, then $x_{i^*} \geq x_{i^*}^0 + 1$,
(2) $x_{i^*} < x_{i^*}^0$, then $x_{i^*} \leq x_{i^*}^0 - 1$.

Since $x \in X \setminus \{x^0\} \subset X$ where X is a cubical set, there exists a smaller face $h^* \in \mathcal{K}(X)$ (in the inclusion sense) such that $x \in h^* := \times_{i \in [\![1,n]\!]} [\lfloor x_i \rfloor, \lceil x_i \rceil]$. Then, $x^0 \in h^*$ iff for each $i \in [\![1, n]\!]$, $x_i^0 \in [\lfloor x_i \rfloor, \lceil x_i \rceil]$. However, since in case (1), $x_{i^*}^0 \leq x_{i^*} - 1 < \lfloor x_{i^*} \rfloor$, and in case (2), $x_{i^*}^0 \geq x_{i^*} + 1 > \lceil x_{i^*} \rceil$, then $x^0 \notin h^*$.

Obviously, $h^* \in \mathcal{K}(X)$: otherwise, all the cubes containing h^* do not belong to $\mathcal{K}(X)$, and then $x \notin X \setminus \{x^0\}$. This way, there exists a cube $h \in \mathcal{K}(X)$ such that $x^0 \notin h$ and $x \in h$.

Second, let us prove that: $\widetilde{X}(x^0) \supseteq \bigcup\{h \in \mathcal{K}(X) \; ; \; x^0 \notin h\}$. Let p be an element of $\bigcup\{h \in \mathcal{K}(X) \; ; \; x^0 \notin h\}$. In other words, $p \in h \in \mathcal{K}(X)$ and $x^0 \notin h$. Since $p \in h \in \mathcal{K}(X)$ and $x^0 \notin h$ then $p \in X \setminus \{x^0\}$. Now, let us write $h = \times_{i \in [\![1,n]\!]}[h_i^{\min}, h_i^{\max}]$, where $h^{\min}, h^{\max} \in \mathbb{Z}^n$. Since $x^0 \notin h$, there exists $i^* \in [\![1,n]\!]$ such that $x_{i^*}^0 \notin [h_{i^*}^{\min}, h_{i^*}^{\max}]$. Furthermore, we have:

(a) either $x_{i^*}^0 \leq h_{i^*}^{\min} - 1$,
(b) or $x_{i^*}^0 \geq h_{i^*}^{\max} + 1$.

Since $p \in h$, for each $i \in [\![1,n]\!]$, we have $p_i \in [h_i^{\min}, h_i^{\max}]$, and then $p_{i^*} \in [h_{i^*}^{\min}, h_{i^*}^{\max}]$. Then, in case (a), $x_{i^*}^0 \leq h_{i^*}^{\min} - 1 \leq p_{i^*} - 1$, which leads to $p_{i^*} - x_{i^*}^0 \geq 1$, and in case (b), $x_{i^*}^0 \geq h_{i^*}^{\max} + 1 \geq p_{i^*} + 1$, which leads to $x_{i^*}^0 - p_{i^*} \geq 1$. In both cases, we obtain that $\|p - x^0\|_\infty \geq 1$. $\qquad\square$

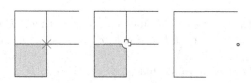

Fig. 3. $X \setminus \{x^0\}$ is homotopy equivalent to $\widetilde{X}(x^0)$: From left to right, a cubical set X (see the location of the central point x^0 in red), X minus its central point x^0 and the new cubical set $\widetilde{X}(x^0)$ homotopy equivalent to $X \setminus \{x^0\}$. (Color figure online)

Now, let us prove that $X \setminus \{x^0\}$ and $\widetilde{X}(x^0)$ are homotopy equivalent (as depicted on Fig. 3).

Proposition 10. *Let X be a cubical set in $\mathbb{R}^n$ and x^0 be a point of $X \cap \mathbb{Z}^n$. Then, $X \setminus \{x^0\}$ is homotopy equivalent to $\widetilde{X}(x^0)$.*

Proof. Let $f : X \setminus \{x^0\} \to \mathbb{R}^n$ be the function defined such as:

$$f(x) := \begin{cases} x & \text{when } \|x - x^0\|_\infty \geq 1, \\ x^0 + \frac{x - x^0}{\|x - x^0\|_\infty} & \text{otherwise,} \end{cases}$$

and let $g : \widetilde{X}(x^0) \to X \setminus \{x^0\}$ be the map from $\widetilde{X}(x^0)$ to $X \setminus \{x^0\}$ such that:

$$\forall x \in \widetilde{X}(x^0), \; g(x) = x,$$

which is possible since $\widetilde{X}(x^0) \subseteq X \setminus \{x^0\}$. Now, let us proceed step by step.

Step 1: $X \setminus \{x^0\}$ *and* $\widetilde{X}(x^0)$ *are topological spaces.* The sets $X \setminus \{x^0\}$ and $\widetilde{X}(x^0)$ are topological spaces since they are subsets of $\mathbb{R}^n$ supplied with the usual Euclidian distance.

***Step 2*:** *f is a map from $X \setminus \{x^0\}$ to $\widetilde{X}(x^0)$.* Let x be an element of $X \setminus \{x^0\}$. When $\|x - x^0\|_\infty \geq 1$, $f(x) = x$. This way, $f(x) \in X \setminus \{x^0\}$ and $\|f(x) - x^0\|_\infty \geq 1$, then $f(x) \in \widetilde{X}(x^0)$. When $\|x - x^0\|_\infty < 1$, $f(x) = x^0 + \frac{x - x^0}{\|x - x^0\|_\infty}$. This way, $\|f(x) - x^0\|_\infty = 1$. Since $x \in X \setminus \{x^0\} \subset X$ with X a cubical set, there exists a cube $h \in \mathcal{K}(X)$ such that $x \in h$. Furthermore, this cube h contains x^0 since all the cubes containing a point of $\times_{i \in [\![1,n]\!]}]x_i^0 - 1, x_i^0 + 1[$ contain also x^0 (the cubes are defined relatively to integral coordinates). Since $h = \times_{i \in [\![1,n]\!]} h_i$, then for each $i \in [\![1,n]\!]$, $x_i \in h_i$, and

$$(f(x))_i = x_i^0 + \frac{x_i - x_i^0}{\|x - x^0\|_\infty}.$$

Let us prove that this last equality shows that $f(x) \in h$. Since n is finite, there exists some $i^* \in [\![1,n]\!]$ such that: $\|x - x^0\|_\infty = |x_{i^*} - x_{i^*}^0|$, then:

$$(f(x))_i = x_i^0 + \frac{x_i - x_i^0}{|x_{i^*} - x_{i^*}^0|} \quad (P1)$$

Let us assume without constraint that $x_{i^*} > x_{i^*}^0$, then $(f(x))_{i^*} = x_{i^*}^0 + 1$. However, $x_{i^*} > x_{i^*}^0$ implies that $h_{i^*} = [x_{i^*}^0, x_{i^*}^0 + 1]$ since h contains x^0. When $i \neq i^*$, since $x_i \in h_i$, and since $h \ni x^0 \in \mathbb{Z}^n$, $h_i = [x_i^0, x_i^0 + 1]$ or $h_i = [x_i^0 - 1, x_i^0]$. Let us assume without constraint that $x_i > x_i^0$, then $h_i = [x_i^0, x_i^0 + 1]$. Because of $(P1)$, it follows easily that $(f(x))_i \in h_i$ since $\frac{x_i - x_i^0}{|x_{i^*} - x_{i^*}^0|} \in [0, 1]$.

Then, we have proven that when $\|x - x^0\|_\infty < 1$, there exists $h \in \mathcal{K}(X)$ such that for any $i \in [\![1,n]\!]$, $(f(x))_i \in h_i$, that is to say,

$$f(x) \in h. \quad (P2)$$

Also, $x \neq x^0$, which is equivalent to $f(x) \neq x^0$. Then, $f(x) \in \widetilde{X}(x^0)$.

***Step 3*:** *$g \circ f$ is homotopic to $\mathrm{Id}_{X \setminus \{x^0\}}$.*

– We can observe that $g \circ f : X \setminus \{x^0\} \to X \setminus \{x^0\}$ is the continuous function defined as:

$$g \circ f = \begin{cases} x & \text{when } \|x - x^0\|_\infty \geq 1, \\ x^0 + \frac{x - x^0}{\|x - x^0\|_\infty} & \text{otherwise.} \end{cases}$$

– Let $H : (X \setminus \{x^0\}) \times [0, 1] \to \mathbb{R}^n$ defined such that for any $x \in X \setminus \{x^0\}$ and any $\lambda \in [0, 1]$,

$$H(x, \lambda) := \lambda x + (1 - \lambda) g \circ f(x),$$

then:
  - H is continuous as a composition of continuous functions,
  - H is a function from $(X \setminus \{x^0\}) \times [0, 1]$ to $X \setminus \{x^0\}$:
    * when $\|x - x^0\|_\infty \geq 1$, $H(x, \lambda) = x \in X \setminus \{x^0\}$,

* when $\|x - x^0\|_\infty < 1$, $H(x, \lambda) = \lambda x + (1 - \lambda)f(x)$. However, we have seen that in this case, cf. $(P2)$, there exists a cube $h \in \mathcal{K}(X)$ such that $f(x) \in h$. Since h is a cube, it is convex, and then $H(x, \lambda) \in h$. Then, $H(x, \lambda) \in X$. Also, we can prove that $H(x, \lambda) \neq x^0$: the cases $\lambda = 0$ and $\lambda = 1$ are obvious; in the case $\lambda \in]0, 1[$, we can see that

$$H(x, \lambda) = \lambda x + (1 - \lambda)\left(x^0 + \frac{x - x^0}{\|x - x^0\|_\infty}\right)$$

and then, by assuming without constraints that $x^0 = 0$ and that for any $i \in [\![1, n]\!]$, $x_i \geq 0$, we obtain that for any $i \in [\![1, n]\!]$:

$$(H(x, \lambda))_i = \lambda x_i + (1 - \lambda)\frac{x_i}{\|x\|_\infty} = (\lambda(\|x\|_\infty - 1) + 1)\frac{x_i}{\|x\|_\infty},$$

then, because $(\|x\|_\infty - 1) < 1$ and $\frac{x_i}{\|x\|_\infty} \geq 0$, $(H(x, \lambda))_i$ is decreasing relatively to λ, and then

$$x_i \leq (H(x, \lambda))_i \leq \frac{x_i}{\|x\|_\infty}.$$

Since $x \neq 0$, there exists i^* such that $x_{i^*} \neq 0$, and then such that $(H(x, \lambda))_{i^*} > 0$ since $x_{i^*} > 0$. Therefore, $H(x, \lambda) \neq 0$, that is, $H(x, \lambda) \neq x^0$. Then, $H(x, \lambda) \in X \setminus \{x^0\}$.

- We can see that $H(x, 0) = g \circ f(x), \forall x \in X \setminus \{x^0\}$,
- We can also observe that $H(x, 1) = x, \forall x \in X \setminus \{x^0\}$.

Then $g \circ f$ is homotopic to $\mathrm{Id}_{X \setminus \{x^0\}}$.

Step 4: $f \circ g$ *is homotopic to* $\mathrm{Id}_{\widetilde{X}(x^0)}$. Since $f \circ g$ is equal to $\mathrm{Id}_{\widetilde{X}(x^0)}$, they are homotopic.

Step 5: *Conclusion.* $X \setminus \{x^0\}$ and $\widetilde{X}(x^0)$ are homotopically equivalent.   □

Corollary 11. *Assuming the notations of Proposition 9, we can compute the homology groups of* $X \setminus \{x^0\}$ *based on the ones of the cubical set* $\widetilde{X}(x^0)$. *Indeed, for each* $i \in \mathbb{Z}$, *we have the following equality:* $\mathbb{H}_i(X \setminus \{x^0\}) = \mathbb{H}_i(\widetilde{X}(x^0))$.

Proof. This follows from Propositions 9 and 10.   □

3.3 Defining the Cubical Set $\varphi_d(\mathcal{X})$

Let us define the mapping $\varphi : \mathbb{Z}/2 \to \mathcal{K}^1$:

$$\forall x \in \mathbb{Z}/2, \quad \varphi(x) := \begin{cases} [x, x+1] & \text{when } x \in \mathbb{Z}, \\ \{x + \frac{1}{2}\} & \text{otherwise.} \end{cases}$$

Then we can define: $\forall x \in (\mathbb{Z}/2)^d$, $\varphi_d(x) := \times_{i=1}^d \varphi(x_i)$ where, as usual, x_i denotes the ith coordinate of x; φ_d is then a bijection between $(\mathbb{Z}/2)^d$ and $\mathcal{K}^d$. Note that the underlying polyhedron of $\varphi_d(\mathcal{X})$ is equal to $\mathrm{CA}(\mathcal{X})$ up to a translation, and in this way they are topologically equivalent.

3.4   Choosing a Particular Point x^0 in the Boundary X of $\varphi_4(\mathcal{X})$

Let us begin with a simple property.

Property 12. *The point $x^0 := (1, 1, 1, 1)$ belongs to the boundary X of $\varphi_4(\mathcal{X})$.*

Proof. Let us recall that $\varphi_4(\mathcal{X})$ is a translation of the set $\mathrm{CA}(\mathcal{X})$ by a vector $v := (\frac{1}{2}, \frac{1}{2}, \frac{1}{2}, \frac{1}{2})$. Also, the point $p := (\frac{1}{2}, \frac{1}{2}, \frac{1}{2}, \frac{1}{2}) \in \mathbb{R}^4$ belongs to $\mathrm{CA}(\mathcal{X})$ and does not belong to $\mathrm{Int}(\mathrm{CA}(\mathcal{X}))$ since there does not exist any topological open ball $B(p, \varepsilon)$, $\varepsilon > 0$, in $\mathbb{R}^4$ centered at p and contained in $\mathrm{Int}(\mathrm{CA}(\mathcal{X}))$ since $B(p, \varepsilon)$ intersects $\mathrm{Int}(\mathrm{CA}(\mathcal{X}^c))$. This way, p belongs to $\mathrm{bdCA}(\mathcal{X})$. Finally, we obtain that the translation $x^0 = p + v$ of p by v belongs to the translation X of $\mathrm{bdCA}(\mathcal{X})$ by v. □

3.5   Computation of $\mathbb{H}(X, X \setminus \{x^0\})$

Let us compute the relative homology groups $\mathbb{H}_i(X, A)$ for each $i \in \mathbb{Z}$.

Obviously, since $\mathcal{C}_k(X, X \setminus \{x^0\}) = 0$ for $k \in \mathbb{Z} \setminus [\![0, 4]\!]$, then $\mathbb{H}_k(X, X \setminus \{x^0\}) = 0$. Also, thanks to Proposition 4, we know that $\mathbb{H}_0(X, X \setminus \{x^0\}) = 0$. To compute the other relative homology groups, we will use the exact homology sequence of $(X, X \setminus \{x^0\})$ discussed in Lemma 6: the sequence

$$\cdots \to \mathbb{H}_{k+1}(X \setminus \{x^0\}) \xrightarrow{\iota_*} \mathbb{H}_{k+1}(X) \xrightarrow{\pi_*} \mathbb{H}_{k+1}(X, X \setminus \{x^0\}) \xrightarrow{\partial_*} \mathbb{H}_k(X \setminus \{x^0\}) \to \cdots$$

is exact, and then by computing the homology groups $\mathbb{H}_k(X)$ and $\mathbb{H}_k(X \setminus \{x^0\})$ and using the First Isomorphism Theorem, we will be able to compute the local homology groups $\mathbb{H}_k(X, X \setminus \{x^0\})$ and to deduce if X is a homology 3-manifold or not.

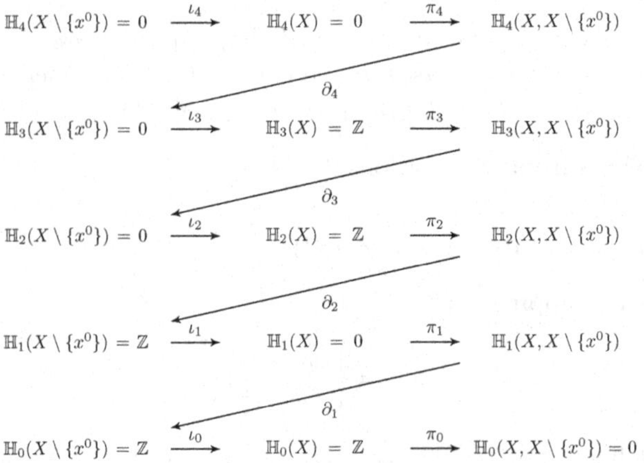

Fig. 4. Long exact sequence of the cubical pair $(X, X \setminus \{x^0\})$.

Using CHomP [8], we compute the local homology groups $\mathbb{H}(\widetilde{X}(x^0))$ and $\mathbb{H}_i(X)$ for $i \in [\![0, 4]\!]$. Using Corollary 11 we obtain $\mathbb{H}_i(X \setminus \{x^0\})$, and replacing this information in the long exact sequence discussed in Lemma 6, we obtain Fig. 4.

Let us compute $\mathbb{H}_4(X, X \setminus \{x^0\})$. By exactness, im $\pi_4 = 0 = \ker \partial_4$ and $\mathbb{H}_4(X, X \setminus \{x^0\})/\ker \partial_4 \simeq \mathrm{im}\, \partial_4 = 0$, then $\mathbb{H}_4(X, X \setminus \{x^0\}) = 0$.

Now, let us compute $\mathbb{H}_3(X, X \setminus \{x^0\})$. By exactness, im $\iota_3 = 0 = \ker \pi_3$, and $\mathbb{Z}/\ker \pi_3 \simeq \mathrm{im}\, \pi_3 \simeq \mathbb{Z}$, then im $\pi_3 = \ker \partial_3 = \mathbb{H}_3(X, X \setminus \{x^0\}) \simeq \mathbb{Z}$.

Concerning $\mathbb{H}_2(X, X \setminus \{x^0\})$, by exactness, im $\iota_2 = 0 = \ker \pi_2$, and $\mathbb{Z}/\ker \pi_2 \simeq \mathrm{im}\, \pi_2 \simeq \mathbb{Z} \simeq \ker \partial_2$. Also, $\ker(\iota_1) = \mathbb{Z} = \mathrm{im}\, \partial_2$ and $\mathbb{H}_2(X, X \setminus \{x^0\})/\ker \partial_2 \simeq \mathrm{im}\, \partial_2$ imply that: $\mathbb{H}_2(X, X \setminus \{x^0\}) \simeq \mathbb{Z}^2$.

Finally, let us compute $\mathbb{H}_1(X, X \setminus \{x^0\})$. By exactness, im $\pi_1 = 0 = \ker \partial_1$. Also, $\ker \pi_0 = \mathbb{Z} \simeq \mathrm{im}\, \iota_0$, and $\mathbb{Z}/\ker \iota_0 \simeq \mathrm{im}\, \iota_0$ imply that $\ker \iota_0 = 0 = \mathrm{im}\, \partial_1$. Then, $\mathbb{H}_1(X, X \setminus \{x^0\}) = 0$.

3.6 Our Final Observation

Fig. 5. Projection in the 3D space of the continuous analog of the 4D counter-example. Each color corresponds to a same (projected) hypercube. Note that the pinch is not observable in 3D.

Since we have $\mathbb{H}_2(X, X \setminus \{x^0\}) \simeq \mathbb{Z}^2 \neq 0$, X is not a homology 3-manifold, and then it is not a topological 3-manifold, which implies that DWCness does not imply CWCness in 4D, which contradicts the conjecture arguing that DWCness and CWCness are equivalent in nD on cubical grids [2]. See Fig. 5 for some 3D projections of the continuous analog of our 4D counter-example. Furthermore, this counter-example shows that a digital set which is *well-composed in the sense of Alexandrov (AWC)* [6,16] is not always CWC, since it has been proven in [5] that AWCness and DWCness are equivalent in nD.

4   Conclusion

The counter-example presented in this paper shows that in 4D, DWCness does not imply CWCness. It shows how much it is important to explicit which flavour of well-composedness we consider when we work with nD discrete images.

Furthermore, two questions arise in a natural way. First, is it possible to find a generic counter-example showing that DWCness does not imply CWCness in any dimension greater than 3 (like the product of the set $\mathcal{X}$ with $\{0,1\}^{n-4}$). Second, does CWCness imply DWCness in nD? This last question seems intuitive but we will show in a future report that it is far from being so simple.

References

1. Alexandrov, P.S.: Combinatorial Topology, vol. 1-3. Graylock, Rochester (1956)
2. Boutry, N.: A study of well-composedness in n-D. Ph.D. thesis, Université Paris-Est, France (2016)
3. Boutry, N., Géraud, T., Najman, L.: How to make nD functions digitally well-composed in a self-dual way. In: Benediktsson, J.A., Chanussot, J., Najman, L., Talbot, H. (eds.) ISMM 2015. LNCS, vol. 9082, pp. 561–572. Springer, Cham (2015). https://doi.org/10.1007/978-3-319-18720-4_47
4. Boutry, N., Géraud, T., Najman, L.: A tutorial on well-composedness. J. Math. Imaging Vis. (2017). https://doi.org/10.1007/s10851-017-0769-6
5. Boutry, N., Najman, L., Géraud, T.: About the equivalence between AWCness and DWCness. Research report, LIGM - Laboratoire d'Informatique Gaspard-Monge; LRDE - Laboratoire de Recherche et de Développement de l'EPITA, October 2016. https://hal-upec-upem.archives-ouvertes.fr/hal-01375621
6. Boutry, N., Najman, L., Géraud, T.: Well-composedness in alexandrov spaces implies digital well-composedness in $\mathbb{Z}^n$. In: Kropatsch, W.G., Artner, N.M., Janusch, I. (eds.) DGCI 2017. LNCS, vol. 10502, pp. 225–237. Springer, Cham (2017). https://doi.org/10.1007/978-3-319-66272-5_19
7. Caselles, V., Monasse, P.: Geometric Description of Images as Topographic Maps. Lecture Notes in Mathematics, vol. 1984. Springer, Heidelberg (2009). https://doi.org/10.1007/978-3-642-04611-7
8. CHomP: http://chomp.rutgers.edu/software/
9. Géraud, T., Carlinet, E., Crozet, S., Najman, L.: A quasi-linear algorithm to compute the tree of shapes of nD images. In: Hendriks, C.L.L., Borgefors, G., Strand, R. (eds.) ISMM 2013. LNCS, vol. 7883, pp. 98–110. Springer, Heidelberg (2013). https://doi.org/10.1007/978-3-642-38294-9_9
10. Kaczynski, T., Mischaikow, K., Mrozek, M.: Computational Homology, vol. 157. Springer, New York (2006). https://doi.org/10.1007/b97315
11. Kelley, J.L.: General Topology. Graduate Texts in Mathematics, vol. 27. Springer, New York (1955)
12. Lachaud, J.O.: Espaces non-euclidiens et analyse d'image: modèles déformables riemanniens et discrets, topologie et géométrie discrète. Ph.D. thesis, Université Sciences et Technologies-Bordeaux I (2006)
13. Lachaud, J.O., Thibert, B.: Properties of Gauss digitized shapes and digital surface integration. J. Math. Imaging Vis. **54**(2), 162–180 (2016)
14. Latecki, L.J.: 3D well-composed pictures. Graph. Models Image Process. **59**(3), 164–172 (1997)
15. Latecki, L.J.: Well-composed sets. Adv. Imaging Electron Phys. **112**, 95–163 (2000)

16. Najman, L., Géraud, T.: Discrete set-valued continuity and interpolation. In: Hendriks, C.L.L., Borgefors, G., Strand, R. (eds.) ISMM 2013. LNCS, vol. 7883, pp. 37–48. Springer, Heidelberg (2013). https://doi.org/10.1007/978-3-642-38294-9_4

17. Ranicki, A., Casson, A., Sullivan, D., Armstrong, M., Rourke, C., Cooke, G.: The hauptvermutung book. Collection of papers by Casson, Sullivan, Armstrong, Cooke, Rourke and Ranicki, K-Monographs in Mathematics 1 (1996)

3D-Array Token Petri Nets Generating Tetrahedral Picture Languages

T. Kalyani[1], K. Sasikala[2], D. G. Thomas[3](✉), Thamburaj Robinson[4],
Atulya K. Nagar[5], and Meenakshi Paramasivan[6]

[1] Department of Mathematics, St. Joseph's Institute of Technology,
Chennai 119, India
kalphd02@yahoo.com
[2] Department of Mathematics, St. Joseph's college of Engineering,
Chennai 119, India
sasikalaveerabadran2013@gmail.com
[3] Department of Science and Humanities, (Mathematics Division),
Saveetha School of Engineering, SIMATS, Chennai 602105, India
dgthomasmcc@yahoo.com
[4] Department of Mathematics, Madras Christian College, Chennai 59, India
robinson@mcc.edu.in
[5] Department of Computer Science, Liverpool Hope University, Hope Park,
Liverpool L169JD, UK
nagara@hope.ac.uk
[6] Department of Computer Science, University of Trier, Trier, Germany
meena_maths@yahoo.com

Abstract. The study of two-dimensional picture languages has a wide application in image analysis and pattern recognition [5,7,13,17]. There are various models such as grammars, automata, P systems and Petri Nets to generate different picture languages available in the literature [1–4,6,8,10–12,14,15,18]. In this paper we consider Petri Nets generating tetrahedral picture languages. The patterns generated are interesting, new and are applicable in floor design, wall design and tiling. We compare the generative power of these Petri Nets with that of other recent models [9,16] developed by our group.

Keywords: Petri Net · Tetrahedral tiles · P systems

1 Introduction

The art of tiling plays an important role in human civilization. A two-dimensional pattern generating model called pasting system was introduced in the literature which glues two square tiles together at the edges. Later on two isosceles right angled triangular tiles are pasted together at the gluable edges and a new pasting system called triangular tile pasting system was introduced in [3].

Petri Nets are mathematical models introduced to model dynamic systems [15]. Tokens represented by black dots are used to simulate the dynamic activity

© Springer Nature Switzerland AG 2020
T. Lukić et al. (Eds.): IWCIA 2020, LNCS 12148, pp. 88–105, 2020.
https://doi.org/10.1007/978-3-030-51002-2_7

of the system. Array token Petri Nets are models which generate array languages [12]. Array Token Petri nets have applications in the following areas namely character recognition, generation and recognition of picture patterns, tiling pattern and kolam patterns. Arrays are used as token. The transitions are associated with catenation rules. Firing of transitions catenate arrays to grow in bigger size.

The area of membrane computing is a new computability model called P system introduced by Gh. Păun inspired by the functioning of living cells. Ceterchi et al. [4] proposed a theoretical model of P-system called Array Rewriting P-system for generating two-dimensional patterns. Motivated by these studies a three-dimensional pattern generating model called tetrahedral tile pasting system and tetrahedral tile pasting P system were introduced in [9] by gluing two tetrahedral tiles at the glueable edges. In the literature the studies on membrane computing generating picture languages is very limited. We have used membrane computing to generate 3D Tetrahedral picture languages, in which we can generate both rectangular and non rectangular 3D pictures like stars, triangles, rhombuses, hexagons, octagons and some kolam patterns which are some of the interesting patterns.

In this paper we introduce 3D-Array token Petri Nets generating three-dimensional tetrahedral picture languages (3D-TetATPN) and this model is compared with K-Tabled Tetrahedral Tile Pasting System (K-TTTPS), Tetrahedral Tile Pasting P System (TetTPPS), Regular Tetrahedral Array Languages (RTAL) and Basic Puzzle Tetrahedral Array Languages (BPTAL) for generative powers. The patterns generated by the Petri Nets are useful in floor design, wall design and tiling.

2 Preliminaries

In this section we recall the notion of tetrahedral tiles, K-TTTPS and TetTPPS
.

Definition 1. *[9] A tetrahedral tile is a polyhedral which has four vertices, four faces and six edges. Each face is an equilateral triangle. f_4 is the base of the tetrahedron(Fig. 1).*

Fig. 1. A Tetrahedron.

We consider tetrahedral tiles of the following four types, named as

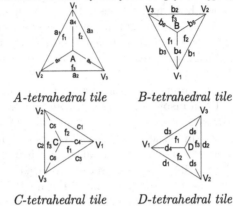

A-tetrahedral tile   B-tetrahedral tile

C-tetrahedral tile   D-tetrahedral tile

f_4 is the base $V_1 V_2 V_3$ of the tetrahedral tile.

Definition 2. *[9] A K-Tabled Tetrahedral Tile Pasting System (K-TTPS) is a 4-tuple $M = (\Gamma, E, P, t_0)$, where Γ is a finite set of tetrahedral tiles of the forms A and B. E is a set of edge labels of base of tetrahedral tiles A and B. P is a finite set of tables $\{T_1, T_2, \ldots T_K\}$ where $T_1, T_2, \ldots T_K$ $(k \geq 1)$ are finite sets of pasting rules. t_0 is the axiom pattern.*

A tiling pattern t_{i+1} is generated from a pattern t_i in k stages. In each stage, the rules of the table T_i $(i = 1, 2, \ldots k)$ are applied in parallel to the boundary edges of the pattern obtained in the previous stage. When all the rules in P are applied one after the other in succession the pattern t_{i+1} is generated from t_i. i.e. $t_i \Rightarrow t_{i+1}$. We write $t_0 \overset{*}{\Rightarrow} t_j$ if $t_0 \Rightarrow t_1 \Rightarrow t_2 \Rightarrow \cdots \Rightarrow t_j$. The collection of all patterns generated by K-TTPS derived from the axiom t_0 using the pasting rules of the system M is denoted by $T(M) = \{t_j \in \Gamma^{***} : t_0 \overset{*}{\Rightarrow} t_j / j \geq 0\}$, where Γ^{***} represents the set of all three-dimensional tetrahedral patterns obtained by gluing tetrahedral tiles of Γ.

The family of all three-dimensional patterns generated by K-TTPS is denoted as $\mathcal{L}(K\text{-}TTPS)$.

Example 1. A one-tabled Tetrahedral Tile Pasting System, generating a sequence of three-dimensional patterns whose boundaries are hexagons and stars alternatively is given below:

$$M = (\Gamma, E, P, t_0) \text{ where } \Gamma = \left\{ \begin{array}{c} a_1 \triangle a_3, b_3 \overset{b_2}{\triangledown} b_1 \\ a_2 \end{array} \right\}$$

$$E = \{a_1, a_2, a_3, b_1, b_2, b_3\}; P = \{T_1\};$$
$$T_1 = \{(a_1, b_1), (a_2, b_2), (a_3, b_3), (b_1, a_1), (b_2, a_2), (b_3, a_3)\}$$

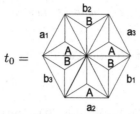

$$t_0 =$$

The first three members of $T(M)$ are shown in Fig. 2.

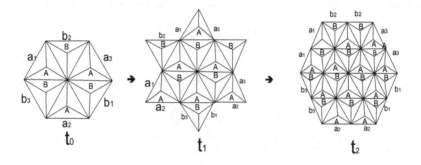

Fig. 2. Hexagon and Star polyhedral.

Definition 3. *[9] A Tetrahedral Tile Pasting P system (TetTPPS) $\Pi = (\Gamma, \mu, F_1, \ldots, F_m, R_1, R_2, \ldots, R_m, i_0)$ where Γ is a finite set of labeled tetrahedral tiles; μ is a membrane structure with m membranes, labeled in an one-to-one way with $1, 2, \ldots m$; $F_1, F_2, \ldots F_m$ are finite sets of three-dimensional picture patterns over tiles of Γ associated with the m regions of μ; $R_1, R_2, \ldots, R_m$ are finite sets of pasting rules of the type $(t_i, (x_i, y_i), tar)$, $1 \leq i \leq n$ associated with the m regions of μ and i_0 is the output membrane which is an elementary membrane.*

The computation process in TetTPPS is defined as, to each 3D-Picture pattern present in the region of the system, the pasting rule associated with the respective region should be applied in parallel to the boundary edges of the base of the tetrahedral tile. Then the resultant tetrahedral 3D-pattern is moved (remains) to another region (in the same region) with respect to the target indicator in_j (here) associated with the pasting rule. If the target indicator is out, then the resultant tetrahedral 3D-pattern is sent immediately to the next outer region of the membrane structure.

The computation is successful only if the pasting rules of each region are applied. The computation stops if no further application of pasting rule is applicable. The result of a halting computation consists of the 3D-picture patterns composed only of tetrahedral tiles from Γ placed in the membrane with label i_0 in the halting configuration.

The set of all such tetrahedral 3D-patterns computed or generated by a TetTPPS Π is denoted by TetPL(Π). The family of all such languages

TetPL(Π) generated by system Π with at most m membranes, is denoted by TetPL_m (TetTPPS).

Example 2. Consider the Tetrahedral Tile Pasting P System, TetTPPS

$$\Pi_1 = (\Gamma, \mu = [_1[_2[_3]_3]_2]_1, F_1, F_2, F_3, R_1, R_2, R_3, 3),$$

which generates a sequence of tetrahedral 3D-picuture patterns whose boundaries are hexagons, μ indicates that the system has three regions one within another i.e. region 1 is the 'skin' membrane which contains region 2, which in turn contains region 3, $i_0 = 3$ indicates that region 3 is the output region.

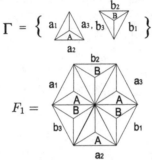

$$R_1 = \{(B, (a_1, b_1), here), (A, (b_1, a_1), here), (B, (a_2, b_2), here),$$
$$(A, (b_3, a_3), here), (B, (a_3, b_3), here), (A, (b_2, a_2), in)\}$$
$$R_2 = \{(B, (a_1, b_1), here), (A, (b_1, a_1), here), (B(a_2, b_2), here),$$
$$(A, (b_3, a_3), here), (B, (a_3, b_3), here), (A, (b_2, a_2), in), (A, (b_2, a_2), out)\}$$
$$R_3 = \emptyset.$$

Beginning with the initial object F_1 in region 1, the pasting rule R_1 is applied, where the rules in R_1 are applied in parallel to the boundary edges of the picture pattern present in region 1. Once the rule $(A, (b_2, a_2), in)$ is applied, the generated 3D-pattern is sent to the inner membrane 2, and in region 2, the rules of R_2 are applied in parallel to the boundary edges of the pattern generated in region 1. If the rule $(A, (b_2, a_2), out)$ is applied, the 3D-pattern generated is sent to region 1, and the process continues. Whereas if the rule $(A, (b_2, a_2), in)$ is applied the 3D-pattern generated is sent to region 3, which is the output region, wherein it is collected in the 3D-picture pattern language formed by TetTPPSΠ_1. TetTPPSΠ_1 is the tetrahedral 3D-Picture language whose boundary is the hexagon.

3   3D-Array Token Petri Nets

In this section we recall some notions of Petri Nets. For more details we refer to [15]. Here we introduce catenation rules and firing rules for 3D-Array token Petri Nets and Tetrahedral 3D-Array token Petri Nets structure.

Definition 4. *[11] A Petri Net structure is a four tuple $C = (P, T, I, O)$ where $P = \{P_1, P_2, \ldots P_n\}$ is a finite set of places, $n \geq 0$, $T = \{t_1, t_2, \ldots t_m\}$ is a finite set of transitions $m \geq 0$, $P \cap T = \emptyset$, $T \to P^\infty$ is the input function from transitions to bags of places and $O : T \to P^\infty$ is the output function from transitions to bags of places.*

Definition 5. *[11] A Petri Net marking is an assignment of tokens to the places of a Petri Net. The tokens are used to define the execution of a Petri Net. The number and position of tokens may change during the execution of a Petri Net. The marking can be defined as an n-vector $\mu = (\mu_1, \mu_2, \mu_3, \ldots \mu_n)$ where μ_i is the number of tokens in the P_i, $i = 1, 2, \ldots, n$. We can also write $\mu(P_i) = \mu_i$.*

Definition 6. *[11] A Petri Net C with initial marking μ is called a marked Petri Net. A marked Petri Net $M = (C, \mu)$ can also be written as $M = (P, T, I, O, \mu)$.*

When a transition is fired one token is removed from its input place and one token is placed in each of its output place. For example when t_1 is fired in the following figure one token from place A is removed and one token is placed in both B & C which are the output places of t_1.

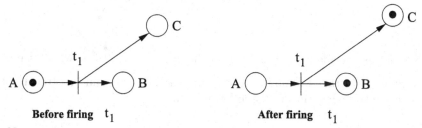

Before firing t_1 **After firing** t_1

Now we turn our attention to define Tetrahedral 3D Array Token Petri Net.

Catenation Rules
The catenation rules which glue any two tetrahedral tiles at the glueable edges are given below:

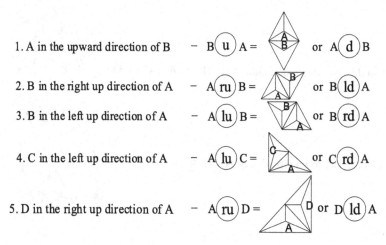

1. A in the upward direction of B $-$ $B\,(\mathbf{u})\,A =$ or $A\,(\mathbf{d})\,B$

2. B in the right up direction of A $-$ $A\,(\mathbf{ru})\,B =$ or $B\,(\mathbf{ld})\,A$

3. B in the left up direction of A $-$ $A\,(\mathbf{lu})\,B =$ or $B\,(\mathbf{rd})\,A$

4. C in the left up direction of A $-$ $A\,(\mathbf{lu})\,C =$ or $C\,(\mathbf{rd})\,A$

5. D in the right up direction of A $-$ $A\,(\mathbf{ru})\,D =$ or $D\,(\mathbf{ld})\,A$

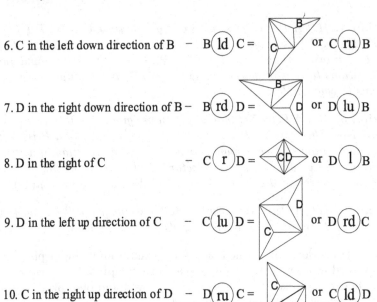

6. C in the left down direction of B   – $B\left(ld\right)C =$ or $C\left(ru\right)B$

7. D in the right down direction of B – $B\left(rd\right)D =$ or $D\left(lu\right)B$

8. D in the right of C   – $C\left(r\right)D =$ or $D\left(l\right)B$

9. D in the left up direction of C   – $C\left(lu\right)D =$ or $D\left(rd\right)C$

10. C in the right up direction of D – $D\left(ru\right)C =$ or $C\left(ld\right)D$

Now let us consider the hexagonal polyhedral $H =$, which is made up of gluing A-tetrahedral and B tetrahedral tiles. This H can be catenated to A and B - tetrahedral tiles in the following manner.

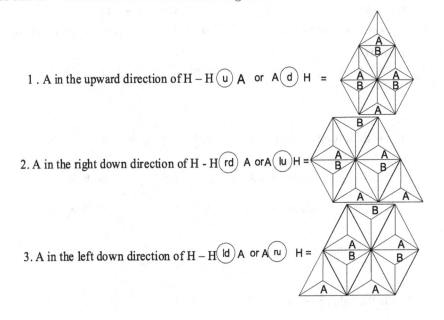

1 . A in the upward direction of $H - H\left(u\right)A$ or $A\left(d\right)H =$

2. A in the right down direction of H - $H\left(rd\right)A$ or$A\left(lu\right)H =$

3. A in the left down direction of $H - H\left(ld\right)A$ or $A\left(ru\right)H =$

4. B in the down ward direction H - H(d) B or B(u) H =

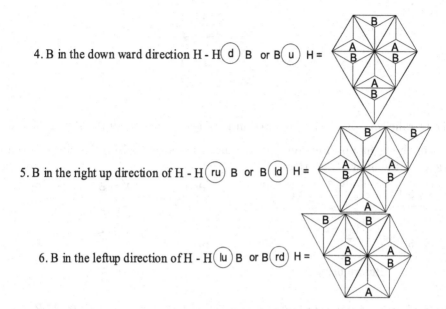

5. B in the right up direction of H - H(ru) B or B(ld) H =

6. B in the leftup direction of H - H(lu) B or B(rd) H =

Firing Rules

The transitions of the Petri Net are associated with the catenation rules of the form P (cat) Q where $P, Q \in \{A, B, C, D, H\}$ and (cat) is any one of the above catenation rules. When transition fires, the array in the input place gets catenated according to the catenation rule and the resultant array is placed in the output place. The transitions will be enabled as per the following conditions.

(i) All the input places will have the same array as token.
(ii) If there is no label for the transition then the same array will be moved to all the output places.
(iii) If there is a label i.e a catenation rule for the transition then the array in the input place gets catenated according to the catenation rule and the resultant array is moved to all the output places.

Example 3. If the input place of transition has the tetrahedral polyhedral

S =

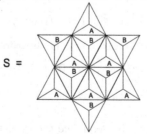

as token and the transition is attached to the rule A (ru) B then after the firing the output places of the transition will have the tetrahedral picture

$$T = $$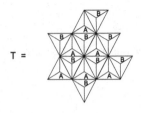

The tetrahedral tile - B is catenated in parallel manner to all the A-tetrahedral tiles in the right up direction. The 3D-array token Petri Net diagram is given below for the above transition.

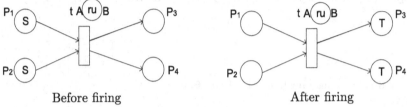

Before firing   After firing

Definition 7. *A 3D Tetrahedral Tile Array Token Petri Net (3D-TetATPN) is a six tuple $N = (\Sigma, C, \mu, S, \sigma, F)$ where Σ is an alphabet of tetrahedral tiles or extended tetrahedral tiles (3D-picture made up of tetrahedral tiles), C is a Petri Net structure, μ is an initial marking of 3D-pictures made up of tetrahedral tiles or extended tetrahedral tiles kept in some places of the net, S is a set of catenation rules, σ is a partial mapping which attaches rules to the various transitions of the Petri Net of the form $\sigma(t_i) = P \textcircled{cat} Q$, F is a subset of the set of places of the Petri Net where the final 3D-tetrahedral picture is stored after all the firing of the various possible transitions of the Petri Net.*

Definition 8. *The language generated by 3D-TetATPN is the set of all 3D-tetrahedral pictures stored in the final places of the Petri Net structure and is denoted by $\mathcal{L}(N)$.*

Example 4. Consider the 3D-TetATPN $N_1 = (\Sigma, C, \mu, S, \sigma, F)$ where $\Sigma = \{R, A, B\}$ where $R = \triangle$, $C = (P, T, I, O)$ where $P = \{P_1, P_2, P_3, P_4, P_5\}$, $T = \{t_1, t_2, t_3, t_4\}$. The initial marking μ is the rhombus polyhedral R in the place of P_1. $S = \{\text{H}\textcircled{ru}\text{B}, \text{H}\textcircled{rd}\text{A}, \text{B}\textcircled{rd}\text{A}, \text{D}\textcircled{ru}\text{C}, \text{C}\textcircled{ru}\text{H}, \text{C}\textcircled{rd}\text{H}, \text{H}\textcircled{rd}\text{H}\}$ σ the mapping from the set of transitions to the set of rules is shown in Fig. 3 and $F = \{P_5\}$.

Starting with R, on firing the sequence $t_1 t_2 t_3 t_4$, the rhombus polyhedral is generated. The first two members of the language are shown in the following Fig. 4.

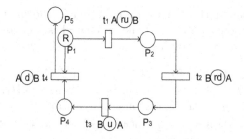

Fig. 3. The 3D - Array token Petri Net generating rhombus polyhedral.

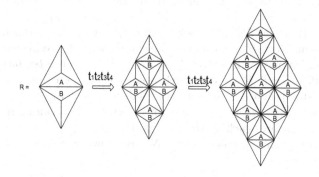

Fig. 4. Rhombus polyhedral.

Example 5. Consider the 3D-TetATPN $N_2 = (\Sigma, C, \mu, S, \sigma, F)$ where, $\Sigma = \{H, A, B, C, D\}$, $C = (P, T, I, O)$, $P = \{P_1, P_2, P_3, \ldots P_8\}$, $T = \{t_1, t_2, t_3, \ldots t_7\}$. The initial marking μ is the hexagonal polyhedral H in the place of P_1.

$$S = \{A \ \textcircled{ru} \ B, B \ \textcircled{rd} \ A, B \ \textcircled{u} \ A, A \ \textcircled{d} \ B\}.$$

σ the mapping from the set of transitions to the set of rules is shown in Fig. 5 and $F = \{P_8\}$

Starting with H on firing the sequence $t_1 t_2 t_3 t_4$ the tetrahedral tiles B, A, D and C are catenated to H according to the catenation rules respectively and the resultant 3D-array is sent out to place P_5. On firing the sequence $t_5 t_6$ Hexagonal polyhedrals are catenated to C-tetrahedral tile in parallel in the right up and right down directions and then firing t_7 hexagonal polyhedrals are catenated to hexagonal polyhedrals in the right down direction in parallel and finally the resultant sequence of hexagonal polyhedral language is sent to the final place P_8.

The first member of the language generated is shown in Fig. 6. In every generation the hexagonal polyhedral tile catenated is increased by one. In the first member two hexagonal polyhedral are catenated twice.

Example 6. Consider the 3D-TetATPN $N_3 = (\Sigma, C, \mu, S, \sigma, F)$ where, $\Sigma = \{H, A, B\}$, $C = (P, T, I, O)$, $P = \{P_1, P_2, P_3, \ldots P_{12}\}$, $T = \{t_1, t_2, t_3, \ldots t_{12}\}$. The initial marking μ is the hexagonal polyhedral H in the place of P_1.

$$S = \left\{ \begin{array}{l} H\,(u)\,A,\ H\,(ru)\,B, H\,(rd)\,A,\ H\,(d)\,B,\ H\,(ld)\,A,\ H\,(lu)\,B,\ A\,(ru)\,B,\ A\,(lu)\ B, \\ A\,(d)\,B, B\,(u)\,A,\ B\,(rd)\,A,\ B\,(ld)\,A \end{array} \right\}$$

σ the mapping from the set of transitions to the set of rules is shown in Fig. 7 and $F = \{P_{13}\}$.

Starting with H on firing the sequence $t_1 t_2 t_3 t_4 t_5 t_6$ the tetrahedral tiles A and B are catenated to H according to the catenation rules respectively and the resultant 3D-array is sent to place P_7 on firing the sequence $t_7 t_8 t_9 t_{10} t_{11} t_{12}$ the tetrahedral tiles A and B are catenated according to the catenation rules and the resultant star polyhedral is generated and it is sent to place P_{13} which is the final place or the sequence of transitions $t_7 t_8 t_9 t_{10} t_{11} t_{12}$ can be repeated any number of times before reaching the final destination P_{13}.

The first two members generated by N_3 are shown in Fig. 8.

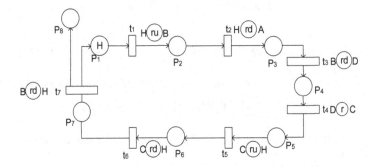

Fig. 5. 3D-Array token Petri Net generating increasing sequence of hexagonal polyhedrals.

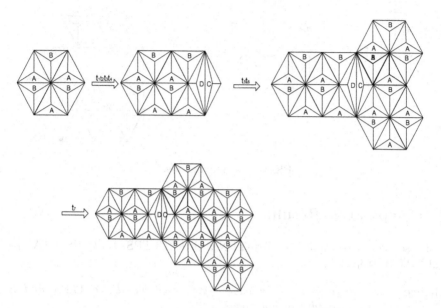

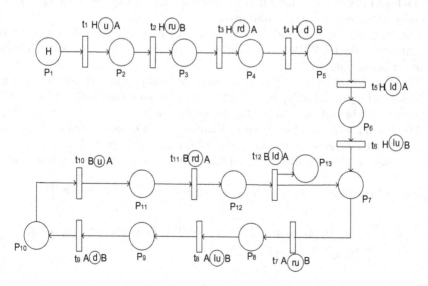

Fig. 6. First member of the language generated by N_2.

Fig. 7. 3D-array token Petri Net generating sequence of star polyhedrals.

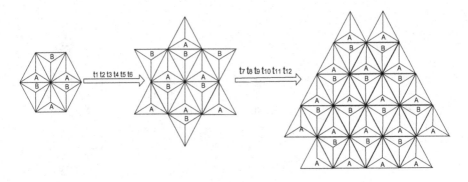

Fig. 8. Star polyhedrals.

4   Comparative Results

In this section, we compare 3D-TetATPN with K-TTTPS, TetTPPS, RTAL [16] and BPTAL [16].

Theorem 1. *The families of languages generated by 3D-TetATPN and K-TTTPS are incomparable but not disjoint.*

Proof. The families of languages generated by K-TTTPS and 3D-TetATPN are by parallel mechanism. The constraint in K-TTTPS is that the pasting rules of the tables are applied in parallel to the pattern obtained in the previous stage. In 3D-TetATPN the catenation rules generate the family of languages where extended tetrahedral tiles are also used.

The language of star and hexagonal polyhedrals in Example 1 cannot be generated by 3D-TetATPN, since the catenation rules are applied in parallel wherever applicable.

The language of increasing sequence of hexagonal polyhedrals given in Example 5 cannot be generated by K-TTTPS, since extended tetrahedral tile, namely hexagonal polyhedral, is used in the catenation rules.

The language of rhombus given in Example 4 can be generated by both systems. A 3-TTTPS generating the family of rhombuses is given below:

Consider a three tabled Tetrahedral Tile Pasting System generating a sequence of rhombuses, $M = (\Gamma, E, P, t_0)$ where $\Gamma = \{\ a_1 \triangle a_3, b_3 \triangledown b_1\ \}$

$$E = \{a_1, a_2, a_3, b_1, b_2, b_3\}, P = \{T_1, T_2, T_3\}, T_1 = \{(a_3, b_3), (b_1, a_1)\},$$
$$T_2 = \{(b_2, a_2), (b_1, a_1)\} T_3 = \{(a_2, b_2)\}.$$

$$t_0 = \diamondsuit$$

□

Theorem 2. *The families of languages generated by 3D-TetATPN and TetTPPS are incomparable but not disjoint.*

Proof. In 3D-TetATPN the catenation rules are applied in parallel to generate the language concerned and extended tetrahedral tiles are also used. In TetTPPS the pasting rules are applied in parallel and the target indications permits the array generated to transit within the regions.

The language of stars and hexagonal polyhedrals given in Example 2 generated by TetTPPS cannot be generated by 3D-TetATPN as the catenation rules are applied in parallel wherever applicable.

The language of increasing sequence of hexagonal polyhedrals given in Example 5 cannot be generated by TetTPPS, since extended tetrahedral tiles, namely hexagonal polyhedral, is used in the catenation rules.

The language of rhombuses given in Example 4 is generated by both systems. TetTPPS generating the language of rhombuses is given below. Consider TetTPPS $\pi_2 = (\Gamma, \mu = [_1[_2[_3]_3]_2]_1, F_1, F_2, F_3, R_1, R_2, R_3, 3)$. μ- indicates that the system has three regions one within the another i.e region 1 is the skin membrane which contains region 2, which in turn contains region 3, $i_0 = 3$ indicates that region 3 is the output region.

$$\Gamma = \left\{ \, a_1 \!\!\!\bigtriangleup\!\!\! a_3, b_3 \!\!\!\bigtriangledown\!\!\! b_1 \, \right\} \, , F_1 = \diamondsuit, F_2 = F_3 = \emptyset$$

$$R_1 = \{(A, (b_2, a_2), here), (B, (a_3, b_3), here), (A, (b_1, a_1), in)\}$$
$$R_2 = \{(B, (a_2, b_2), in), (B, (a_2, b_2), out)\}, R_3 = \emptyset.$$

Beginning with the initial object F_1 in region 1, the pasting rule R_1 is applied, where the rules in region 1 are applied in parallel to the boundary edges of the pattern present in region 1. Once the rule $(B, (a_3, b_3), here)$ is applied, the tetrahedral tile B is catenated to A and then, when the rule $(A, (b_1, a_1), in)$ is applied, the picture pattern generated is sent to the inner region 2. In region 2, when the rule $(B, (a_2, b_2), in)$ is applied the generated picture pattern is sent to region 3, which is the output region, where there is no rule exits and the language of rhombus is collected in region 3. Whereas if the rule $(B, (a_2, b_2), out)$ is applied, the generated picture pattern is sent to region 1 and the process continues. $\square$

Theorem 3. $\mathcal{L}(3D - TetATPN) - RTAL \neq \emptyset$.

Proof. We consider a tetrahedral language whose boundary is an equilateral triangle. This language cannot be generated by any Regular Tetrahedral Array Grammar (RTAG) [16]. Since the rules in RTAG are of the following forms:

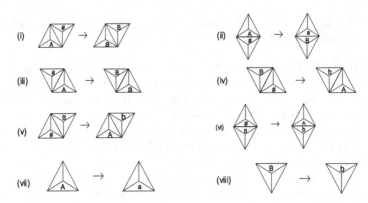

Similar rules can be given for the other two tetrahedral tiles C and D, where A and B are non terminal symbols and a and b are terminal symbols. Starting with a tetrahedral tile A, RTAG can generate at most three connected tiles. So it cannot generate an equilateral triangle of the form ⧋ but this language can be generated by the following 3D-TetATPN.

Consider a 3D-TetATPN $N_4 = (\Sigma, C, \mu, S, \sigma, F)$, where $\Sigma = \{A, B\}$, $C = (P, T, I, O)$ where $P = \{P_1, P_2, P_3, P_4\}$, $T = \{t_1, t_2, t_3\}$. The initial marking μ is the tetrahedral tile A in place P_1.

$$S = \{A \; \text{(ru)} \; B, B \; \text{(u)} \; A, B \; \text{(rd)} \; A \}$$

σ the mapping from the set of transitions to the set of rules is shown in Fig. 9, $F = \{P_4\}$ and the language generated by N_4 is shown in Fig. 10.   □

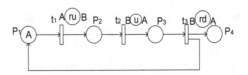

Fig. 9. 3D-TetATPN of the language of equilateral triangle tetrahedral.

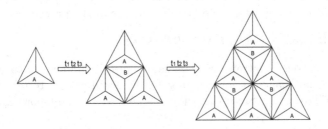

Fig. 10. Language of equilateral triangle tetrahedral.

Theorem 4. $\mathcal{L}(3DTetATPN)$ *and BPTAL are incomparable but not disjoint.*

Proof. Consider a tetrahedral language whose boundary is an equilateral triangle of size 2. This language is generated by both systems Basic Puzzle Tetrahedral Array Grammar (BPTAG) [16] as well as by 3D-TetATPN.

Consider a BPTAG, $G = (\{$ 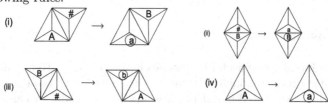 $\}, \{$ $\}, \mathbf{P}, \mathbf{S})$ where P consists of the following rules:

(i)   →   (ii)   →

(iii)   →   (iv)   →

The language generated by G is an equilateral triangle tetrahedral of size 2 which is shown in Fig. 11.

This language can be generated by the following 3D-TetATPN:

Consider a 3DTetATPN $N_5 = (\Sigma, C, \mu, S, \sigma, F)$, where $\Sigma = \{A, B\}$, $C = (P, T, I, O)$ where $P = \{P_1, P_2, P_3, P_4\}$, $T = \{t_1, t_2, t_3\}$. The initial marking μ is the tetrahedral tile A in place $P1$. $S = \{ A\,\textcircled{ru}\,B, B\,\textcircled{u}\,A, B\,\textcircled{rd}\,A \}$, σ the mapping from the set of transitions to the set of rules is shown in Fig. 12 and $F = \{P_4\}$

Equilateral triangle tetrahedral of size more than 2 cannot be generated by BPTAG, whereas it can be generated by 3D-TetATPN (as in Theorem 3). On the other hand the sequence of overlapping equilateral triangle tetrahedral can be generated by the above BPTAG, whereas it cannot be generated by any 3D-TetATPN as the catenation rules are applied in parallel wherever possible. □

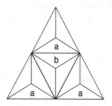

Fig. 11. Equilateral triangle tetrahedral picture of size 2.

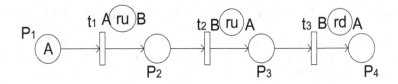

Fig. 12. 3D-TetATPN generating equilateral triangle tetrahedral of size 2.

5 Conclusion

This model is found to be useful in generating interesting patterns and is compared with other recent models in terms of their generative powers. P systems are by definition distributed parallel computing devices and they can solve computationally hard problems in a feasible time. There are NP hard problems in picture languages also. We will analyze whether these problems can be studied using membrane computing. Also we propose to work on a more powerful model of 3D-TetATPN using a concept called 'inhibitor arc' to generate further useful patterns and closure properties of 3D-TetATPN. This is our future work.

References

1. Alhazov, A., Fernau, H., Freund, R., Ivanov, S., Siromoney, R., Subramanian, K.G.: Contextual Array Grammars with matrix control, regular control languages and tissue P systems control. Theor. Comput. Sci. **682**, 5–21 (2017)
2. Immanuel, B., Usha, P.: Array-token Petri nets and 2d grammars. Int. J. Pure Appl. Math. **101**(5), 651–659 (2015)
3. Bhuvaneswari, K., Kalyani, T., Lalitha, D.: Triangular tile pasting P system and array generating Petri nets. Int. J. Pure Appl. Math. **107**(1), 111–128 (2016)
4. Ceterchi, R., Mutyam, M., Păun, G., Subramanian, K.G.: Array - rewriting P Systems. Nat. Comput. **2**, 229–249 (2003)
5. Giammarresi, D., Restivo, A.: Two-dimensional languages. In: Rozenberg, G., Salomaa, A. (eds.) Handbook of Formal Languages, pp. 215–267. Springer, Heidelberg (1997). https://doi.org/10.1007/978-3-642-59126-6_4
6. Siromoney, G., Siromoney, R., Krithivasan, K.: Picture language with array rewrite rules. Inf. Control **22**(5), 447–470 (1973)
7. Fernau, H., Paramasivan, M., Schmid, M.L., Thomas, D.G.: Simple picture processing based on finite automata and regular grammars. J. Comput. Syst. Sci. **95**, 232–258 (2018)
8. Venkat, I., Robinson, T., Subramanian, K.G., de Wilde, P.: Generation of kolam-designs based on contextual array P systems. In: Chapman, P., Stapleton, G., Moktefi, A., Perez-Kriz, S., Bellucci, F. (eds.) Diagrams 2018. LNCS (LNAI), vol. 10871, pp. 79–86. Springer, Cham (2018). https://doi.org/10.1007/978-3-319-91376-6_11
9. Kalyani, T., Raman, T.T., Thomas, D.G.: Tetrahedral tile pasting P system for 3D patterns. Math. Eng. Sci. Aerosp. **11**(1), 255–263 (2020)
10. Kamaraj, T., Lalitha, D., Thomas, D.G.: A formal study on generative power of a class of array token Petrinet structure. Int. J. Syst. Assur. Eng. Manag. **9**(3), 630–638 (2018)

11. Lalitha, D., Rangarajan, K.: Characterisation of pasting system using array token Petri nets. Int. J. Pure Appl. Math. **70**(3), 275–284 (2011)
12. Lalitha, D., Rangarajan, K., Thomas, D.G.: Rectangular arrays and Petri nets. In: Barneva, R.P., Brimkov, V.E., Aggarwal, J.K. (eds.) IWCIA 2012. LNCS, vol. 7655, pp. 166–180. Springer, Heidelberg (2012). https://doi.org/10.1007/978-3-642-34732-0_13
13. Anselmo, M., Giammarresi, D., Madonia, M.: A common framework to recognize two-dimensional languages. Fundam. Inform. **171**(1–4), 1–17 (2020)
14. Păun, G.: Computing with Membranes: An Introduction. Springer, Berlin (2002). https://doi.org/10.1007/978-3-642-56196-2
15. Peterson, J.L.: Petri Net Theory and Modeling of Systems. Prentice Hall Inc., Englewood Cliffs (1981)
16. Raman, T.T., Kalyani, T., Thomas, D.G.: Tetrahedral array grammar system. Math. Eng. Sc. Aerosp. **11**(1), 237–254 (2020)
17. Rosenfeld, A.: Picture Languages (Formal Models for Picture Recognition). Academic Press, New York (1979)
18. Subramanian, K.G., Sriram, S., Song, B., Pan, L.: An overview of 2D picture array generating models based on membrane computing. In: Adamatzky, A. (ed.) Reversibility and Universality. ECC, vol. 30, pp. 333–356. Springer, Cham (2018). https://doi.org/10.1007/978-3-319-73216-9_16

Simulating Parallel Internal Column Contextual Array Grammars Using Two-Dimensional Parallel Restarting Automata with Multiple Windows

Abhisek Midya[1]([✉]) [iD], Frits Vaandrager[1] [iD], D. G. Thomas[2] [iD]
, and Chandrima Ghosh[3] [iD]

[1] Institute for Computing and Information Sciences, Radboud University,
Nijmegen, The Netherlands
abhisekmidyacse@gmail.com, F.Vaandrager@cs.ru.nl
[2] Department of Science and Humanities, Saveetha School of Engineering,
Chennai, India
dgthomasmcc@yahoo.com
[3] Computer Science and Engineering, Presidency University, Bangalore, India
chandrimaghosh211098@gmail.com

Abstract. The connection between picture languages and restarting automata has been established in Otto (2014). An interesting class of picture languages generated by parallel contextual array grammars was studied with application in image generation and analysis in Subramanian et al. (2008). In this paper, we introduce a variant of two dimensional restarting automata that accepts a subclass of parallel internal contextual array languages. We show that these automata can simulate parallel internal column contextual array grammars in reverse order.

Keywords: Parallel internal column contextual array grammars ·
Membership problem · Restarting automaton

1 Introduction

Syntactic approaches, on account of their structure-handling capability, have played an important role in the problem of description of picture patterns considered as connected digitized, finite arrays of symbols. Using the techniques of formal string language theory, various types of picture or array grammars have been introduced and investigated in [3,9,10,24,25]. Most of the array grammars are based on Chomskian string grammars. Some recent results on picture languages can be found in [2,8,18]. Another interesting class of string grammars, called the class of contextual grammars, was proposed by Marcus in [16]. A contextual grammar defines a string language by starting from a given finite set of strings and adjoining iteratively pairs of strings (called contexts) associated

ⓒ Springer Nature Switzerland AG 2020
T. Lukić et al. (Eds.): IWCIA 2020, LNCS 12148, pp. 106–122, 2020.
https://doi.org/10.1007/978-3-030-51002-2_8

to sets of words (called selectors), to the strings already obtained. These contextual grammars [6,23] are known to provide new approaches for a number of basic problems in formal language theory. Recently, extension of string contextual grammars to array structures has been attempted in [1,7,15,23]. A new method of description of pictures of digitized rectangular arrays, through parallel contextual array grammars, was introduced [4,26]. In this paper, we establish a relationship between *two dimensional restarting automata* and *parallel internal column contextual array grammars*.

The concept of restarting automaton was introduced in [14], in order to model the 'analysis by reduction', which is a technique used in linguistics to analyze sentences of natural languages. Analysis by reduction consists of stepwise simplifications (reductions) of a given (lexically disambiguated) extended sentence unless a correct simple sentence is obtained. A word is accepted until an error is found - the process continues until either the automaton accepts or an error is detected. Each simplification replaces a short part of the sentence by an even shorter one. The one dimensional restarting automaton contains a finite control unit, a head with a look-ahead window attached to a tape.

It has been shown in [12], that restarting automaton with delete (simply, DRA) can represent the analyzer for characterizing the class of *contextual grammars with regular selector* (CGR). Also [13] showed that restarting automata recognize a family of languages which can be generated by certain type of contextual grammars, called *regular prefix contextual grammars with bounded infix* (RPCGBI).

Here, we focus on two dimensional parallel restarting automata as we are dealing with rectangular picture languages and bring the concept of multiple windows in order to capture the parallel application of rules of parallel internal column contextual array grammars. A two dimensional parallel restarting automaton can delete adjoined sub-arrays in a cycle and followed by restart (*DEL-RST*). We exploit the *DEL-RST* operation to reverse the adjoining contexts that take place in a derivation of a parallel internal column contextual array grammar. We use two dimensional parallel restarting automaton with multiple windows to simulate parallel internal column contextual array grammars in reverse order.

The *membership problem* for a language asks whether, for a given grammar G and a string w, w belongs to the language generated by G or not.

The remainder of this paper is organized as follows. Section 2 describes the basic classes of contextual grammars in more detail which is followed by an example in Subsect. 2.1. Section 3 presents the new variant of two dimensional parallel restarting automata with multiple windows. In Sect. 4, we describe the connection between parallel internal column contextual array grammars and two dimensional parallel restarting automata with multiple windows, also an example is given in Subsect. 4.1 for better understanding. Subsection 4.2 presents some interesting properties of the proposed automata and in Subsect. 4.3 we discuss about the complexity of membership problem for parallel internal column contextual array languages, also we introduce some new definitions. Section 5 concludes the work and shows a future direction of work.

2   Preliminaries

Let V be a finite alphabet. We write V^* for the set of all finite strings over V, which includes the empty string λ. An *image* or *picture* over V is a rectangular $m \times n$ array of elements of V or in short $[a_{ij}]_{m \times n}$. The set of all images over V is denoted by V^{**}. A *picture language* or *two dimensional language* over V is a subset of V^{**}. We define $V^{m,n} = \{A \in V^{**} \mid A \text{ has } m \text{ rows and } n \text{ columns}\}$. If $a \in V$, then $[a^{m,n}]$ is the array over $\{a\}$ with m rows and n columns. In this paper, Λ denotes any empty array. The notion of *column concatenation* is defined as follows: if A and B are two arrays where

$$A = \begin{bmatrix} a_{1,j} & \cdots & a_{1,k} \\ a_{2,j} & \cdots & a_{2,k} \\ \cdots & \cdots & \cdots \\ a_{l,j} & \cdots & a_{l,k} \end{bmatrix}, B = \begin{bmatrix} b_{1,m} & \cdots & b_{1,n} \\ b_{2,m} & \cdots & b_{2,n} \\ \cdots & \cdots & \cdots \\ b_{l,m} & \cdots & b_{l,n} \end{bmatrix} \text{ then } A\Phi B = \begin{bmatrix} a_{1,j} & \cdots & a_{1,k} & b_{1,m} & \cdots & b_{1,n} \\ a_{2,j} & \cdots & a_{2,k} & b_{2,m} & \cdots & b_{2,n} \\ \cdots & \cdots & \cdots & \cdots & \cdots & \cdots \\ a_{l,j} & \cdots & a_{l,k} & b_{l,m} & \cdots & b_{l,n} \end{bmatrix}.$$

If L_1, L_2 are two picture languages over an alphabet V, the *column concatenation* $L_1 \Phi L_2$ of L_1, L_2 is defined by $L_1 \Phi L_2 = \{A\Phi B \mid A \in L_1, B \in L_2\}$. Column concatenation is only defined for pictures that have the same number of rows. Note that operation Φ is associative. If X is an array, the set of all sub-arrays of X is denoted by $sub(X)$. We now recall the notion of column array context [4,26].

Definition 1. *Let V be an alphabet. A* column array context c *over V is of the form*

$$c = \begin{bmatrix} u_1 \\ u_2 \end{bmatrix} \psi \begin{bmatrix} v_1 \\ v_2 \end{bmatrix} \in V^{**} \psi V^{**},$$

where u_1, u_2 are arrays of sizes $1 \times p$, and v_1, v_2 are arrays of sizes $1 \times q$, for some $p, q \geq 1$, and ψ is a special symbol not in V.

The next definition deals with the parallel internal column contextual operation.

Definition 2. *Let V be an alphabet, C a finite set of column array contexts over V, and $\varphi : V^{**} \to 2^C$ a mapping, called choice mapping. For an array* $A = \begin{bmatrix} a_{1,j} & \cdots & a_{1,k} \\ a_{2,j} & \cdots & a_{2,k} \\ \cdots & \cdots & \cdots \\ a_{l,j} & \cdots & a_{l,k} \end{bmatrix}$, $j \leq k, a_{ij} \in V$, *we define* $\hat{\varphi} : V^{**} \to 2^{V^{**} \psi V^{**}}$ *such that* $L\psi R \in \hat{\varphi}(A)$, *where*

$$L = \begin{bmatrix} u_1 \\ u_2 \\ \vdots \\ u_l \end{bmatrix}, R = \begin{bmatrix} v_1 \\ v_2 \\ \vdots \\ v_l \end{bmatrix},$$

and $c_i = \begin{bmatrix} u_i \\ u_{i+1} \end{bmatrix} \psi \begin{bmatrix} v_i \\ v_{i+1} \end{bmatrix} \in \varphi \begin{bmatrix} a_{i,j} & \cdots a_{i,k} \\ a_{i+1,j} & \cdots a_{i+1,k} \end{bmatrix}$, with $c_i \in C, (1 \leq i \leq l-1)$, not all need to be distinct. Given an array $X = [a_{ij}]$ of size $m \times n$, $a_{ij} \in V, X = X_1 \Phi X_2 \Phi X_3$ where

$$X_1 = \begin{bmatrix} a_{1,1} & \cdots & a_{1,p-1} \\ a_{2,1} & \cdots & a_{2,p-1} \\ \vdots & \vdots & \vdots \\ a_{m1} & \cdots & a_{m,p-1} \end{bmatrix}, X_2 = \begin{bmatrix} a_{1,p} & \cdots & a_{1,q} \\ a_{2,p} & \cdots & a_{2,q} \\ \vdots & \vdots & \vdots \\ a_{m,p} & \cdots & a_{m,q} \end{bmatrix}, X_3 = \begin{bmatrix} a_{1,q+1} & \cdots & a_{1,n} \\ a_{2,q+1} & \cdots & a_{2,n} \\ \vdots & \vdots & \vdots \\ a_{m,q+1} & \cdots & a_{m,n} \end{bmatrix}$$

and $1 \leq p \leq q \leq n$, we write $X \Rightarrow_{in} Y$ if $Y = X_1 \Phi L \Phi X_2 \Phi R \Phi X_3$ such that $L\psi R \in \hat{\varphi}(X_2)$. Here L and R are called left and right contexts respectively.

We say that Y is obtained from X by parallel internal column contextual operation ($\Rightarrow$).

Now we consider the notion of parallel internal column contextual array grammar [4,26].

Definition 3. *A parallel internal column contextual array grammar (PICCAG) is an ordered system $G = (V, \mathbf{A}, C, \varphi)$, where V is an alphabet, $\mathbf{A}$ is a finite subset of V^{**} called the axiom set, C is a finite set of column array contexts over V, $\varphi : V^{**} \to 2^C$ is the choice mapping which performs the parallel internal column contextual operation. When φ is omitted we call G a parallel internal contextual array grammar without choice.*
We already discussed the notion of $X \Rightarrow_{in} Y$ in the previous definition. Here we denote by $\Rightarrow^*_{in}$ the reflexive transitive closure of $\Rightarrow_{in}$. The parallel internal column contextual array language (PICCAL) generated by G is defined as the set $L_{in}(G) = \{Y \in V^{**} \mid \exists X \in \mathbf{A} \text{ such that } X \Rightarrow^*_{in} Y\}$.

2.1 Example

Let $G = (V, \mathbf{A}, C, \varphi)$ be a parallel internal column contextual array grammar (*PICCAG*) where $V = \{a, b\}$, $\mathbf{A} = \left\{ B = \begin{bmatrix} a & a & b & b \\ a & a & b & b \\ b & b & a & a \\ b & b & a & a \end{bmatrix} \right\}$,

$C = \{[\begin{smallmatrix} a \\ b \end{smallmatrix}] \psi [\begin{smallmatrix} b \\ a \end{smallmatrix}], [\begin{smallmatrix} a \\ a \end{smallmatrix}] \psi [\begin{smallmatrix} b \\ b \end{smallmatrix}], [\begin{smallmatrix} b \\ b \end{smallmatrix}] \psi [\begin{smallmatrix} a \\ a \end{smallmatrix}]\}$, φ is a choice mapping satisfying

$$\varphi \begin{bmatrix} a & b \\ b & a \end{bmatrix} = [\begin{smallmatrix} a \\ b \end{smallmatrix}] \psi [\begin{smallmatrix} b \\ a \end{smallmatrix}], \varphi \begin{bmatrix} a & b \\ a & b \end{bmatrix} = [\begin{smallmatrix} a \\ a \end{smallmatrix}] \psi [\begin{smallmatrix} b \\ b \end{smallmatrix}], \varphi \begin{bmatrix} b & a \\ b & a \end{bmatrix} = [\begin{smallmatrix} b \\ b \end{smallmatrix}] \psi [\begin{smallmatrix} a \\ a \end{smallmatrix}].$$

Then,

$$L_{in}(G) = \left\{ \begin{bmatrix} (a^n \ b^n)_m \\ (b^n \ a^n)_m \end{bmatrix}_a \mid n \geq 2, m = 2 \right\}, \text{ where } a^n = aa...a \ (n \text{ times}) \text{ and}$$

$a_m = \begin{smallmatrix} : \\ a \end{smallmatrix}$, with m rows. A simple derivation of a member of $L_{in}(G)$ is as follows:

$$B = \begin{bmatrix} a & a & b & b \\ a & a & b & b \\ b & b & a & a \\ b & b & a & a \end{bmatrix} \Rightarrow \begin{bmatrix} a & a & a & b & b & b \\ a & a & a & b & b & b \\ b & b & b & a & a & a \\ b & b & b & a & a & a \end{bmatrix} = \begin{bmatrix} (a^3 \ b^3)_2 \\ (b^3 \ a^3)_2 \end{bmatrix} \in L_{in}(G).$$

Now, if we consider $a =$ white box and $b =$ black box, we get a nice rectangular picture, see Fig. 1 and Fig. 2.

3 Two Dimensional Parallel Restarting Automata with Multiple Windows

It is interesting to find the connection between picture languages with deterministic two-dimensional three-way ordered restarting automata (det-2D-3W-ORWW) and deterministic two-dimensional extended two-way ordered restarting automata (det-2D-x2W-ORWW) in [19,20].

In this section we present a variant of two dimensional restarting automaton called a two dimensional parallel restarting automaton with multiple windows

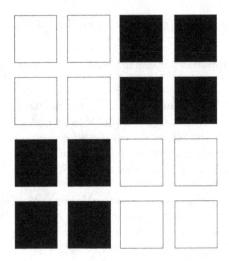

Fig. 1. A rectangular picture of size 4 × 4

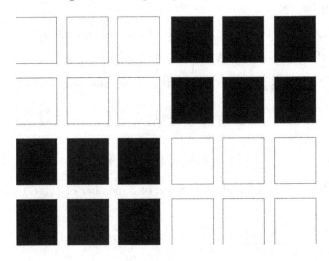

Fig. 2. A rectangular picture of size 4 × 6

(2D-PRA-W$_m$) in order to simulate *PICCAL* in reverse order. Here we introduce multiple windows to deal with the parallel application of rules of parallel internal column contextual array grammars. This automaton works when *PICCAG* is given.

We describe the basic working nature of 2D-PRA-W$_m$. It contains finite control unit and multiple tapes and each tape is associated with individual head and they work in a parallel way. At several points, it cuts-off sub-arrays from each sub-window using DEL operation followed by *restart* (RST) operation, that is, DEL-RST. Here in each sub-window the same number of columns are deleted, this happens in exactly the same positions.

All the heads move together right along the individual tape until it takes any DEL-RST operation. RST implies that the restarting automaton places all the windows over the left border of the individual tape and it completes one *cycle*. After performing a DEL-RST operation, the restarting automaton is unable to remember any step of computation that was performed already.

Let W be an array of size $m \times n$, with $m \geq 2$. Assume

$$W = \begin{bmatrix} a_{1,1} & \cdots & a_{1,p-1} \\ a_{2,1} & \cdots & a_{2,p-1} \\ \vdots & \vdots & \vdots \\ a_{m1} & \cdots & a_{m,p-1} \end{bmatrix}$$

Now W can be viewed as $[W]^{m,n} = \begin{bmatrix} [W_i]^{2,k} \\ [W_{i+1}]^{2,k} \\ \vdots \\ [W_f]^{2,k} \end{bmatrix}$ *where*

$$[W_1]^{2,n} = \begin{bmatrix} a_{1,1} & \cdots & a_{1,n} \\ a_{2,1} & \cdots & a_{2,n} \end{bmatrix}, [W_2]^{2,n} = \begin{bmatrix} a_{2,1} & \cdots & a_{2,n} \\ a_{3,1} & \cdots & a_{3,n} \end{bmatrix}, [W_3]^{2,n} = \begin{bmatrix} a_{3,1} & \cdots & a_{3,n} \\ a_{4,1} & \cdots & a_{4,n} \end{bmatrix},$$

$$[W_{m-1}]^{2,n} = \begin{bmatrix} a_{m-1,1} & \cdots & a_{m-1,n} \\ a_{m,1} & \cdots & a_{m,n} \end{bmatrix}$$

Now we present the concept of super window and sub-window.

$$[W]^{f+1,k} = \begin{bmatrix} [W_i]^{2,k} \\ [W_{i+1}]^{2,k} \\ \vdots \\ [W_f]^{2,k} \end{bmatrix}$$

where $[W]^{f+1,k}$ is called the super window (array) of size $((f+1) \times k)$ which contains sub-windows (arrays) $[W_i]^{2,k}$ of sizes $(2 \times k)$ where $[W_i]^{2,k} = ([\triangleleft^{2,1}]\varPhi V^{2,k-1}) \cup (V^{2,k}) \cup (V^{2,k-1}\varPhi[\triangleright^{2,1}]) \cup ([\triangleleft^{2,1}]\varPhi V^{2,k-2}\varPhi[\triangleright^{2,1}])$. Here f denotes the number of sub-windows. The second row of each ith sub-window $[W_i]^{2,k}$ overlaps with the first row of each $(i+1)$th sub-window $[W_{i+1}]^{2,k}$. Now we show the transition function δ of 2D-PRA-W_m for set of sub-windows. Here $[\triangleleft^{2,1}], [\triangleright^{2,1}]$ denote one column and 2 rows of $\triangleleft, \triangleright$ marker respectively.

Suppose $W = \begin{bmatrix} a & a & a & a & b & b & b & b \\ a & a & a & a & b & b & b & b \\ b & b & b & b & a & a & a & a \\ b & b & b & b & a & a & a & a \end{bmatrix}$.

Now W can be viewed in the following way with the help of sub window and super window. The size of each sub window depends on the given grammar. Interestingly, size of super window depends on the number of sub window. (Discussed in detail in Theorem 1). We have shown below sub window $[W_1], [W_2], [W_3]$ and the super window $[W]$ which contains $[W_1], [W_2], [W_3]$.

$$[W_1] = \begin{bmatrix} a & a & a & a & b & b \\ a & a & a & a & b & b \end{bmatrix}, [W_2] = \begin{bmatrix} a & a & a & a & b & b \\ b & b & b & b & a & a \end{bmatrix}, [W_3] = \begin{bmatrix} b & b & b & b & a & a \\ b & b & b & b & a & a \end{bmatrix},$$

$$[W] = \begin{bmatrix} [W_1] \\ [W_2] \\ [W_3] \end{bmatrix},$$

Let $G = PICCAG = (V, \mathbf{A}, C, \varphi)$.

Definition 4. *A two dimensional parallel restarting automaton with multiple window (2D-PRA-W_m), is given through a 6-tuple, $M = (Q, V, \triangleleft, \triangleright, q_0, \delta)$ where*

- Q *is a finite set of states,*
- V *is the input alphabet,*
- $\triangleleft, \triangleright$ *are left border, right border markers respectively,*
- $q_0 \in Q$ *is the initial state,*

- $\delta : Q \times [W_i]^{2,k} \rightarrow 2^{((Q \times \{MVR, DEL-RST\}) \cup \{Accept, Reject\})}$ *is the transition function. This function describes four different types of transition steps:*
  - *MVR: $(q, MVR) \in \delta(q, ([W_i]^{2,k}, [W_{i+1}]^{2,k}, ..., [W_f]^{2,k}))$. Thus each sub-window of M sees the sub-array of size $2 \times k$. Applying the transition function δ, each sub-window of M moves through left to right using MVR until it takes DEL-RST or $\triangleright \notin [W_i]^{2,k}$.*
  - *DEL-RST: $(q_0, DEL - RST) \in \delta(q, ([W_i]^{2,k}, [W_{i+1}]^{2,k}, ..., [W_f]^{2,k}))$: For possible contents of each sub-window, it deletes a subarray and causes M to move its sub-window to the left border marker $\triangleleft$ and re-enters into the initial state q_0.*
  - *ACCEPT: Accept $\in \delta(q, ([W_i]^{2,k}, [W_{i+1}]^{2,k}, ..., [W_f]^{2,k}))$, it gets into an accepting state.*
  - *REJECT: Reject $\in \delta(q, ([W_i]^{2,k}, [W_{i+1}]^{2,k}, ..., [W_f]^{2,k})) = \emptyset$ (i.e., when δ is undefined), then M will reject the input.*
- *Let $P \in V^{r,s}$ is accepted by $2D - PRA - W_m$ M, if there is a computation, which starts from the initial configuration $q_0[\triangleleft^{r,1}]\Phi P\Phi[\triangleright^{r,1}]$, and reaching the Accept state. By $L(M)$, we denote the language consisting of all arrays accepted by M. In formal notation $L(M) = \{P \in V^{r,n} \mid q_0[\triangleleft^{r,1}]\Phi P^{r,s}\Phi[\triangleright^{r,1}] \vdash^* Accept\}$. Here $[\triangleleft^{r,1}], [\triangleright^{r,1}]$ denote one column and r rows of $\triangleleft, \triangleright$ marker respectively.*

In general, the 2D-PRA-W_m is nondeterministic, that is, there can be two or more instructions with the same left-hand side. If this is not the case, the automaton is deterministic.

Proposition 1 *(Error preservation of 2D-PRA-W_m). If $[W]^{f+1,k} \vdash^*_M [W']^{f+1,k'}$ and $[W]^{f+1,k} \notin L(M)$ then $[W']^{f+1,k'} \notin L(M)$ where $[W]^{f+1,k}, [W']^{f+1,k'} \in V^{**}, k > k'$.*

4   2D-PRA-W_m and PICCAG

Before we analyze the relationship between 2D-PRA-W_m and *PICCAG*, which is the objective of this section, we first need to understand the relationship of DRA with string contextual grammars [12].

External contextual grammars were introduced by Marcus in 1969 [16]. Internal contextual grammars [22] produce strings starting from an *axiom* and in each step *left context* and *right context* are adjoined to the string based on certain string called selector present as a sub-string in the derived string. u, v are called *left context* and *right context* respectively. For more details on contextual grammars, we refer to [21].

- The selector in a contextual grammar can be of arbitrary type in nature, like regular, context-free etc, but the strings u, v are finite.
- Normal DRA works in the opposite way of contextual grammars in accepting strings [12]. In a normal DRA M, w is given as an input, it checks the items of the window with the contextual grammar G that any given rule has been used or not.
- If it finds that any rule has been used then the automaton deletes the left and right context u, v and takes the RST operation, otherwise takes MVR and checks whether any rule in G can be applied.
- In this way, the automaton simulates the derivation of contextual grammar in reverse order and if the input string can be reduced back to the axiom B^1, it implies that the string w can be generated using the given grammar G, thus $w \in L(G)$.
- Here the *size* of the tape of the automaton M is same as the size of the array w. Step by step, the automaton M only deletes subarrays of w, so the size of the tape becomes smaller and smaller.

In this paper, we adapt the working nature of DRA to solve membership problem for *PICCAL*. We show that the membership problem for *PICCAL* is solvable by the introduced 2D-PRA-W$_m$. The paradigm of this version of 2D-PRA-W$_m$ is closely related to *PICCAG*. A *PICCAG* works just in the *opposite direction* of 2D-PRA-W$_m$. The connection is established based on the following observation. For a *PICCAG* rule $\varphi[x_i] = L_i \psi R_i$, now if we present that in two-dimensional form-

$$\varphi[x_i] = \begin{bmatrix} l_{i,1} & \cdots & l_{i,m} \\ l_{i+1,1} & \cdots & l_{i+1,m} \end{bmatrix} \psi \begin{bmatrix} R_{i,1} & \cdots & R_{i,n} \\ R_{i+1,1} & \cdots & R_{i+1,n} \end{bmatrix}$$

where 2D-PRA-W$_m$ has to delete the left context L_i and right context R_i, that is, $\varphi[x_i] = L_i \psi R_i$ is occurred as a subarray in the given input array. In that case, we informally say that a *PICCAG* rule is found in the window as a subarray.

Let M be 2D-PRA-W$_m$. A reduction system induced by M is $RS(M) = (V^{**}, \vdash_M)$. For each *PICCAG* G, we define a reduction system induced by G as $RS(G) = (V^{**}, \Rightarrow_G^{-1})$ where $([W]^{f+1,k} \vdash_M^{-1} [W']^{f+1,k'})$ iff $[W']^{f+1,k'} \Rightarrow_G [W]^{f+1,k}$.

With the above detail we will construct a 2D-PRA-W$_m$ in such a way that if $B \Rightarrow_G^* P$ then $P \vdash_M^* B$ for $P, B \in V^{**}, B \in \mathbf{A}$, thus $RS(G) = RS(M)$. Also $\mathcal{L}$2D-PRA-W$_m$ denotes the class of languages accepted by 2D-PRA-W$_m$.

Theorem 1. *For a PICCAG G, a 2D-PRA-W$_m$ automaton M can be constructed in such a way that $RS(G) = RS(M)$ and $L_{in}(G) = L(M)$.*

Proof. Given a *PICCAG* $G = (V, \mathbf{A}, C, \varphi)$ we have to construct a 2D-PRA-W$_m$ automaton $M = (Q, V, \triangleleft, \triangleright, q_0, \delta)$, that accepts $L_{in}(G)$ where

- $Q = \{q_0, q, Accept, Reject\}$
- V is the input alphabet

---

[1] We consider $\mathbf{A}$ is a singleton axiom set and $B \in A$.

- $\triangleleft, \triangleright$ are left and right borders respectively and $\triangleleft, \triangleright \notin V$
- q_0 is the initial state.

$$[W]^{f+1,k} = \begin{bmatrix} [W_i]^{2,k} \\ [W_{i+1}]^{2,k} \\ \vdots \\ [W_f]^{2,k} \end{bmatrix}$$

- Here number of columns in each window of M will be $k = max(Rule_{max}, k_b + 2)$ where $Rule_{max}$ is the maximum size given rule - $Rule_{max} = max\{|Rule_1|_c, |Rule_2|_c, ..., |Rule_n|_c\}$ where $|Rule_i|_c$ denotes the number of columns in the ith rule and $1 \le i \le n, n \ge 1$.
- Let k_b be axiom size. 2 is added there for the left border $\triangleleft$ and the right border $\triangleright$. The reason for 2 is added with k_b is to satisfy the accepting condition - ACCEPT - $Accept \in \delta(q, ([W_i]^{2,k}, [W_{i+1}]^{2,k}, ..., [W_f]^{2,k}))$ where $[W]^{f+1,k} = [\triangleleft^{f+1,1}]\Phi B\Phi[\triangleright^{f+1,1}]$ where $B \in \mathbf{A}$.
- If the rule is $Rule_i = (\varphi[x_i] = [L_i]\psi[R_i])$ where x_i, L_i, R_i are arrays of size $2 \times k, k \ge 1$, then we define $|Rule_i|_c = |x_i|_c + |L_i|_c + |R_i|_c$, $Rule_{max} = max\{|Rule_1|_c, |Rule_2|_c, ..., |Rule_n|_c\}$

Lemma 1. *If the input is $P^{m,n}$, then number of windows will be $m - 1$.*

Proof. Each window will take care of each rule in a parallel way. According to Definition 2, we know that if the input is P of size $m \times n$ then the number of parallel rules will be $m - 1$, from this fact we can conclude this. (see example for better understanding)

- *DEL-RST*: The DEL-RST instruction of the 2D-PRA-W$_m$ for solving membership problem of *PICCAL*, works in the following manner:
  - Now M works in a parallel way on, $(q, ([W_i]^{2,k}, [W_{i+1}]^{2,k}, ..., [W_f]^{2,k}))$ where $i \ge 1$, and applies DEL-RST on each sub-window from state q to arrive at $((q_0, [W_i']^{2,k'}), (q_0, [W_{i+1}']^{2,k'}), ..., (q_0, [W_f']^{2,k'}))$ and eventually reaching $(q_0, ([W_i']^{2,k'}, [W_{i+1}']^{2,k'}, ..., [W_f']^{2,k'}))$ where $[W']^{f+1,k'}, [W_i']^{2,k'}$ are scattered sub-array of $[W]^{f+1,k}, [W_i]^{2,k}$ respectively and $k > k'$, immediately followed by a RST instruction: RST $\in \delta(q, [W_i]^{2,k})$ for any possible contents $[W_i]^{2,k}$ of the window. If no *PICCAG* rule does belong to window as a subarray and $\triangleright$ does not belong to window ($\triangleright \notin [W_i]^{2,k}$) then the automaton takes *MVR* operation.
- ACCEPT: $Accept \in \delta(q, ([W_i]^{2,k}, [W_{i+1}]^{2,k}, ..., [W_f]^{2,k}))$ where $[W]^{f+1,k} = [\triangleleft^{f+1,1}]\Phi B\Phi[\triangleright^{f+1,1}]$ where $B \in \mathbf{A}$. Here $[\triangleleft^{f+1,1}], [\triangleright^{f+1,1}]$ denote one column of $\triangleleft, \triangleright$ marker respectively, here we deal with singleton axiom set.
- REJECT: $Reject \in \delta(q, ([W_i]^{2,k}, [W_{i+1}]^{2,k}..., [W_f]^{2,k})) = \emptyset$. That is when δ is undefined. In other words, when $2D - PRA - W_m$ is unable to take any of the DEL-RST or MVR operation, then the transition becomes undefined.

2D-PRA-W$_m$ simulates the derivation of *PICCAG* in reverse order, in case of any *PICCAG* rule it deletes the left and right contexts using DEL-RST instruction which is defined already. For *PICCAG*, the derivation starts from the axiom to the generated array, the automaton starts the reduction from the generated array to the axiom. If $B \Rightarrow_G^* P$ then $P \vdash_M^* B$ where $P, B \in V^{**}, B \in$ **A** is axiom, thus $RS(G) = RS(M)$.

Corollary 1. *The membership problem for PICCAL can be solved by 2D-PRA-W$_m$.*

Proof. We conclude this important result from Theorem 1.

4.1   Example

Consider the *PICCAG G* given in Example 2.1. Suppose $P = \begin{bmatrix} a & a & a & a & b & b & b & b \\ a & a & a & a & b & b & b & b \\ b & b & b & b & a & a & a & a \\ b & b & b & b & a & a & a & a \end{bmatrix}$ is given as an input and we note that $P \in L_{in}(G)$. Now we can construct a 2D-PRA-W$_m$ automaton $M = (Q, V, \triangleleft, \triangleright, q_0, \delta)$, that accepts P where

- $Q = \{q_0, q, Accept, Reject\}$
- V is the input alphabet
- $\triangleleft, \triangleright$ are left and right borders respectively and $\triangleleft, \triangleright \notin V$
- q_0 is the initial state
- The number of columns in each window is $k = 6$ and the number of windows is 3 respectively.
- In the first cycle, rule and $\triangleleft$ are not found in the window and so it takes MVR: $(q, MVR) \in \delta(q, ([W_1]^{2,6}, [W_2]^{2,6}, [W_3]^{2,6}))$, where

$$[W_1]^{2,6} = \begin{bmatrix} \triangleleft & a & a & a & a & b \\ \triangleleft & a & a & a & a & b \end{bmatrix}, [W_2]^{2,6} = \begin{bmatrix} \triangleleft & a & a & a & a & b \\ \triangleleft & b & b & b & b & a \end{bmatrix}, [W_3]^{2,6} = \begin{bmatrix} \triangleleft & b & b & b & b & a \\ \triangleleft & b & b & b & b & a \end{bmatrix},$$

- After taking the MVR operation the elements of windows get changed and M takes $DEL-RST$: $(q_0, ([W_1']^{2,4}, [W_2']^{2,4}, [W_3']^{2,4})) \in \delta(q, [W_1]^{2,6}, [W_2]^{2,6}, [W_3]^{2,6}))$, where $[W_i']^{2,4}$ is the scattered sub-array of $[W_i]^{2,6}, i \geq 1$ and

$$[W]^{4,6} = \begin{bmatrix} [W_1]^{2,6} \\ [W_2]^{2,6} \\ [W_3]^{2,6} \end{bmatrix}, [W']^{4,4} = \begin{bmatrix} [W_1']^{2,4} \\ [W_2']^{2,4} \\ [W_3']^{2,4} \end{bmatrix}$$

$$[W_1]^{2,6} = \begin{bmatrix} a & a & a & a & b & b \\ a & a & a & a & b & b \end{bmatrix}, [W_1']^{2,4} = \begin{bmatrix} a & a & a & b \\ a & a & a & b \end{bmatrix}$$

$$[W_2]^{2,6} = \begin{bmatrix} a & a & a & a & b & b \\ b & b & b & b & a & a \end{bmatrix}, [W_2']^{2,4} = \begin{bmatrix} a & a & a & b \\ b & b & b & a \end{bmatrix}$$

$$[W_3]^{2,6} = \begin{bmatrix} b & b & b & b & a & a \\ b & b & b & b & a & a \end{bmatrix}, [W_3']^{2,4} = \begin{bmatrix} b & b & b & a \\ b & b & b & a \end{bmatrix}$$

- In the next cycle, again M can take $DEL-RST$: $(q_0, ([W_1']^{2,4}, [W_2']^{2,4}, [W_3']^{2,4})) \in \delta(q_0, [W_1]^{2,6}, [W_2]^{2,6}, [W_3]^{2,6}))$ where $[W_i']^{2,4}$ is the scattered sub-array of $[W_i]^{2,6}, i \geq 1$ and

$$[W]^{4,6} = \begin{bmatrix} [W_1]^{2,6} \\ [W_2]^{2,6} \\ [W_3]^{2,6} \end{bmatrix}, [W']^{4,4} = \begin{bmatrix} [W_1']^{2,4} \\ [W_2']^{2,4} \\ [W_3']^{2,4} \end{bmatrix}$$

$$[W_1]^{2,6} = \begin{bmatrix} \lhd a\ a\ a\ a\ b\ b \\ \lhd a\ a\ a\ a\ b\ b \end{bmatrix}, [W_1']^{2,4} = \begin{bmatrix} \lhd a\ a\ b \\ \lhd a\ a\ b \end{bmatrix}$$

$$[W_2]^{2,6} = \begin{bmatrix} a\ a\ a\ a\ b\ b \\ b\ b\ b\ b\ a\ a \end{bmatrix}, [W_2']^{2,4} = \begin{bmatrix} \lhd a\ a\ b \\ \lhd b\ b\ a \end{bmatrix}$$

$$[W_3]^{2,6} = \begin{bmatrix} b\ b\ b\ b\ a\ a \\ b\ b\ b\ b\ a\ a \end{bmatrix}, [W_3']^{2,4} = \begin{bmatrix} \lhd b\ b\ a \\ \lhd b\ b\ a \end{bmatrix}$$

– In the next cycle, $Accept \in \delta(q_0, [W_1]^{2,6}, [W_2]^{2,6}, [W_3]^{2,6})$ where

$$[W]^{4,6} = \begin{bmatrix} \lhd a\ a\ b\ b\ \rhd \\ \lhd a\ a\ b\ b\ \rhd \\ \lhd b\ b\ a\ a\ \rhd \\ \lhd b\ b\ a\ a\ \rhd \end{bmatrix}$$

In this way, every member of $L_{in}(G)$ is accepted by M. Now we consider the input $P' = \begin{bmatrix} a\ a\ a\ a\ b\ b\ b \\ a\ a\ a\ a\ b\ b\ b \\ b\ b\ b\ b\ a\ a\ a \\ b\ b\ b\ b\ a\ a\ a \end{bmatrix}$ and we note that $P' \notin L_{in}(G)$.

In the first cycle, rule and $\lhd$ are not found in the window, so it takes MVR: $(q, MVR) \in \delta(q, ([W_1]^{2,6}, [W_2]^{2,6}, [W_3]^{2,6}))$, where

$$[W_1]^{2,6} = \begin{bmatrix} \lhd a\ a\ a\ a\ b \\ \lhd a\ a\ a\ a\ b \end{bmatrix}, [W_2]^{2,6} = \begin{bmatrix} \lhd a\ a\ a\ a\ b \\ \lhd b\ b\ b\ b\ a \end{bmatrix}, [W_3]^{2,6} = \begin{bmatrix} \lhd b\ b\ b\ b\ a \\ \lhd b\ b\ b\ b\ a \end{bmatrix},$$

– Now M takes $DEL - RST$: $(q_0, ([W_1']^{2,4}, [W_2']^{2,4}, [W_3']^{2,4})) \in \delta(q, ([W_1]^{2,6}, [W_2]^{2,6}, [W_3]^{2,6}))$, where $[W_i']^{2,4}$ is the scattered sub-array of $[W_i]^{2,6}, i \geq 1$ and

$$[W]^{4,6} = \begin{bmatrix} [W_1]^{2,6} \\ [W_2]^{2,6} \\ [W_3]^{2,6} \end{bmatrix}, [W']^{4,4} = \begin{bmatrix} [W_1']^{2,4} \\ [W_2']^{2,4} \\ [W_3']^{2,4} \end{bmatrix}$$

$$[W_1]^{2,6} = \begin{bmatrix} a\ a\ a\ a\ b\ b \\ a\ a\ a\ a\ b\ b \end{bmatrix}, [W_1']^{2,4} = \begin{bmatrix} a\ a\ a\ b \\ a\ a\ a\ b \end{bmatrix}$$

$$[W_2]^{2,6} = \begin{bmatrix} a\ a\ a\ a\ b\ b \\ b\ b\ b\ b\ a\ a \end{bmatrix}, [W_2']^{2,4} = \begin{bmatrix} a\ a\ a\ b \\ b\ b\ b\ a \end{bmatrix}$$

$$[W_3]^{2,6} = \begin{bmatrix} b\ b\ b\ b\ a\ a \\ b\ b\ b\ b\ a\ a \end{bmatrix}, [W_3']^{2,4} = \begin{bmatrix} b\ b\ b\ a \\ b\ b\ b\ a \end{bmatrix}$$

In the next cycle, again M can take $DEL - RST$: $(q_0, ([W_1']^{2,4}, [W_2']^{2,4}, [W_3']^{2,4})) \in \delta(q_0, ([W_1]^{2,6}, [W_2]^{2,6}, [W_3]^{2,6}))$ where $[W_i']^{2,4}$ is the scattered sub-array of $[W_i]^{2,6}, i \geq 1$ and

$$[W]^{4,6} = \begin{bmatrix} [W_1]^{2,6} \\ [W_2]^{2,6} \\ [W_3]^{2,6} \end{bmatrix}, [W']^{4,4} = \begin{bmatrix} [W_1']^{2,4} \\ [W_2']^{2,4} \\ [W_3']^{2,4} \end{bmatrix}$$

$$[W_1]^{2,6} = \begin{bmatrix} \lhd a\ a\ a\ b\ b \\ \lhd a\ a\ a\ b\ b \end{bmatrix}, [W_1']^{2,4} = \begin{bmatrix} \lhd a\ a\ b \\ \lhd a\ a\ b \end{bmatrix}$$

$$[W_2]^{2,6} = \begin{bmatrix} a\ a\ a\ a\ b\ b \\ b\ b\ b\ b\ a\ a \end{bmatrix}, [W_2']^{2,4} = \begin{bmatrix} \lhd a\ a\ b \\ \lhd b\ b\ a \end{bmatrix}$$

$$[W_3]^{2,6} = \begin{bmatrix} b\ b\ b\ b\ a\ a \\ b\ b\ b\ b\ a\ a \end{bmatrix}, [W_3']^{2,4} = \begin{bmatrix} \lhd b\ b\ a \\ \lhd b\ b\ a \end{bmatrix}$$

In the next cycle, it rejects because this time transition function is undefined. $Reject \in \delta(q_0, ([W_1]^{2,6}, [W_2]^{2,6}, [W_3]^{2,6}))$ where

$$[W]^{4,6} = \begin{bmatrix} \lhd a\ a\ b\ \rhd \\ \lhd a\ a\ b\ \rhd \\ \lhd b\ b\ a\ \rhd \\ \lhd b\ b\ a\ \rhd \end{bmatrix}$$

4.2   Properties of 2D-PRA-W$_m$

In this section, we introduce some important properties of 2D-PRA-W$_m$.

Lemma 2. *The language class* $\mathcal{L}(2D\text{-}PRA\text{-}W_m)$ *is closed under* $180°$ *rotation.*

Proof. Let 2D-PRA-W$_m$ be $M = (Q, V, \triangleleft, \triangleright, q_0, \delta)$, that accepts a language $L \subseteq V^{*,*}$ where $Q = \{q_0, q, Accept, Reject\}$, V is the input alphabet, $\triangleleft, \triangleright$ are left and right borders respectively and $\triangleleft, \triangleright \notin V$, q_0 is the initial state, $Accept \in \delta(q, ([W_i]^{2,k}, [W_{i+1}]^{2,k}, ..., [W_f]^{2,k}))$, $Reject \in \delta(q, ([W_i]^{2,k}, [W_{i+1}]^{2,k}, ..., [W_f]^{2,k})) = \emptyset$ (i.e., when δ is undefined), then M will reject.

Now, from M we can construct M_R (after $180°$ rotation of M) where $M_R = (Q, V, \triangleleft, \triangleright, q_0, \delta)$, that accepts a language $L \subseteq V^{*,*}$ where $Q = \{q_0, q, Accept', Reject\}$, V is the input alphabet, $\triangleleft, \triangleright$ are left and right borders respectively and $\triangleleft, \triangleright \notin V$, q_0 is the initial state, $Accept' \in \delta(q, ([W_i]_R^{2,k}, [W_{i+1}]_R^{2,k}, ..., [W_f]_R^{2,k}))$ where $[W_i]_R^{2,k}$ is the ith sub-window after $180°$ rotation of $[W_i]^{2,k}$ and $1 \leq i \leq f$, $Reject \in \delta(q, ([W_i]_R^{2,k}, [W_{i+1}]_R^{2,k}, ..., [W_f]_R^{2,k})) = \emptyset$ (i.e., when δ is undefined), then M will reject the input. $\qquad\blacksquare$

Lemma 3. *The language class* $\mathcal{L}(2D\text{-}PRA\text{-}W_m)$ *is closed under complement.*

Proof. Let M be a 2D-PRA-W$_m$, that accepts a language $L \subseteq V^{*,*}$. Now, from M we can construct M_c (complement of M) by interchanging undefined and accepting transitions. $\qquad\blacksquare$

Lemma 4. *The language class* $\mathcal{L}(2D\text{-}PRA\text{-}W_m)$ *is closed under column concatenation.*

Proof. Let M_1 be a 2D-PRA-W$_m$, on V, that accepts a language $L_1 \subseteq V^{*,*}$ where $Accept \in \delta(q, ([W_i]^{2,k}, [W_{i+1}]^{2,k}, ..., [W_f]^{2,k}))$. Consider M_2 be another 2D-PRA-W$_m$ which accepts a language $L_2 \subseteq V^{*,*}$ where $Accept \in \delta(q, ([W_i'']^{2,k}, [W_{i+1}'']^{2,k}, ..., [W_f'']^{2,k}))$. Now, we can construct M which can accept $L_1 \Phi L_2$ by modifying the accepting state, that is, $Accept \in \delta(q, ([W_i]^{2,k} \Phi [W_i'']^{2,k}, [W_{i+1}]^{2,k} \Phi [W_{i+1}'']^{2,k}, ..., [W_f]^{2,k} \Phi [W_f'']^{2,k}))$. $\qquad\blacksquare$

Lemma 5. *The language class* $\mathcal{L}(2D\text{-}PRA\text{-}W_m)$ *with auxiliary special symbol is closed under intersection.*

Proof. Let 2D-PRA-W$_m$ be a $M_1 = (Q_1, V, \Gamma_1, \triangleleft, \triangleright, q_0', \delta_1)$ with auxiliary special symbols. Let $M_2 = (Q_2, V, \Gamma_2, \triangleleft, \triangleright, q_0'', \delta_2)$ be another 2D-PRA-W$_m$ with special symbols where $\Gamma_1, \Gamma_2 \supseteq V$. Now we can construct $M = (Q, V, \Gamma, \triangleleft, \triangleright, q_0, \delta)$ such that $L(M) = L(M_1) \bigcap L(M_2)$. Essentially, M will work as follows:

- M first simulates M_1, that is, it behaves exactly like M_1. If M_1 should get stuck on the given input, that is, M_1 does not accept, then neither does M. If, however, M_1 accepts, then instead of accepting, M marks the position (i, j) at which M_1 accepts, using a special symbol.

– Now only M should start simulating M_2. So, it is understood that we need to mark the last position by special symbol and because of that we introduced $\Gamma = V \cup T$ is the tape alphabet, $\Gamma \supseteq V$.

Lemma 6. *The language class* $\mathcal{L}(2D\text{-}PRA\text{-}W_m)$ *is not closed under transposition.*

Proof. Let $G = (V, A, C, \varphi)$ be a parallel internal column contextual array grammar where $V = \{a, b\}$,

$$A = \left\{ B = \begin{bmatrix} a & a & a & a & b \\ a & b & a & b & b \\ a & b & a & b & b \end{bmatrix} \right\}, C = \left\{ \begin{bmatrix} a \\ b \end{bmatrix} \psi \begin{bmatrix} a \\ b \end{bmatrix}, \begin{bmatrix} b \\ b \end{bmatrix} \psi \begin{bmatrix} b \\ b \end{bmatrix} \right\},$$

where $B \in A$, φ is a choice mapping,

$$\varphi \begin{bmatrix} a & a & a \\ b & a & b \end{bmatrix} = \begin{bmatrix} a \\ b \end{bmatrix} \psi \begin{bmatrix} a \\ b \end{bmatrix}, \varphi \begin{bmatrix} b & a & b \\ b & a & b \end{bmatrix} = \begin{bmatrix} b \\ b \end{bmatrix} \psi \begin{bmatrix} b \\ b \end{bmatrix}$$

We can construct 2D-PRA-W_m such that $L(M) = L(G)$ where the configuration of acceptance , ACCEPT- $[W]^{3,7} = [\triangleleft]^{3,1} B [\triangleright]^{3,1}$. where $B \in A$. Now, if we consider the transposition of B, we obtain

$$B_T = \begin{bmatrix} a & a & a \\ a & b & b \\ a & a & a \\ a & b & b \\ b & b & b \end{bmatrix}.$$

Clearly, $B_T \notin L(M)$ because in this case we cannot construct M_T. If $B_T \in L(M)$ then the content of each sub-window w_j in each cycle c_i should be transposed to w_{j_T} such that $\forall i \forall j \; Transpose(w_j, w_{j_T})$. In order to do that, the working procedure of our M for given G, needs to be changed.

4.3 Complexity of Membership Problem for PICCAL

In this section, we discuss the complexity of solving the membership problem for *PICCAL*. Let us start with *internal contextual string languages with finite choice* $(ICSL(FIN))$. $ICSL(FIN)$ is contained in the family of languages generated by *growing context-sensitive grammars* $(GCSG)$, and from this scenario we will comment on the time complexity of solving membership problem for *PICCAL*. So here, first we recall the definition of $ICSL(FIN)$ and $GCSG$.

Definition 5 [17]. *For an alphabet* Σ, *we denote by* Σ^* *the free monoid generated by* Σ, *by* λ *its identity, and* $\Sigma^+ = \Sigma^* - \{\lambda\}$. *The family of finite languages is denoted by* FIN. *Contextual grammar is a construct,* $G = (\Sigma, A, (sel_1, C_1), (sel_2, C_2), ..., (sel_k, C_k))$, *for some* $k \geq 1$, *where* Σ *is an alphabet,* $A \subset \Sigma^*$ *is a finite set, called the axiom set,* $sel_i \subseteq \Sigma^*, 1 \leq i \leq k$, *are the sets of selectors, and* $C_i \subset \Sigma^* \times \Sigma^*$ *where* $1 \leq i \leq k$, *and* C_i *is a finite set of contexts. There are two basic modes of derivation, the internal mode of derivation as follows. For two words* $x, y \in \Sigma^*$, *we have the internal mode of derivation:*

$x \Longrightarrow_{in} y$ iff $x = x_1 x_2 x_3, y = x_1 u x_2 v x_3, x_2 \in sel_i, (u, v) \in C_i$, for some $1 \leq i \leq k$.

The language generated by internal mode of derivation is: $L_{in}(G) = \{w \in \Sigma^* \mid x \in A, x \Longrightarrow_{in}^* w\}$, where $\Longrightarrow_{in}^*$ denotes the reflexive - transitive closure of $\Longrightarrow_{in}$.

If the sets $sel_1, sel_2, ..., sel_k$ are languages in a given family FIN, then G is said to be with FIN choice. The family of languages generated by contextual grammars with FIN choice in the internal mode of derivation is denoted by $ICSL(FIN)$.

Now we recall the definition from [11].

Definition 6. A context-sensitive grammar (CSG) is a tuple $G = (V, T, P, S)$, where V is a set of alphabets, T is a finite set of terminal symbols, P is a finite set of production rules, and S is the starting symbol. We say that G is growing if S does not appear on the right and $|\alpha| < |\beta|$ for any $(\alpha \rightarrow \beta)$, with $\alpha \neq S$, from P.

Definition 7. A CSG $G = (V, T, P, S)$ is QGCSG if there exists a function $f : (V \cup T)^* \mapsto \mathbb{Z}^+$ such that, for all $p \in P$, $f(\alpha) > f(\beta)$.

Lemma 7. $ICSL(FIN) \subset GCSG$

Proof. $G = (\Sigma, A, (sel_1, C_1), (sel_2, C_2), ..., (sel_k, C_k))$ be $ICSL(FIN)$. We can assume that $(\lambda, \lambda) \notin C_i$ for all $1 \leq i \leq k$. The problem in developing a $QGCSG$, is to simulate an insertion step $x \Rightarrow_{in} y$ if $x = x_1 x_3, \lambda \in S_i, y = x_1 u v x_3$, and $(u, v) \in C_i$ for some $1 \leq i \leq k$. In order to avoid this, we do as follows:
We define homomorphism $h : \Sigma^* \rightarrow \Sigma'^*$ where $\Sigma' = \{a' \mid a \in \Sigma\}$ such that $\Sigma \cap \Sigma' = \phi$ and $h(a) = a'$ for $a \in \Sigma$. Now we are ready to construct $QGCSG$ $G' = (\Sigma' \cup S, \Sigma, P, S)$ where $S \notin \Sigma'$, the P is given below.
$P = \{S \rightarrow h(x) \mid x \in A\} \cup \{S \rightarrow h(u, v) \mid \lambda \in A \cap S_i, (u, v) \in C_i, 1 \leq i \leq k\} \cup \{h(x) \rightarrow h(uxv) \mid x \in S_i \setminus \{\lambda\}, (u, v) \in C_i, 1 \leq i \leq k\} \cup \{h(a) \rightarrow h(uva), h(a) \rightarrow h(auv) \mid a \in \Sigma, \lambda \in S_i, (u, v) \in C_i, 1 \leq i \leq k\} \cup \{h(a) \rightarrow a \mid a \in \Sigma\}$ with the valuation $f(S) = 1$ and $f(h(a)) = 2, f(a) = 3$, if $a \in \Sigma$. So, here the constructed grammar is $QGCSG$ and $L(G') = L_{in}(G)$.

Since $QGCSG = GCSG$, we can state that $ICSL(FIN) \subset GCSG$. Here the inclusion is strict because the cross-dependency language $L_{cross-dependency} = \{a^n b^m c^n d^m \mid n, m \geq 1\} \notin ICSL(FIN)$ but $L_{cross-dependency} \in GCSL$.

Lemma 8 [11]. The membership problem for internal contextual string languages with finite choice ($ICSL(FIN)$) is $LOG(CFL) - hard$.

Proof. From Lemma 7, we concluded that $ICSL(FIN) \subset GCSG$. In [5], it is shown that $GCSL$ family of languages, is contained in $LOG(CFL)$. This shows that the upper bound for membership problem for $ICSL(FIN)$ is $LOG(CFL)$.

Lemma 9. The membership problem for parallel internal column contextual array languages (PICCAL) is contained in NP.

Proof. Let $VERIFIER(W, C)$ be a procedure where W, C are given inputs and denote a word and certificate respectively. Here C is a certificate, i.e., a derivation. The procedure $VERIFIER(W, C)$ returns "YES" if the given certificate C is correct, otherwise "NO". In other words, $VERIFIER(W, C)$ verifies the correctness of C. Moreover the running time of $VERIFIER(W, C)$ is bounded by a polynomial in $|W|$ where $|W|$ denotes the size of W. See Algorithm 1.

Algorithm 1. Polynomial Time Verifier

1: **procedure** VERIFIER(W, C)
2:      Initialize $w_i \leftarrow Axiom$ ▷ w_i stores the Axiom
3:      Initialize $k \leftarrow |W|$ ▷ k stores the length of W
4:      Initialize $N \leftarrow |C|$ ▷ Nstores the length of C
5:      **for** $i = 1$ to N **do**
6:          $w_i \Rightarrow_{ithstep} w_{i+1}$
7:          **if** $w_{i+1} == W$ **then**   ▷ ith step of the derivation
8:              print YES **return**   ▷ C is correct
9:          **else**
10:             $i \leftarrow i + 1$
11:         **end if**
12:     **end for**
13:     **if** $|w_i| > k$ **then**
14:         Print NO   ▷ C is incorrect
15:     **end if**
16: **end procedure**

Corollary 2. *The membership problem for PICCAL is at least $LOG(CFL) -$ hard and is contained in NP.*

Proof. From Lemma 8 and Lemma 9, we can easily conclude this Corollary 2.

5   Conclusion and Future Work

In this paper, we have introduced a non-deterministic 2D-PRA-W_m to solve the membership problem of *PICCAL*. Here we have introduced multiple windows in order to capture the parallel application of the parallel column contextual array rules. Also we discussed some of the important properties of 2D-PRA-W_m and commented on the complexity of membership problem for *PICCAL*.

Here our focus was on column concatenation only. We can extend our work to take care of row concatenation too. In terms of future direction of work, it could be also interesting, if we can define a powerful subclass of *PICCAG* and solve the membership problem using deterministic 2D-PRA-W_m.

Acknowledgement. Supported by NWO TOP project 612.001.852 Grey-box learning of Interfaces for Refactoring Legacy Software (GIRLS).

References

1. Alhazov, A., Fernau, H., Freund, R., Ivanov, S., Siromoney, R., Subramanian, K.: Contextual array grammars with matrix control, regular control languages, and tissue P systems control. Theor. Comput. Sci. **682**, 5–21 (2017)
2. Anselmo, M., Giammarresi, D., Madonia, M.: A common framework to recognize two-dimensional languages. Fundamenta Informaticae **171**(1–4), 1–17 (2020)
3. Bunke, H., Sanfeliu, A.: Syntactic and Structural Pattern Recognition: Theory and Applications, vol. 7. World Scientific, New York (1990)
4. Chandra, P.H., Subramanian, K., Thomas, D.: Parallel contextual array grammars and languages. Electron. Notes Discrete Math. **12**, 106–117 (2003)
5. Dahlhaus, E., Warmuth, M.K.: Membership for growing context-sensitive grammars is polynomial. J. Comput. Syst. Sci. **33**(3), 456–472 (1986)
6. Ehrenfeucht, A., Păun, G., Rozenberg, G.: Contextual grammars and formal languages. In: Rozenberg, G., Salomaa, A. (eds.) Handbook of Formal Languages, pp. 237–293. Springer, Heidelberg (1997). https://doi.org/10.1007/978-3-662-07675-0_6
7. Fernau, H., Freund, R., Siromoney, R., Subramanian, K.: Non-isometric contextual array grammars and the role of regular control and local selectors. Fundamenta Informaticae **155**(1–2), 209–232 (2017)
8. Fernau, H., Paramasivan, M., Schmid, M.L., et al.: Simple picture processing based on finite automata and regular grammars. J. Comput. Syst. Sci. **95**, 232–258 (2018)
9. Firschein, O.: Syntactic pattern recognition and applications. Proc. IEEE **71**(10), 1231–1231 (1983)
10. Giammarresi, D., Restivo, A.: Two-dimensional languages. In: Rozenberg, G., Salomaa, A. (eds.) Handbook of Formal Languages, pp. 215–267. Springer, Heidelberg (1997). https://doi.org/10.1007/978-3-642-59126-6_4
11. Holzer, M.: On fixed and general membership for external and internal contextual languages. In: Developments in Language Theory: Foundations, Applications, and Perspectives, pp. 351–361. World Scientific, New York (2000)
12. Janar, P., Mráz, F., Páltek, M., Procházka, M., Vogel, J.: Deleting automata with a restart operation. In: 3rd International Conference Developments in Language Theory, DLT, pp. 191–202 (1997)
13. Jancar, P., Mraz, F., Plátek, M., Procházka, M., Vogel, J.: Restarting automata, marcus grammars and context-free languages. In: Developments in Language Theory, pp. 102–111. World Scientific, New York (1995)
14. Jančar, P., Mráz, F., Plátek, M., Vogel, J.: Restarting automata. In: Reichel, H. (ed.) International Symposium on Fundamentals of Computation Theory, pp. 283–292. Springer, Heidelberg (1995). https://doi.org/10.1007/3-540-60249-6_60
15. Krithivasan, K., Balan, M.S., Rama, R.: Array contextual grammars. In: Recent Topics in Mathematical and Computational Linguistics, pp. 154–168 (2000)
16. Marcus, S.: Contextual grammars. In: International Conference on Computational Linguistics Coling 1969: Preprint No. 48 (1969)
17. Midya, A., Thomas, D., Malik, S., Pani, A.K.: Polynomial time learner for inferring subclasses of internal contextual grammars with local maximum selectors. In: Hung, D., Kapur, D. (eds.) International Colloquium on Theoretical Aspects of Computing, pp. 174–191. Springer, Cham (2017). https://doi.org/10.1007/978-3-319-67729-3_11

18. Mráz, F., Průša, D., Wehar, M.: Two-dimensional pattern matching against basic picture languages. In: Hospodar, M., Jiraskova, G. (eds.) International Conference on Implementation and Application of Automata, pp. 209–221. Springer, Cham (2019). https://doi.org/10.1007/978-3-030-23679-3_17

19. Otto, F.: Restarting automata for picture languages: a survey on recent developments. In: Holzer, M., Kutrib, M. (eds.) International Conference on Implementation and Application of Automata, pp. 16–41. Springer, Cham (2014). https://doi.org/10.1007/978-3-319-08846-4_2

20. Otto, F., Mráz, F.: Extended two-way ordered restarting automata for picture languages. In: Dediu, A.H., Martin-Vide, C., Sierra-Rodriguez, J.L., Truthe, B. (eds.) International Conference on Language and Automata Theory and Applications. pp. 541–552. Springer, Cham (2014). https://doi.org/10.1007/978-3-319-04921-2_44

21. Paun, G.: Marcus Contextual Grammars, vol. 67. Springer, Dordrecht (2013). https://doi.org/10.1007/978-94-015-8969-7

22. Paun, G., Nguyen, X.M.: On the inner contextual grammars. Rev. Roum. Math. Pures Appl. **25**(4), 641–651 (1980)

23. Rama, R., Smitha, T.: Some results on array contextual grammars. Int. J. Pattern Recogn. Artif. Intell. **14**(04), 537–550 (2000)

24. Rosenfeld, A.: Picture Languages: Formal Models for Picture Recognition. Academic Press, Cambridge (2014)

25. Rosenfeld, A., Siromoney, R.: Picture languages: a survey. Lang. Design **1**(3), 229–245 (1993)

26. Subramanian, K., Van, D.L., Chandra, P.H., Quyen, N.D.: Array grammars with contextual operations. Fundamenta Informaticae **83**(4), 411–428 (2008)

Grayscale Uncertainty of Projection Geometries and Projections Sets

László G. Varga[✉], Gábor Lékó, and Péter Balázs

Department of Image Processing and Computer Graphics, University of Szeged,
Szeged, Hungary
{vargalg,leko,pbalazs}@inf.u-szeged.hu

Abstract. In some cases of tomography, the projection acquisition process has limits, and thus one cannot gain enough projections for an exact reconstruction. In this case, the low number of projections leads to a lack of information, and uncertainty in the reconstructions. In practice this means that the pixel values of the reconstruction are not uniquely determined by the measured data and thus can have variable values. In this paper, we provide a theoretically proven uncertainty measure that can be used for measuring the variability of pixel values in grayscale reconstructions. The uncertainty values are based on linear algebra and measure the slopes of the hyperplane of solutions in the algebraic formulation of tomography. The methods can also be applied for any linear equation system, that satisfy a given set of conditions. Using the uncertainty measure, we also derive upper and lower limits on the possible pixel values in tomographic reconstructions.

Keywords: Uncertainty · Computed tomography · Algebraic reconstruction

1 Introduction

In Computed Tomography (CT) [4], X-ray radiation is used to produce the projections of an object. The projections themselves represent the attenuation of the beams passing through the object, giving information about the density of the material. By gathering these projections from different angles one can reconstruct the interior of the subject of investigation. In contrast to CT, Discrete Tomography (DT) [5,6] uses the prior information that the cross-section image to be reconstructed contains only a few different intensities which are known in advance. Binary Tomography (BT) is a more restricted variant of discrete tomography. In this case every single pixel of the image to be reconstructed can take only two different intensities. In practice, objects corresponding to these images must be made (or must consist) of a homogeneous material.

In an ideal case, having the reconstruction is performed using a large number of projections. This, however, is not always possible and the low number of projections turn the reconstruction task into an ill-posed problem, having various

© Springer Nature Switzerland AG 2020
T. Lukić et al. (Eds.): IWCIA 2020, LNCS 12148, pp. 123–138, 2020.
https://doi.org/10.1007/978-3-030-51002-2_9

possible solutions. This lack of information produces errors in the reconstructions, because we do not have enough information for determining exact pixel values.

Two research groups simultaneously investigated the theory of pixel uncertainty in discrete reconstructed images. They gave two different approaches for measuring the variability of reconstructions [2,11]. While [11] could be used in binary cases, [2] presents a way to apply uncertainty in Discrete Tomography. Unfortunately, these previous measures do not consider the uncertainty of a reconstructed image in continuous CT.

This field has a wide range of applications. It was shown that entropy based uncertainty could be efficiently used in projection selection [3,8,9]. [8] adapts the global uncertainty presented in [11] to select the most informative angles through an offline method. [1] presents an improved pixel update strategy by introducing a probability map that measures the classification accuracy of each pixel based on its grey value evolution throughout the iterations, which can be applied also in the case of Discrete Tomography, but it still does not give information about the uncertainty of the projection sets and their geometry.

Although several different types of uncertainty descriptors have been invented during the last decades, they mostly focus on discrete tomography, or the direct application of descriptors. The reconstruction uncertainty of the projection geometry and the variability pixels on grayscale reconstructions - to the best of our knowledge - has never been discussed yet. Such tools could be used to analyse complex CT sinograms and reconstructed CT slices to tell how stable or accurate their reconstructions are.

The aim of this paper is to define grayscale uncertainty and to examine its behavior with different projection sets. First, we provide a measure for the variability of the reconstructed pixels based only on the projection geometry. Then we used this variability measure to give theoretically proven upper and lower bounds for the pixel values in the reconstructions.

The structure of the paper is the following. In Sect. 2 we give a brief explanation of the reconstruction problem and its algebraic formulation. In Sect. 3 we propose our new findings in three theorems and in the same time prove them. In Sect. 4 we give details about the experimental frameset used for investigating the practical properties of our methods, while in Sect. 5 we present the experimental results. Finally, Sect. 6 is for the conclusions.

2   The Reconstruction Problem

We use the algebraic formulation of computed tomography (see, e.g., chapter 7 of [7]), and assume that the object to be reconstructed is represented in a two dimensional image. The idea is to describe the connections between projections and pixels using equations. Assuming that the size of the image to be reconstructed is $n \times n$ (N will denote the $n \times n$ product), the reconstruction problem can be described as a system of equations

$$\mathbf{Ax} = \mathbf{b}, \quad \mathbf{A} \in \mathbb{R}_{\geq 0}^{m \times N}, \quad \mathbf{x} \in \mathbb{R}_{\geq 0}^{N}, \quad \mathbf{b} \in \mathbb{R}_{\geq 0}^{m}, \tag{1}$$

where $\mathbf{x}$ is the vector of all n^2 unknown image pixels, m is the total number of projection lines used, $\mathbf{b}$ is the vector of all m measured projection values and $\mathbf{A}$ describes the projection geometry with all $a_{i,j}$ elements giving the length of the line segment of the i-th projection line through the j-th pixel (see Fig. 1 for illustration).

We must note, that the basic formulation is given for the 2D case of tomography. However, the results are directly applicable to the 3D case of tomography as well.

One can note, that the stability of the reconstruction is connected to the condition number of the $\mathbf{A}$ matrix. Unfortunately, classical tools for analyzing the condition number are hard to carry out due to the large size of projection matrices, therefore alternative approaches are needed.

Some of our proposed methods strongly rely on that the projection matrix, the projection values and the reconstructed pixels are all non-negative. This assumption can be made in transmission tomography since there are no materials with negative density or projection rays with line segments of negative lengths. We also use that the plane of solutions is a linear hyperplane in $\mathbb{R}_{\geq 0}^N$.

There are various methods for tomographic reconstructions. When we needed an actual reconstruction, we approximated a solution of Eq. (1) with the Bounded version of the Simultaneous Iterative Reconstruction Technique (SIRT) [4,7,10] and the Conjugate Gradient Least Squares (CGLS) technique [10]. Both SIRT and CGLS are iterative processes capable of approximating the correct reconstruction by iteratively subtracting the back-projected error of the intermediate state from itself. In general, these methods give continuous reconstructions with real pixel values and (with the proper setup) can produce images which are the closest one (in the Euclidean sense) to the initial image. We used the SIRT and CGLS methods because their advantages and drawbacks made them suitable for various purposes. The CGLS for example has a faster convergence then the SIRT method (see, e.g., Theorem 6.23 of [10]) that means in practice it gives an accurate unconstrained solution for a general reconstruction in significantly less iterations. The SIRT on the other hand is easy to modify into a projected gradient method and it can be used for solving the reconstruction problem with a $x_i \geq 0$ lower bound.

3 Proposed Methods

Before getting to the formulation of the uncertainty measure we need some definitions.

Definition 1. *Let* $\mathbf{1}_N$ *be a column vector such that*

$$(\mathbf{1}_N)_k = 1 \ , \ \forall k \in \{1, \ldots, N\}. \tag{2}$$

Definition 2. *Let* $\mathbf{0}_N$ *be a column vector such that*

$$(\mathbf{0}_N)_k = 0 \ , \ \forall k \in \{1, \ldots, N\}. \tag{3}$$

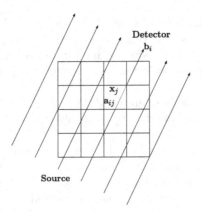

Fig. 1. Equation system-based representation of the parallel-beam projection geometry on a discrete image.

Definition 3. *Let* $\mathbf{e}_{N,i}$ *be a column vector such that*

$$(\mathbf{e}_{N,i})_k = \begin{cases} 1, & \text{if } k = i\,, \\ 0, & \text{if } k \neq i \end{cases} \quad \forall k \in \{1, \dots, N\}\,. \tag{4}$$

When it does not lead to confusion we will omit the size of $\mathbf{0}_N$ and $\mathbf{1}_N$ by just writing, $\mathbf{0}$ and $\mathbf{1}$.

Now we can define the core concept of the uncertainty measure.

Definition 4. *For any* $\mathbf{A}$ *projection matrix let the* i-*th perturbed reconstruction of* $\mathbf{A}$ *be*

$$\mathcal{P}_{\mathbf{A}}(i) = \arg\min_{\mathbf{x}} \left(\|\mathbf{x} - \mathbf{e}_{N,i}\| \mid \mathbf{A}\mathbf{x} = \mathbf{0}_N,\ \mathbf{x} \in \mathbf{R}^N \right). \tag{5}$$

In practice, the $\mathcal{P}_{\mathbf{A}}(i)$ can be calculated by the SIRT algorithm, or the CGLS method.

Technically, the $\mathbf{A}\mathbf{x} = \mathbf{0}_N$ is the $\mathbf{A}\mathbf{x} = \mathbf{b}$ hyperplane of reconstructions shifted into the origin. The $\mathcal{P}_{\mathbf{A}}(i)$ point is a reconstruction on this plane, that is the closest to $\mathbf{e}_{N,i}$. These perturbed reconstructions will be used to calculate the uncertainty. In the following sections we will highlight some properties of $\mathcal{P}_{\mathbf{A}}(i)$.

For a convenience we will denote by $\mathcal{N}$ and $\mathcal{M}$ the sets

$$\mathcal{N} = \{1, 2, \dots, N\}\,, \tag{6}$$

and

$$\mathcal{M} = \{1, 2, \dots, m\}\,, \tag{7}$$

Definition 5. *Let* $\mathcal{G}_{\mathbf{A}}(i)$ *be the largest gradient of the solutions of* $\mathbf{A}\mathbf{x} = \mathbf{0}_N$ *with respect to* x_i, *i.e.*,

$$\mathcal{G}_{\mathbf{A}}(i) = \max(x_i \mid \mathbf{A}\mathbf{x} = \mathbf{0}_N,\ \sum_{k \in \mathcal{N} \setminus \{i\}} (x_k^2) = 1)\,. \tag{8}$$

In a more intuitive manner, the $\mathcal{G}_\mathbf{A}(i)$ is a point that gives us information on the slope of the hyperplane of solutions on the x_i axis. This gives a vector along which the x_i value can change the fastest within the hyperplane of solutions.

3.1   The Variability of the Reconstructed Pixels

Theorem 1. *For any* $\mathbf{A}$ *projection matrix if* $(\mathcal{P}_\mathbf{A}(i))_i \neq 0$ *and* $(\mathcal{P}_\mathbf{A}(i))_i \neq 1$, *then there is a* $c \in \mathbb{R}^+$ *constant such that*

$$\mathcal{G}_\mathbf{A}(i) = c \cdot \mathcal{P}_\mathbf{A}(i) \ . \tag{9}$$

Proof. Given a fixed i pixel index, denote

$$\mathbf{p} = \mathcal{P}_\mathbf{A}(i) \tag{10}$$

We stated that $p_i \neq 0$. We can also state that $p_i > 0$ because if $\mathbf{Ax} = \mathbf{0}_N$ then

$$\mathbf{A}(-\mathbf{p}) = -\mathbf{0}_N = \mathbf{0}_N \ , \tag{11}$$

and also if $x_i < 0$ then

$$(1 - x_i)^2 > (1 - (-x_i))^2 \ , \tag{12}$$

leading to

$$\begin{aligned}
\|\mathbf{e}_i - \mathbf{p}\|_2 &= \sqrt{\sum_{k \in \mathcal{N} \setminus \{i\}} (0 - p_k)^2 + (1 - p_i)^2} \\
&= \sqrt{\sum_{k \in \mathcal{N} \setminus \{i\}} p_k^2 + (1 - p_i)^2} \\
&> \sqrt{\sum_{k \in \mathcal{N} \setminus \{i\}} p_k^2 + (1 - (-p_i))^2} \\
&= \sqrt{\sum_{k \in \mathcal{N} \setminus \{i\}} (0 - (-p_k))^2 + (1 - (-p_i))^2} \\
&= \|\mathbf{e}_i - (-\mathbf{p})\|_2 \ .
\end{aligned} \tag{13}$$

This would mean that $\|\mathbf{e}_i - \mathbf{p}\|_2 > \|\mathbf{e}_i - (-\mathbf{p})\|_2$ that is a contradiction with the definition of $\mathbf{p}$.

We can also state that $0 < p_i < 1$. This is because the $\mathbf{p}$, $\mathbf{0}_N$ and $\mathbf{e}_{N,i}$ points form a right-angled triangle in space having the $(\mathbf{0}_N, \mathbf{e}_{N,i})$ hypotenuse of length 1. For an illustration see Fig. 2.

Now let δ be

$$\delta = \sqrt{\sum_{k \in \mathcal{N} \setminus \{i\}} p_k^2} \ . \tag{14}$$

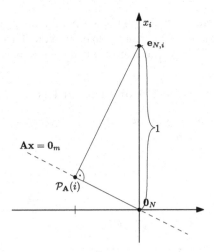

Fig. 2. Illustration of the $\mathcal{P}_\mathbf{A}(i)$ perturbed reconstruction.

Furthermore, we will denote $\mathbf{v}$ the vector such that

$$\mathbf{v} = \delta \cdot \mathcal{G}_\mathbf{A}(i) \tag{15}$$

This way, we know that

$$\sqrt{\sum_{k\in\mathcal{N}\backslash\{i\}} p_k^2} = \delta = \sqrt{\sum_{k\in\mathcal{N}\backslash\{i\}} \delta^2 \cdot (\mathcal{G}_\mathbf{A}(i))_k^2} = \sqrt{\sum_{k\in\mathcal{N}\backslash\{i\}} v_k^2} \tag{16}$$

Also, from the definition of $\mathcal{G}_\mathbf{A}(i)$ we get that

$$v_i = \max(x_i \mid \mathbf{A}\mathbf{x} = \mathbf{0}, \sum_{k\in\mathcal{N}\backslash\{i\}} x_k^2 = \delta^2) \ . \tag{17}$$

Assume, that $v_i > p_i$ and there is an $\alpha \in \mathbb{R}_{>0}$ such that

$$v_i = p_i + \alpha \ . \tag{18}$$

Now we have two cases. (For an illustration see, Fig. 3.)

1. First, if $0 < v_i < 1$ (left side of Fig. 3) then we have

$$v_i = p_i + \alpha > p_i \ . \tag{19}$$

This gives us

$$\|e_i - v\|_2 = \sqrt{\sum_{k\in\mathcal{N}\backslash\{i\}} (0 - v_k)^2 + (1 - v_i)^2} = \sqrt{\sum_{k\in\mathcal{N}\backslash\{i\}} v_k^2 + (1 - v_i)^2}$$
$$= \sqrt{\delta^2 + (1 - (p_i + \alpha))^2} < \sqrt{\delta^2 + (1 - p_i)^2} \tag{20}$$
$$= \|e_i - p\|_2 \ .$$

that is a contradiction with the definition of $\mathbf{p} = \mathcal{P}_\mathbf{A}(i)$.

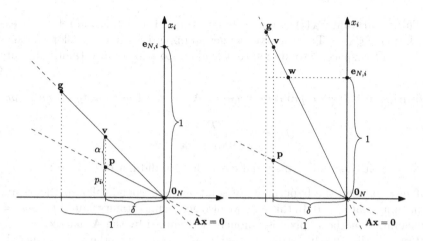

Fig. 3. Two cases of $v_i > p_i$. On the left: $0 < v_i < 1$; On the right: $v_i \geq 1$.

2. In the second case, assume, that $v_i \geq 1$. In this case let

$$\mathbf{w} = \mathbf{v}/v_i \ . \tag{21}$$

Now we have

$$\sqrt{\sum_{k \in \mathcal{N} \setminus \{i\}} w_k^2} = \sqrt{\sum_{k \in \mathcal{N} \setminus \{i\}} \left(\frac{v_k}{v_i}\right)^2} = \frac{\sqrt{\sum_{k \in \mathcal{N} \setminus \{i\}} v_k^2}}{v_i} = \frac{\delta}{v_i} \leq \delta \tag{22}$$

and by definition $w_i = 1$ therefore

$$1 - w_i = 0 \ . \tag{23}$$

This leads to

$$\begin{aligned}
\|e_i - w\|_2 &= \sqrt{\sum_{k \in \mathcal{N} \setminus \{i\}} (0 - w_k)^2 + (1 - w_i)^2} \\
&= \sqrt{\sum_{k \in \mathcal{N} \setminus \{i\}} w_k^2 + (1 - 1)^2} = \sqrt{\left(\frac{\delta}{v_i}\right)^2 + 0^2} \\
&< \sqrt{\delta^2 + (1 - p_i)^2} = \|e_i - p\|_2 \ ,
\end{aligned} \tag{24}$$

This also contradicts the definition of **p**.

Furthermore, $\mathbf{p} = \mathcal{P}_\mathbf{A}(i)$ is unique and $\mathcal{G}_\mathbf{A}(i)$ is also unique if the conditions of the theorem hold. We also have that p_i cannot be greater than v_i (as it would contradict the definition of v_i), therefore, we have that $p_i = v_i$. Together with (16), this means that (9) can only hold if $\mathbf{v} = \mathbf{p}$ and $c = 1/\delta$. ∎

This means that $\mathcal{P}_{\mathbf{A}}(i)$ can directly be used for the simple and relatively easy calculation of $\mathcal{G}_{\mathbf{A}}(i)$. This way, we can get an upper bound for the slope of the x_i variable in the space of solutions, that leads us to a local uncertainty measure. ∎

Definition 6. *Given a projection matrix* $\mathbf{A}$, *let the Uncertainty of the* x_i *pixel be*

$$U_{\mathbf{A}}(i) = \frac{\|\mathcal{P}_{\mathbf{A}}(i)\|_2}{\|\mathbf{e}_{N,i} - \mathcal{P}_{\mathbf{A}}(i)\|_2} . \tag{25}$$

Note, that there are two special cases that we did not discuss yet:

- If $p_i = 0$, then the hyperplane of solutions is perpendicular to the x_i axis, meaning that $v_i = 0$. In this case we also get an uncertainty value of $U_{\mathbf{A}}(i) = 0$ meaning that the x_i pixel is uniquely determined by the $\mathbf{A}$ matrix.
- If $p_i = 1$, then the hyperplane of solutions is parallel to the x_i axis, meaning that either $v_i = \infty$, or v_i does not exist. In this case we get an uncertainty value of $U_{\mathbf{A}}(i) = \infty$ meaning that the x_i pixel can take any values independent of the other variables.

Using the above results we can also find an interesting connection between $U_{\mathbf{A}}(i)$ and the $\mathcal{G}_{\mathbf{A}}(i)$ vectors.

Lemma 1. *Let* $\mathbf{g}$ *be a vector for a fixed* i *such that*

$$\mathbf{g} = \mathcal{G}_{\mathbf{A}}(i) , \tag{26}$$

then,

$$U_{\mathbf{A}}(i) = g_i . \tag{27}$$

Proof. Let $\hat{\mathbf{g}}$ be a vector such that

$$\hat{g}_k = \begin{cases} g_k & \text{if } i \neq k \\ 0 & \text{if } i = k \end{cases} \tag{28}$$

By definition $\|\hat{\mathbf{g}}\|_2 = 1$. Also, let $\mathbf{p} = \mathcal{P}_{\mathbf{A}}(i)$ as in (10). This way we can get two triangles. The first is determined by the $(\mathbf{0}_n, \mathbf{g}, \hat{\mathbf{g}})$ points, and the other one is determined by $(\mathbf{e}_{N,i}, \mathbf{p}, \mathbf{0}_n)$ (for an illustration, see Fig. 4).

The $(\mathbf{0}_n, \mathbf{g}, \hat{\mathbf{g}})$ and $(\mathbf{e}_{N,i}, \mathbf{p}, \mathbf{0}_n)$ triangles are similar, and we get

$$U_{\mathbf{A}}(i) = \frac{\|\mathcal{P}_{\mathbf{A}}(i)\|_2}{\|\mathbf{e}_{N,i} - \mathcal{P}_{\mathbf{A}}(i)\|_2} = \frac{\|\mathbf{p}\|_2}{\|\mathbf{e}_{N,i} - \mathbf{p}\|_2} = \frac{g_i}{1} = g_i . \tag{29}$$

∎

According the the previous results, the $U_{\mathbf{A}}(i)$ is an upper bound for the slope of space of solutions on the x_i axis. It was also shown, that the uncertainty values can be calculated in three different ways, i.e.,

$$U_{\mathbf{A}}(i) = \frac{\|\mathcal{P}_{\mathbf{A}}(i)\|_2}{\|\mathbf{e}_{N,i} - \mathcal{P}_{\mathbf{A}}(i)\|_2} = \frac{p_i}{\delta} = g_i . \tag{30}$$

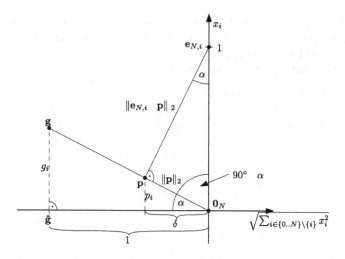

Fig. 4. Connection between the $U_\mathbf{A}(i)$ uncertainty value and the $\mathcal{G}_\mathbf{A}(i)$ vector.

This means that there are more possible formulations and either $\mathcal{P}_\mathbf{A}(i)$ or $\mathcal{G}_\mathbf{A}(i)$ could be used for calculating the uncertainty. The reason of the current definition of $U_\mathbf{A}(i)$ is that this form is less effected by the quantization error of floating-point numbers when the uncertainty is small. We must also note, that $\mathcal{G}_\mathbf{A}(i)$ was only used for theoretical purposes, as it is useful for showing the properties of the uncertainty. In practice the calculation of the $\mathcal{P}_\mathbf{A}(i)$ is computationally more effective, and sufficient for determining the uncertainty.

There is also an important property of the uncertainty coming from the fact that the space of solutions is a linear hyperplane.

Corollary 1. *Let* $\mathbf{A}$ *be a projection matrix,* $\mathbf{b}$ *a set of projection values, and* $\mathbf{x}$ *and* $\mathbf{y}$ *two vectors such that*

$$\mathbf{Ax} = \mathbf{b} , \quad \mathbf{Ay} = \mathbf{b} , \quad \mathbf{x}, \mathbf{y} \in \mathbb{R}_{\geq 0}^N . \tag{31}$$

If

$$\sqrt{\sum_{k\in\mathcal{N}\backslash\{i\}} (x_k - y_k)^2} = 1 , \tag{32}$$

then

$$|x_i - y_i| \leq U_\mathbf{A}(i) . \tag{33}$$

This later Corollary only says that the upper bound on the slopes of $\mathbf{Ax} = \mathbf{0}_N$ also apply for $\mathbf{Ax} = \mathbf{b}$.

3.2   Bounds of Pixel Values

Using the bound on the slopes of the hyperplane of solutions we can give bounds on the pixel values in the reconstructions. In this section, we will give two possible bounds. The first one is only based on the equation system of the projections, while the second one on using the first bound and the uncertainty together for giving upper and lower limits on the pixel values.

Let us start with the first upper limit.

Lemma 2. *For any x_k variable*

$$x_i \leq \min_{j \in \mathcal{M}} \frac{b_j}{a_{j,i}} \ . \tag{34}$$

Proof. For any $(i,j) \in (\{1,..,m\},\{1,..,N\})$

$$0 \leq x_j \ , \quad 0 \leq a_{i,j} \ , \tag{35}$$

therefore

$$0 \leq a_{i,j}x_j \ . \tag{36}$$

■

We also know that for any $i \in \mathcal{N}$ and $j \in \mathcal{M}$

$$x_i = \frac{b_j - \sum_{k \in \mathcal{N}\setminus\{i\}}(a_{j,k}x_j)}{a_{j,i}} \tag{37}$$

and from this we have

$$x_i = \frac{b_j - \sum_{k \in \mathcal{N}\setminus\{i\}}(a_{j,k}x_j)}{a_{j,i}} \leq \frac{b_j - \sum_{k \in \mathcal{N}\setminus\{i\}} 0}{a_{j,i}} = \frac{b_j}{a_{j,i}} \ , \tag{38}$$

Taking the minima of this upper bound for all the equations we have the upper bound for the theorem.

■

Now using this limit, we can get a different bound on the pixel values by also using the uncertainty values.

Definition 7. *Let $L_\mathbf{A}(i)$ be a value such that*

$$L_\mathbf{A}(i) = \min_{j \in \mathcal{M}} \frac{b_j}{a_{j,i}} \ . \tag{39}$$

Definition 8. *Let $D_\mathbf{A}(i)$ be a value that is*

$$D_\mathbf{A}(i) = \sqrt{\sum_{k \in \mathcal{N}\setminus\{i\}}(L_\mathbf{A}(k))^2} \tag{40}$$

Theorem 2. *Given an $\mathbf{A}$ projection matrix, a $\mathbf{b}$ set of projections, and an $\hat{\mathbf{x}} \in \mathbb{R}^N_{\geq 0}$ reconstruction such that*

$$\mathbf{A}\hat{\mathbf{x}} = \mathbf{b} \ , \quad \hat{\mathbf{x}} \in \mathbb{R}^n_{\geq 0} \ . \tag{41}$$

For any $\mathbf{y} \in \mathbb{R}^n_{\geq 0}$ reconstruction and any $i \in \mathcal{N}$ index

$$\hat{x}_i - U_\mathbf{A}(i) \cdot D_\mathbf{A}(i) \ \leq \ y_i \ \leq \ \hat{x}_i + U_\mathbf{A}(i) \cdot D_\mathbf{A}(i) \ . \tag{42}$$

Proof. Let us have a fixed i value. Let γ be

$$\gamma = \sqrt{\sum_{k \in \mathcal{N} \setminus \{i\}} (x_k - y_k)^2} \tag{43}$$

and define $\mathbf{z}$ as a vector such that

$$\mathbf{z} = \frac{\mathbf{y} - \mathbf{x}}{\gamma} . \tag{44}$$

Form Lemma 2 we have that if $\mathbf{Ax} = \mathbf{b}$ then for any $i \in \mathcal{N}$

$$0 \le x_i \le L_{\mathbf{A}}(k) = \min_{j \in M} \frac{b_j}{a_{j,k}} . \tag{45}$$

This also means that the space of solutions is an N-dimensional hyperrectangle with and for any $\mathbf{x}, \mathbf{y}$ pair of solutions

$$\gamma = \sqrt{\sum_{k \in \mathcal{N} \setminus \{i\}} z_k^2} = \sqrt{\sum_{k \in \mathcal{N} \setminus \{i\}} (y_k - x_k)^2}$$
$$\le \min_{j \in \{1..m\}} \sqrt{\sum_{k \in \{1..M\} \setminus i} (L_{\mathbf{A}}(k))^2} = D_{\mathbf{A}}(k) \tag{46}$$

We also have that $U_{\mathbf{A}}(i) = (\mathcal{G}_{\mathbf{A}}(i))_i$. By definition

$$\sqrt{\sum_{k \in \mathcal{N} \setminus \{i\}} z_k^2} = \sqrt{\sum_{k \in \mathcal{N} \setminus \{i\}} (x_k - y_k)^2} = 1 , \tag{47}$$

Therefore, from Corollary 1 we have

$$z_i = y_i - x_i \le U_{\mathbf{A}}(i) . \tag{48}$$

This gives us

$$y_i = x_i + \gamma \cdot z_i \le x_i + D_{\mathbf{A}}(k) z_i \le x_i + D_{\mathbf{A}}(k) \cdot U_{\mathbf{A}}(i) , \tag{49}$$

and as the hyperplane of solutions is symmetric we also have

$$y_i \ge x_i - D_{\mathbf{A}}(k) \cdot U_{\mathbf{A}}(i) . \tag{50}$$

∎   ∎

Regarding the relation between the upper bounds we can say the following. As the upper bounds in Lemma 2 rely on mostly the upper limit on the projection values, in practice, this bound will get to its minima after a few projections and will not get significantly lower by increasing the number of projections.

On the other hand, the bounds in Theorem 2 work differently. The $D_{\mathbf{A}}$ value is practically a huge constant in the upper bounds giving very loose limits with a low number of projections. However, when increasing the number of projections, the $U_{\mathbf{A}}$ values in the bound will reach towards zero making the bound more-and-more strict.

This means that in practice the bounds of Lemma 2 are more useful with a low projection number, and Theorem 2 is more strict with higher projection counts.

4   Test Data and Experimental Settings

To further investigate the uncertainty and the limits from a practical point of view, we performed experimental tests on a set of different images. Our database consisted of 7 phantoms with different structural complexity, each with a size of 64×64 pixels and an intensity range of $[0,1]$. These images can be seen on Fig. 5. We used parallel beam geometry for the acquisitions of the projections. In every projection we set the distance of the beams and detector elements to 1 pixel and used $64 \cdot \sqrt{2}$ of them to cover the whole image. The rotation center was placed into the center of the image. The reconstructions were performed using 7 different number of projections: 2, 4, 8, 16, 32, 64 and 128.

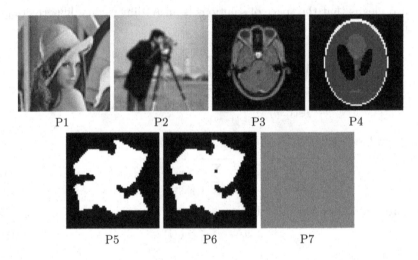

Fig. 5. Images used for testing. P1: Lena, P2: Cameraman, P3: a head CT slice, P4: Shepp-Logan phantom, P5: a random binary shape, P6: the same phantom with a hole, P7: homogeneous image with value 0.5 in every pixel.

The Bounded SIRT was implemented in C++, using the CUDA sdk[1], with GPU acceleration. The reconstructions were performed on a machine powered with 4 NVIDIA Tesla K10 K2 8 GB GPUs. During the reconstruction process, the iterations of SIRT terminated if the difference between two consecutive iterations was less than 0.000001 or the number of iterations reached 1000000. In the case of CGLS, we used a MATLAB implementation. The number of iterations was set to 10000.

5   Results

Let there be a given $\Omega_{\mathbf{A},\mathbf{b}}$ set of reconstructions such that

$$\Omega_{\mathbf{A},\mathbf{b}} = \left\{ \mathbf{x} \mid \mathbf{x} \in \mathbb{R}^n_+, \mathbf{A}\mathbf{x} = \mathbf{b} \right\} . \tag{51}$$

---

[1] https://www.developer.nvidia.com/cuda-zone.

Let $E(\Omega_{\mathbf{A},\mathbf{b}}, i)$ be the expected value of x_i in $\Omega_{\mathbf{A},\mathbf{b}}$, i.e.,

$$E(\Omega_{\mathbf{A},\mathbf{b}}, i) = \frac{\sum_{\mathbf{x} \in \Omega_{\mathbf{A},\mathbf{b}}} (x_i)}{|\Omega_{\mathbf{A},\mathbf{b}}|} \qquad (52)$$

Let $\sigma(\Omega_{\mathbf{A},\mathbf{b}}, i)$ be the standard deviation of x_i in $\Omega_{\mathbf{A},\mathbf{b}}$, i.e.,

$$\sigma(\Omega_{\mathbf{A},\mathbf{b}}, i) = \sqrt{\frac{\sum_{\mathbf{x} \in \Omega_{\mathbf{A},\mathbf{b}}} |x_i - E(\Omega_{\mathbf{A},\mathbf{b}}, i)|^2}{|\Omega_{\mathbf{A},\mathbf{b}}| - 1}} \qquad (53)$$

Furthermore, let $Min(\Omega_{\mathbf{A},\mathbf{b}}, i)$ and $Max(\Omega_{\mathbf{A},\mathbf{b}}, i)$ be the minimum and maximum of x_i in $\Omega_{\mathbf{A},\mathbf{b}}$,

$$Min(\Omega_{\mathbf{A},\mathbf{b}}, i) = \min_{\mathbf{x} \in \Omega_{\mathbf{A},\mathbf{b}}} (x_i) \qquad (54)$$

$$Max(\Omega_{\mathbf{A},\mathbf{b}}, i) = \max_{\mathbf{x} \in \Omega_{\mathbf{A},\mathbf{b}}} (x_i) \qquad (55)$$

First, we investigated the correlation between $\sigma(\Omega_{\mathbf{A},\mathbf{b}}, i)$ and $U_{\mathbf{A}}(i)$, created with CGLS. The Pearson correlations and the Fisher z-transformations with 95% confidence interval can be seen in Table 1. In both cases, we concatenated all the standard deviation and uncertainty maps (belonging to a given phantom) with different number of projections into one matrix and calculated the correlation between them. This way, 7 comparisons were made (because of the 7 test images). The concatenated uncertainty map belonging to P1 can be seen in Fig. 6.

Table 1. The correlation between the concatenated $\sigma(\Omega_{\mathbf{A},\mathbf{b}}, i)$ and $U_{\mathbf{A}}(i)$ values generated from all the different number of projections using the test images in Fig. 5.

Phantoms	Pearson Corr.	Fisher z-trans.
P1	0.8062	[0.8021, 0.8102]
P2	0.8062	[0.8021, 0.8102]
P3	0.8068	[0.8028, 0.8108]
P4	0.8062	[0.8021, 0.8102]
P5	0.8064	[0.8023, 0.8104]
P6	0.8063	[0.8022, 0.8103]
P7	0.8067	[0.8016, 0.8098]

The correlation values around 0.8 indicate a reliable correspondence between the uncertainty measure and the deviation of pixel values in reconstructions. We argue that the correlation is not higher only because the pixel values in our random reconstructions are not from uniform distribution. As the generation of random reconstructions is not the goal of this paper, an extended evaluation could be subject to further studies.

Using one random reconstruction from all the phantoms with different number of projections (created with Bounded SIRT), we also analyzed the range of

Fig. 6. $U_{\mathbf{A}}(i)$ of P1 in case of different number of projections.

the bounds given in Lemma 2 and Theorem 2. In Fig. 7, one can see the profiles of the upper limits $(L_{\mathbf{A}}(i))$ from Lemma 2. It can clearly be seen that as the projection number increases, the limits are getting closer and closer to the x-axis. The distance between the averages can also give us information about the intensity ranges of the phantoms.

Figure 8 shows the profiles of the upper and the lower limits defined in Theorem 2. In the case of a low number of projections the bounds are far from each other and as the projection number increases the bounds are approximating each other.

Taking under consideration the results of Figs. 7 and 8, in the case of small amount of projections, using the upper bound of Lemma 2 seems to be a good choice to limit the reconstruction. As we increase the number of projections it may be increasingly useful to apply the bounds provided by Theorem 2, as they converge to a unique reconstruction.

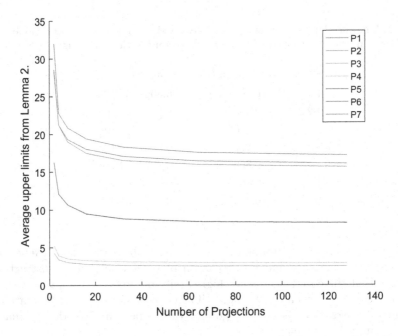

Fig. 7. Average upper limits from Lemma 2 in the case of different number of projections.

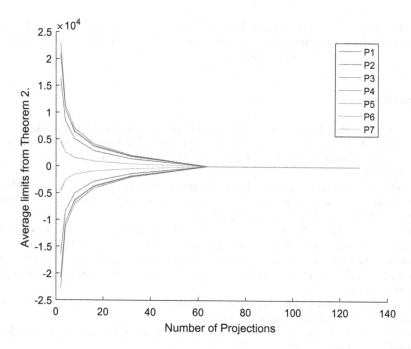

Fig. 8. Average limits from Theorem 2 in the case of different number of projections.

6 Conclusions

We provided a measure for the variability of the reconstructed pixels based only on the projection geometry. Then we used this variability measure to give theoretically proven upper and lower bounds for the pixel values in the reconstructions. We gave a practical analysis to the results evaluating the correlations between the local uncertainties given in Theorem 1 and the standard deviation of a set of reconstructed images. The bounds of Lemma 2 and Theorem 2 are also investigated by comparing them to each other with randomly selected reconstructed images.

As a future work, we are planning to further examine the practical usability of our methods and to evaluate our concepts on real-world data. Another possibility is giving a sophisticated extension of the local uncertainty measure into a global one. Also, the connection of the present work to other numerical tools – like the condition number of matrices – would give interesting insight into the analysis of reconstructions.

Acknowledgements. Gábor Lékó was supported by the UNKP-19-3 New National Excellence Program of the Ministry of Human Capacities. This research was supported by the project "Integrated program for training new generation of scientists in the fields of computer science", no. EFOP-3.6.3-VEKOP16-2017-00002. This research was supported by grant TUDFO/47138-1/2019-ITM of the Ministry for Innovation and Technology, Hungary. The authors would like to thank Bence Savanya for aiding by discovering numerical concepts for the formulas.

References

1. Frenkel, D., Beenhouwer, J.D., Sijbers, J.: An adaptive probability map for the Discrete Algebraic Reconstruction Technique. In: 10th Conference on Industrial Computed Tomography (iCT), (iCT 2020) Wels, Austria, 4–7 February 2020 pp. 1–10 (2020)
2. Frost, A., Renners, E., Hötter, M., Ostermann, J.: Probabilistic evaluation of three-dimensional reconstructions from x-ray images spanning a limited angle. Sensors (Basel, Switzerland) **13**, 137–51 (12 2012)
3. Haque, M., Ahmad, M.O., Swamy, M., Lee, S.: Adaptive projection selection for computed tomography. IEEE Trans. Image Process. **22**, 5085–5095 (2013)
4. Herman, G.T.: Fundamentals of Computerized Tomography: Image Reconstruction from Projections, 2nd edn. Springer, London (2009). https://doi.org/10.1007/978-1-84628-723-7
5. Herman, G.T., Kuba, A.: Discrete Tomography: Foundations, Algorithms, and Applications. Birkhäuser, Basel (1999)
6. Herman, G.T., Kuba, A.: Advances in Discrete Tomography and Its Applications. Birkhäuser, Basel (2007)
7. Kak, A.C., Slaney, M.: Principles of Computerized Tomographic Imaging. IEEE Press, New York (1988)
8. Lékó, G., Balázs, P., Varga, L.G.: Projection selection for binary tomographic reconstruction using global uncertainty. In: Campilho, A., Karray, F., ter Haar Romeny, B. (eds.) ICIAR 2018. LNCS, vol. 10882, pp. 3–10. Springer, Cham (2018). https://doi.org/10.1007/978-3-319-93000-8_1
9. Placidi, G., Alecci, M., Sotgiu, A.: Theory of adaptive acquisition method for image reconstruction from projections and application to EPR imaging. J. Magnet. Resonance Ser. B **108**(1), 50–57 (1995)
10. van der Sluis, A., van der Vorst, H.: Sirt- and cg-type methods for the iterative solution of sparse linear least-squares problems. Linear Algebra Appl. **130**, 257–303 (1990)
11. Varga, L., Nyúl, L., Nagy, A., Balazs, P.: Local and global uncertainty in binary tomographic reconstruction. Comput. Vis. Image Underst. **129**, 52–62 (2014)

Finding the Maximum Empty
Axis-Parallel Rectangular Annulus

Raina Paul[1]([⊠]), Apurba Sarkar[1], and Arindam Biswas[2]

[1] Department of Computer Science and Technology,
Indian Institute of Engineering Science and Technology, Shibpur, India
rainapaul22@gmail.com, as.besu@gmail.com
[2] Department of Information Technology,
Indian Institute of Engineering Science and Technology, Shibpur, India
barindam@gmail.com

Abstract. An annulus is basically a ring-shaped region between two concentric disks on the same plane. However, it can be defined on any other geometrical shapes, for example, a rectangular annulus is defined as the area between two rectangles with one rectangle enclosing the other. The area of the annulus is the area of the region between the two shapes. An axis-parallel rectangular annulus is an annulus where the sides of the rectangles are parallel to the co-ordinate axes. This paper presents a combinatorial technique to find the largest empty axis-parallel rectangular annulus from a given set of n points and runs in $O(n \log n)$ time. It uses two balanced binary search trees to store the points and reduces the complexity of the existing algorithm in the literature.

Keywords: Annulus · Rectangular annulus · Axis-parallel annulus

1 Introduction

Geometric covering is a well known problem in computational geometry and a special sub-case of this is finding the largest empty annulus of different shapes. Annulus of different shapes has importance in many domains such as field of modern industrial design, VLSI design, robotics etc. Details about the potential applications of largest empty annulus of different shapes can be found in [3,7]. In 2002, Banez et al. [4], proposed an algorithm to locate the largest width annulus in between two concentric circles. In their work, they mentioned two variations of the problem, the first variation does not allow the annulus to contain any point and the second variation restricts $k \in \mathcal{O}(n)$ number of points in the inner circle. The first variation takes $\mathcal{O}(n^3 \log n)$ time and $\mathcal{O}(n)$ space, the second variation also requires $\mathcal{O}(n^3 \log n)$ time and $\mathcal{O}(n)$ space but if the value of k is small, i.e, a fixed constant, then it takes $\mathcal{O}(n \log n)$ time and $\mathcal{O}(n)$ space. There are few more works on finding square/rectangular annulus in the literature. P. Mahapatra [5] presented an algorithm to find axis parallel maximum width empty rectangular annulus from a given set of n points. The proposed algorithm

© Springer Nature Switzerland AG 2020
T. Lukić et al. (Eds.): IWCIA 2020, LNCS 12148, pp. 139–146, 2020.
https://doi.org/10.1007/978-3-030-51002-2_10

is simple in nature and runs in $\mathcal{O}(n^2)$ time. Mukherjee et al. [6] proposed an algorithm to find the rectangular annulus of arbitrary orientation from a given set of n points in $\mathcal{O}(n^2 \log n)$ time and $\mathcal{O}(n)$ space. Sang Won Bae [2] studied the problem of finding minimum width square or rectangular annulus problem with k outliers, i.e, the annulus contains at least $n - k$ points out of n points. The k points which are excluded in the process are called outliers. He reported that the k-SquareAnnulus problem can be solved in $O(k^2 n \log n + k^3 n)$ time, and the k-RectAnnulus problem in $\mathcal{O}(nk^2 \log k + k^4 \log^3 k)$ time. Bae [1] also proposed an algorithm to compute minimum width square annulus in arbitrary orientation that runs in $\mathcal{O}(n^3 \log n)$ time. He showed that when the orientation $\theta \in [0, \pi/2)$ and the center, $c \in R^2$, a unique minimum width θ-aligned square annulus containing the points can be obtained. However, the algorithm proposed in this work is an improvement over Mahapatra's algorithm. The proposed algorithm runs in $\mathcal{O}(n \log n)$ time as compared to Mahapatra's algorithm which runs in $\mathcal{O}(n^2)$ time. The improvement in the proposed work is achieved with the use of balanced binary tree to store the points. Figure 2(a) shows a set of point, for which the maximum empty annulus is shown in Fig. 2(b).

The rest of the paper is organized as follows. Section 2 presents a few definitions and preliminaries that are related to the work. The proposed method along with algorithms and running time and proof of correctness are described in Sect. 3. Finally, the paper is concluded in Sect. 4 with future directions.

2   Definitions and Preliminaries

This section contains few definitions and observations that are required to explain the proposed work.

Definition 1 *(Annulus). Mathematically an annulus is defined as a plane figure consisting of the region between a pair of concentric circles.*

For example, the shaded region as shown in Fig. 1(a) is an annulus.

Definition 2 *(Rectangular Annulus). It is defined to be the closed region between two rectangles on the same plane with one enclosing the other.*

As shown in Fig. 1(b), the region bounded by outer rectangle R_1 and inner rectangle R_2 is the rectangular annulus. It is to be noted that unlike circular annulus the rectangular annular region may not be uniform meaning that it may be of different width along four different directions.

Definition 3 *(Axis-Parallel Annulus). Axis-parallel annulus is a rectangular annulus where the sides of the rectangles defining the annulus are axis parallel. The width of a rectangular annulus is the maximum distance between corresponding sides of rectangles in question. An axis parallel empty rectangular annulus of maximum width is said to be maximum empty axis parallel rectangular annulus.*

Figure 1(c) represents an maximum empty axis parallel rectangular annulus with w as the width of the annulus.

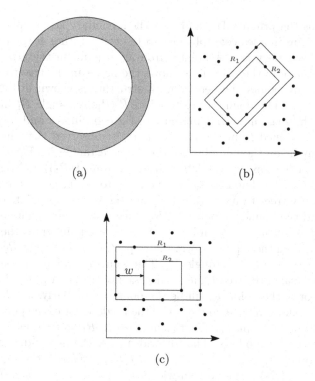

(a)   (b)

(c)

Fig. 1. (a) Annulus, (b) Rectangular annulus, (c) Maximum-width axis parallel annulus.

Definition 4 *(Bounding Rectangle). Bounding Rectangle (BR) R is defined as the rectangle of a fixed orientation in a plane that encloses all the points of the given point set P and there is no other rectangle R' of minimum area that can enclose all the points of P.*

3   Proposed Method

This section discusses the procedure to find the maximum empty axis-parallel annulus from a given set of points. As defined, an annulus is the closed region between two axis-parallel rectangles R_1 and R_2 such that $R_2 \subseteq R_1$ and the closed region does not contain any points. So, the algorithm needs to compute the aforementioned rectangles R_1 and R_2. Initially, the bounding rectangle (BR) that encloses the entire point set is the first potential outer rectangle R_1. The inner rectangle R_2 corresponding to this initial R_1 would give us the first annulus. To find the width of this particular annulus, its top width (t_w), bottom width (b_w), left width (l_w) and right width (r_w) needs to be calculated and the maximum of these four widths would give us the width of the annulus. To perform the above mentioned operation efficiently, the points are stored in two AVL trees T_x and T_y. The tree T_x stores the point in a lexicographically sorted order with x as the primary key and y as the secondary key, similarly the tree T_y stores the point in a lexicographically sorted order with y as the primary key and x as the secondary key. A pointer from each point in the array points to corresponding node in the tree. This allows the nodes in the tree to be directly accessed from any points. The entire data structure is shown in Fig. 3 for the point set in Fig. 2. The bounding rectangle BR (the rectangle R_1 in this case) is formed by the maximum and minimum of x-coordinates and y-coordinates which can be obtained from the tree. It is to be observed that, maximum of four points may lie on the boundary of the outer rectangle R_1. Let p_t, p_l, p_b, p_r be the points on the top ($t(R_1)$), left ($l(R_1)$), bottom ($b(R_1)$) and right ($r(R_1)$) boundary (side) of R_1. The nodes corresponding to p_t, p_l, p_b, p_r are marked by a flag in both the trees to mean that they are the boundary points of R_1. This helps us to find the boundary points of R_2 since, these points can not form the boundary of R_2. Let q_t, q_l, q_b and q_r be the nearest point from the p_t, p_l, p_b, p_r respectively, then these points form the respective boundary ($t(R_2)$), ($l(R_2)$), ($b(R_2)$) and ($r(R_2)$) of the rectangle R_2. The points q_t and q_b can be found out by inorder traversal of the tree T_y, as predecessor of the point p_t will be the point q_t and successor of the point p_b will be the point q_b. Similarly, inorder traversal of the tree T_x would give us q_l and q_r; q_l is the predecessor of the point p_l and q_r is the successor of the point p_r. The region between R_1 and R_2 defines an annulus whose width are as follows. The top-width t_w is the perpendicular distance between the top of R_1 and R_2, i.e. $|t_w| = |d(t(R_1) - t(R_2))|$, the right-width is $|r_w| = |d(r(R_1) - r(R_2))|$, the bottom-width is $|b_w| = |d(b(R_1) - b(R_2))|$, and the left-width $|l_w| = |d(l(R_1) - rl(R_2))|$. This distance can be calculated in constant time as the coordinates of points passing through each of the boundaries are known. The maximum distance between corresponding sides of inner rectangle R_2 and outer rectangle R_1 defines the width of an annulus. Thus the width of the above annulus is $|w_i| = max(t_w, r_w, b_w, l_w)$ as shown in Fig. 1(c). The above steps are repeated $n - 4$ times excluding one point from any boundary of R_1. The node corresponding to the excluded point is also deleted from the tree to keep the tree updated and the location of the point in the array is marked deleted instead of actually deleting it from the array for efficiency. In each iteration the outer rectangle R_1 and the corresponding inner rectangle R_2 are constructed and width of the annulus thus formed by them is calculated. Maximum of all w_i for $i = 1, 2 \ldots n - 4$ would give us the maximum width annulus of the given set of points.

3.1   Observation and Theorem

Observation 1. *Atleast four distinct points are required to form the annulus.*

It is to be observed that minimum two points are required to form a rectangle. This happens when points are in diagonal position. So to form the outer and inner rectangle minimum of four points are required as shown in Fig. 4.

Lemma 1. *There will be maximum of $n - 4$ annuli.*

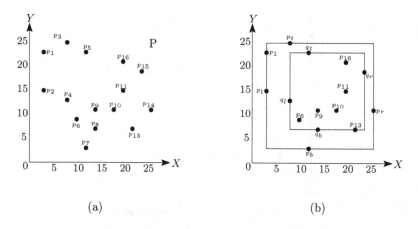

Fig. 2. (a) A sample point set P (b) Maximum Annulus for the point set P.

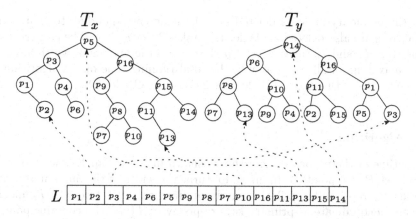

Fig. 3. AVL tree T_x and T_y for the point set in Fig. 2.

Proof. From Observation 1 minimum four points are required to get an annulus. Every time a point is deleted a new rectangle R_1 and its corresponding R_2 can be constructed giving a new annulus. So, there can be maximum of n_4 annuli.

Theorem 1. *Maximum Width Empty Annulus can be calculated in $\mathcal{O}(n \log n)$ time.*

Proof. From Lemma 1, it is proved that there can be a maximum of $n - 4$ annuli and each annulus is a result of a unique outer rectangle R_1 and corresponding inner rectangle R_2. In each iteration, algorithm constructs R_1 by finding the boundary points from the AVL trees T_x and T_y which takes $4 \cdot \log n$ time. The inner rectangle R_2 corresponding to R_1 are formed by searching the AVL trees mentioned above which also takes $4 \cdot \log n$ time. It is to be noted that one point

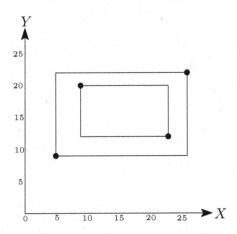

Fig. 4. Annulus formed by 4 points.

from the boundary of R_1 is deleted and the node corresponding to that point in the trees are also deleted. This deletion takes $2 \cdot \log n$ time. The construction of the trees T_x and T_y takes $2 \cdot n \log n$ time. Since the maximum width annulus is the maximum of all $n - 4$ annuli, the total running time of the algorithm is $2 \cdot n \log n + (n - 4)(4 \cdot \log n + 4 \cdot \log n + 2 \cdot \log n) \simeq \mathcal{O}(n \log n)$.

3.2   Algorithm

The outline of the proposed algorithm is shown in Algorithm 1. The algorithm takes a set P of n points as input and returns the width of the maximum width annulus. In step 1 and step 2, the algorithm draws two AVL trees viz. T_x and T_y on x and y coordinate as primary keys respectively. $L[index].p$ contains pointer pointing to the nodes of both the trees. In step 6, function $FindOuR()$ finds the outer rectangle R_1 (explained in Sect. 3). Function $FindOuR()$ takes the updated trees set as input in each iteration. In Step 7, $FindInR()$ finds the inner rectangle corresponding to the outer rectangle R_1 (explained in Sect. 3). The outer rectangle R_1 and remaining points are sent as parameter to the function $FindInR()$. Maximum width of the annulus is calculated in step 8, which is the maximum of the widths viz. top width (t_w), left width (l_w), bottom width (b_w) and right width (r_w). After this a point lying on the boundary of R_1 is deleted and T_x, T_y and L is updated in Step 11. Steps 5–12 are repeated until cardinality of P is less than or equal to $n - 4$. Finally, Step 12 returns the maximum of all the widths.

Algorithm 1: FIND $MAX\_Annulus$

Input: P: set of n points
Output: Maximum Width Annulus

1 $T_x \leftarrow P$; //AVL tree on x
2 $T_y \leftarrow P$; //AVL tree on y
3 $L[index].p \rightarrow T_x, T_y$
4 $max = 0$;
5 **while** $|P| \geq (n - 4)$ **do**
6 $\quad$ $R_1 \leftarrow FindOuR()$;
7 $\quad$ $R_2 \leftarrow FindInR()$;
8 $\quad$ $max_w = max\ (t_w, l_w, b_w, r_w)$;
9 $\quad$ **if** $max_w > max$ **then**
10 $\quad\quad$ $max = max_w$;
11 $\quad$ DeletePoint(T_x, T_y, L);
12 **return** max;

3.3 Proof of Correctness

To prove the correctness of the algorithm, we have to prove that the algorithm constructs all the possible annuli, and then returns the maximum annulus after comparing all the annuli. Lemma 1 proves that there can be at most $n - 4$ annuli. In each iteration, the algorithm constructs one of the annuli and reports its width. The maximum width of these $n - 4$ annuli is the resultant maximum width annulus. The iteration terminates when all of the $n - 4$ annuli are considered and hence the proof.

4 Conclusion

This paper presents a simple algorithm to find the maximum empty axis-parallel rectangular annulus, that runs in $\mathcal{O}(n \log n)$ time. Maximum empty annulus has it's application in the field of VLSI design, robotics, industrial design etc. This algorithm is a modification of the existing algorithm by Mahapatra [5] that runs in $\mathcal{O}(n^2)$ time. Our algorithm uses two AVL trees T_x and T_y and runs in $\mathcal{O}(n \log n)$ time. It is to be noted that the annulus constructed is not unique, i.e, there may be more than one maximum empty annulus, the proposed algorithm reports one of the maximum annulus.

References

1. Bae, S.W.: Computing a minimum-width square annulus in arbitrary orientation. In: Kaykobad, M., Petreschi, R. (eds.) International Workshop on Algorithms and Computation WALCOM, pp. 131–142. Springer, Cham (2016). https://doi.org/10. 1007/978-3-319-30139-6_11

2. Bae, S.W.: Computing a minimum-width square or rectangular annulus with out-liers. Comput. Geom. **76**, 33–45 (2019)
3. de Berg, M., Cheong, O., van Kreveld, M., Overmars, M.: Computational Geometry: Algorithms and Applications. Springer-Verlag, Heidelberg (1997). https://doi.org/10.1007/978-3-540-77974-2
4. Díaz-Báñez, J.M., Hurtado, F., Meijer, H., Rappaport, D., Sellares, T.: The largest empty annulus problem. In: Sloot, P.M.A., Hoekstra, A.G., Tan, C.J.K., Dongarra, J.J. (eds.) ICCS 2002. LNCS, vol. 2331, pp. 46–54. Springer, Heidelberg (2002). https://doi.org/10.1007/3-540-47789-6_5
5. Mahapatra, P.R.S.: Largest empty axis-parallel rectangular annulus. J. Emerg. Trends Comput. Inform. Sci. **3**(6) (2012)
6. Mukherjee, J., Mahapatra, P.R.S., Karmakar, A., Das, S.: Minimum-width rectangular annulus. Theoret. Comput. Sci. **508**, 74–80 (2013). http://www.sciencedirect.com/science/article/pii/S0304397512001934. Frontiers of Algorithmics
7. Preparata, F.P., Shamos, M.I.: Computational Geometry: An Introduction. Springer-Verlag, Heidelberg (1985). https://doi.org/10.1007/978-1-4612-1098-6

Parallel Contextual Array Insertion Deletion Grammar and (Context-Free : Context-Free) Matrix Grammar

S. Jayasankar$^{1(\boxtimes)}$, D. G. Thomas^{2}, S. James Immanuel^{2},
Meenakshi Paramasivan^{3}, T. Robinson^{4}, and Atulya K. Nagar^{5}

^{1} Department of Mathematics, Ramakrishna Mission Vivekananda College,
Chennai 600004, India
ksjayjay@gmail.com
^{2} Department of Science and Humanities (Mathematics Division),
Saveetha School of Engineering, Chennai 602 105, India
dgthomasmcc@yahoo.com, james_imch@yahoo.co.in
^{3} Department of Computer Science, University of Trier, Trier, Germany
meena_maths@yahoo.com
^{4} Department of Mathematics, Madras Christian College, Chennai 600 059, India
robinson@mcc.edu.in
^{5} Department of Mathematics and Computer Science, Liverpool Hope University,
Liverpool L16 9JD, UK
nagara@hope.ac.uk

Abstract. Siromoney et al. introduced Siromoney matrix grammars (1973) which are of sequential-parallel type in the sense that first a horizontal string of nonterminals is derived sequentially by applying the horizontal production rules and then vertical productions are applied in parallel to get the intended two-dimensional picture. In 1999, Radhakrishnan et al. introduced and studied a variant of Siromoney matrix grammars called (X:Y)MG where $X, Y \in \{Context - Free(CF), Regular(R)\}$. James et al. in 2018 introduced Parallel Contextual Array Insertion Deletion Grammar (PCAIDG) to generate two-dimensional array languages using insertion and deletion operations and parallel contextual mappings. In this paper, we prove that this family of languages generated by PCAIDGs properly includes the family (CF : CF) ML.

Keywords: Rectangular array · Parallel contextual array grammar · Insertion · Deletion

1 Introduction

Study of two-dimensional picture languages has picked pace from 1970s for solving problems arising in the framework of pattern recognition and image processing by extending the theory of one dimensional string languages to two dimensional picture languages. Siromoney et al. [16] pioneered study of picture languages by introducing several classes of grammars which are sequential

© Springer Nature Switzerland AG 2020
T. Lukić et al. (Eds.): IWCIA 2020, LNCS 12148, pp. 147–163, 2020.
https://doi.org/10.1007/978-3-030-51002-2_11

and parallel type in that order in the sense that first a horizontal string of non-terminals is derived and then vertical production rules are applied in parallel to get the desired pictures. A. Rosenfeld [14] introduced array grammars that generate or parse set of connected arrays with an emphasis on their relationship to array acceptors. Pattern recognition and image processing field was surveyed by King-Sun Fu and A. Rosenfeld [9] in 1976 and the areas to which these disciplines have been applied include business (e.g., character recognition), medicine (diagnosis, abnormality detection), automation (robot vision), military intelligence, communications (data compression, speech recognition), and many others.

Contextual grammars of Marcus [11] generate a language in an iterative manner by juxtaposing new strings to the current strings based on the pair of strings (called contexts) associated to a set of strings (called selectors), beginning with a finite set of strings (called axiom set). Several variants of contextual grammars were not only available but also studied from mathematical perspective. Many attempts have been made to extend contextual string grammars to the grammars of two-dimensional picture languages [5,19]. Of these, the parallel contextual array grammars capitalize on the use of both row as well as column contexts alternatively. Inspired by these ideas, Thomas et al. [6] have combined the contextual notion of Marcus and insertion deletion idea of Kari [8] and introduced a new grammar called Parallel Contextual Array Insertion Deletion Grammar (PCAIDG) and have proved that PCAIDG has more generative power by showing that the family of languages generated by PCAIDG properly includes the families of languages like LOC [3,4], REC [3,4], and $\mathfrak{L}(CSML)$ [15].

So far as the string languages are concerned there is a definite hierarchy in place in the form of Chomskian hierarchy whereas for two-dimensional picture languages there is a dearth of definite hierarchy. Pictures generated by PCAIDG have applications in floor designs and kolam pattern generation [17]. Motivated by these, an attempt has been made to know the position of the family of languages generated by PCAIDG in two-dimensional languages.

Kamala et al. [12] have introduced a variant of Siromoney matrix grammars $(X:Y)MG$ where X, Y $\in \{Context - free(CF), Regular(R)\}$ and went on to prove that emptiness problem for $(CF : CF)ML$ is decidable and membership testing problem for $(CF : CF)MG$ are decidable.

In this paper it is proved that the family of languages generated by PCAIDGs [6] properly includes the family (CF:CF)ML. This paper is organized as follows: In Sect. 2, necessary preliminaries are given. In Sect. 3, definition of PCAIDG is given along with examples of picture languages generated by PCAIDG which are also generated by the grammars $(X:Y)MG$ where X, Y $\in \{Context - free(CF), Regular(R)\}$. Finally, in Sect. 4, it is proved that $(CF : CF)ML \subsetneq \mathfrak{L}(PCAIDG)$.

Through out this paper the terms matrix, image, picture all mean rectangular array of terminals. This allows very naturally for description of color, texture etc.

2   Preliminaries

In this section we recall some notions related to formal language theory [1] and matrix grammars [12].

Definition 1. *Let V be an alphabet set - a finite non-empty set of symbols. A matrix (or an image) over V is an $m \times n$ rectangular array of symbols from V where $m, n \geq 0$. The set of all matrices over V including Λ is denoted by V^{**} and $V^{++} = V^{**} \setminus \{\Lambda\}$, where Λ is the empty image.*

Definition 2. *Let V^* denote the set of horizontal sequences of letters from V and $V^+ = V^* \setminus \{\lambda\}$, where λ is the identity element (of length zero). V_* denotes the set of all vertical sequences of letters over V, and $V_+ = V_* \setminus \{\lambda\}$. A string s can be a horizontal or vertical sequence of letters, let $|s|$ denote its length. It can also be interpreted as matrix: if $s \in V^+$ then its length coincides with $C(s)$ the number of columns of the matrix, and if $s \in V_+$ then $|s| = R(s)$ the number of its rows.*

For strings x and y, $x = a_1 \dots a_n$, $y = b_1 \dots b_m$, the concatenation (product) of x and y is defined by $x \cdot y = a_1 \dots a_n b_1 \dots b_m$. For matrices we define two types of concatenations, namely, column concatenation and row concatenation (column product and row product).

Definition 3. *The column concatenation of* $A = \begin{bmatrix} a_{11} & \cdots & a_{1p} \\ \vdots & \ddots & \vdots \\ a_{m1} & \cdots & a_{mp} \end{bmatrix}$ *and* $B =$

$\begin{bmatrix} b_{11} & \cdots & b_{1q} \\ \vdots & \ddots & \vdots \\ b_{n1} & \cdots & b_{nq} \end{bmatrix}$ *is defined only when $m = n$ and is given by*

$A \oplus B = \begin{bmatrix} a_{11} & \cdots & a_{1p} & b_{11} & \cdots & b_{1q} \\ \vdots & \ddots & \vdots & \vdots & \ddots & \vdots \\ a_{m1} & \cdots & a_{mp} & b_{n1} & \cdots & b_{nq} \end{bmatrix}$. *Similarly, the row concatenation of A and*

B, *defined only when $p = q$, is given by* $A \ominus B = \begin{bmatrix} a_{11} & \cdots & a_{1p} \\ \vdots & \ddots & \vdots \\ a_{m1} & \cdots & a_{mp} \\ b_{11} & \cdots & b_{1q} \\ \vdots & \ddots & \vdots \\ b_{n1} & \cdots & b_{nq} \end{bmatrix}$. *The empty*

matrix Λ acts as the identity for column and row concatenation of matrices of arbitrary dimensions.

Definition 4. *For any matrix x (also called image) defined over V, we define its horizontal and vertical products iteratively for any natural $n \geq 1$ as follows:*

$$(x)^1 = x, (x)_1 = x, \text{ and } (x)^{n+1} = (x)^n \oplus x, (x)_{n+1} = (x)_n \ominus x$$

Definition 5. *For any string $s = a_1a_2 \ldots a_n \in V^+$, χ is defined by a mapping*

$$\chi : V^+ \rightarrow V_+ \text{ such that } \chi(s) = \begin{matrix} a_1 \\ a_2 \\ \vdots \\ a_n \end{matrix} \text{ i.e., } \chi \text{ transforms a horizontal string to a}$$

vertical string.
Formally if $s = a_1a_2 \ldots a_n \in V^+$, $\chi(s) = a_1 \ominus a_2 \ominus \ldots a_n$.

Definition 6. *Let $c_1, c_2, \ldots c_n \in V^+$ be strings of same length. Then $I = c_1 \odot c_2 \odot \cdots \odot c_n$ is the matrix (or an image) represented by the image $\chi(c_1) \oplus \chi(c_2) \oplus \ldots \oplus \chi(c_n)$.*

We now recall the notion of $(CF : CF)MG$ [12].

Definition 7 (Matrix Grammars). *Let $M = \langle G, G' \rangle$ where $G = \langle N, T, P, S \rangle$ is a grammar in Chomskian hierarchy, $T = \{A_1, A_2, \ldots, A_k\}$, $G' = \{G_1, G_2, \ldots, G_{k-1}, G_k\}$ where each G_i is a Chomskian grammar corresponding to the symbol A_i, defined on Σ, the alphabet set of M. A grammar M is said to be $(CF : CF)MG$ if both G and every $G_i \in G'$ are context free grammars. Other combinations are defined in a similar way.*

Let $I = c_1 \odot c_2 \odot \ldots, \odot c_n$ be an image defined over Σ. $I \in L(M)$ if and only if there exists $A_{x_1} A_{x_2} \ldots A_{x_n} \in L(G)$ such that $c_j \in L(G_{x_j}), 1 \leq j \leq n$. The string $A_{x_1} A_{x_2} \ldots A_{x_n}$ is said to be an intermediate string deriving I with respect to M. It is to be noted that there can be more than one intermediate string deriving I. The family of languages generated by $(X : Y)MG$ is denoted as $(X : Y)ML$ where $X, Y \in \{CF, R\}$, where CF stands for context free and R stands for regular.

Remark 1. The nonterminals in each $G_i, i = 1.2, \ldots k$ are assumed to be disjoint. Hence the choice of a vertical production rule for a particular column at any stage is fixed by the initial symbol of that column determined by the horizontal string. In other words, if the number of columns is n, the vertical production is chosen from the k CFG grammars determined by the horizontal string.

Remark 2. The rules in the k CFG grammars are assumed to be of the form $A \rightarrow BCD$ or $A \rightarrow X, B, D, X \in T_i, i = 1, 2, \ldots k$ so that at each stage, a single terminal is generated in each column.

Remark 3. For every column, a rule from one of $G_i's$ in $G', i = 1, 2, \ldots k$ (which are assumed to be λ free) must be applied at each stage. In other words, no cell in any column is blank or empty. Thus, combining with Remark 2, it means that one and only one element is generated in each column at every stage and all the columns terminate simultaneously.

Example 1. Consider the array language $L=$

$$
\left\{
\begin{array}{llll}
& X\;X\;X\;X\;X & & X\;X\;X\;X\;X\;X\;X \\
X\;X\;X\;X\;X & X\;\bullet\;X\;\bullet\;X & X\;X\;X\;X\;X\;X\;X & X\;\bullet\;\bullet\;X\;\bullet\;\bullet\;X \\
X\;\bullet\;X\;\bullet\;X & X\;\bullet\;X\;\bullet\;X & X\;\bullet\;\bullet\;X\;\bullet\;\bullet\;X & X\;\bullet\;\bullet\;X\;\bullet\;\bullet\;X \\
X\;X\;X\;X\;X, & X\;X\;X\;X\;X, & X\;X\;X\;X\;X\;X\;X, & X\;X\;X\;X\;X\;X\;X, \\
X\;\bullet\;X\;\bullet\;X & X\;\bullet\;X\;\bullet\;X & X\;\bullet\;\bullet\;X\;\bullet\;\bullet\;X & X\;\bullet\;\bullet\;X\;\bullet\;\bullet\;X \\
X\;X\;X\;X\;X & X\;\bullet\;X\;\bullet\;X & X\;X\;X\;X\;X\;X\;X & X\;\bullet\;\bullet\;X\;\bullet\;\bullet\;X \\
& X\;X\;X\;X\;X & & X\;X\;X\;X\;X\;X\;X
\end{array}
\right.
$$

$$
\begin{array}{lll}
X\;X\;X\;X\;X\;X & & X\;X\;X\;X\;X\;X\;X\;X\;X \\
X\;\bullet\;\bullet\;X\;\bullet\;\bullet\;X & X\;X\;X\;X\;X\;X\;X\;X\;X & X\;\bullet\;\bullet\;\bullet\;X\;\bullet\;\bullet\;\bullet\;X \\
X\;\bullet\;\bullet\;X\;\bullet\;\bullet\;X & X\;\bullet\;\bullet\;\bullet\;X\;\bullet\;\bullet\;\bullet\;X & X\;\bullet\;\bullet\;\bullet\;X\;\bullet\;\bullet\;\bullet\;X \\
X\;\bullet\;\bullet\;X\;\bullet\;\bullet\;X & X\;\bullet\;\bullet\;\bullet\;X\;\bullet\;\bullet\;\bullet\;X & X\;\bullet\;\bullet\;\bullet\;X\;\bullet\;\bullet\;\bullet\;X \\
X\;X\;X\;X\;X\;X, & X\;X\;X\;X\;X\;X\;X\;X\;X, & X\;X\;X\;X\;X\;X\;X\;X\;X,\ldots \\
X\;\bullet\;\bullet\;X\;\bullet\;\bullet\;X & X\;\bullet\;\bullet\;\bullet\;X\;\bullet\;\bullet\;\bullet\;X & X\;\bullet\;\bullet\;\bullet\;X\;\bullet\;\bullet\;\bullet\;X \\
X\;\bullet\;\bullet\;X\;\bullet\;\bullet\;X & X\;\bullet\;\bullet\;\bullet\;X\;\bullet\;\bullet\;\bullet\;X & X\;\bullet\;\bullet\;\bullet\;X\;\bullet\;\bullet\;\bullet\;X \\
X\;\bullet\;\bullet\;X\;\bullet\;\bullet\;X & X\;X\;X\;X\;X\;X\;X\;X\;X & X\;\bullet\;\bullet\;\bullet\;X\;\bullet\;\bullet\;\bullet\;X \\
X\;X\;X\;X\;X\;X & & X\;X\;X\;X\;X\;X\;X\;X\;X
\end{array}
\Bigg\}.
$$

L is a $(CF : CF)ML$. The corresponding $(CF : CF)MG$ is $M = (G, G')$ where $G = \langle N, T, P, S \rangle$ with

- $N = \{S, A, B\}$
- $T = \{S_1, S_2\}$
- $P = \{S \to S_1 A S_1,\ A \to S_2 B S_2,\ B \to S_1,\ B \to S_2 B S_2\}$

and

$G' = \{G_1, G_2\}$ with $G_1 = \langle N_1, T_1, P_1, S_1 \rangle$ and $G_2 = \langle N_2, T_2, P_2, S_2 \rangle$. Here

- $N_1 = \{S_1\}$
- $T_1 = \{X\}$
- $P_1 = \{S_1 \to X S_1 X, S_1 \to X\}$

and

- $N_2 = \{S_2, C, D\}$
- $T_2 = \{\bullet, X\}$
- $P_1 = \{S_2 \to XCX, C \to \bullet D\bullet,\ D \to \bullet D\bullet,\ D \to X\}$

A sample derivation of a picture given in Example 1 using the above $(CF : CF)MG$ is given as follows:

$$S \Rightarrow S_1 A S_1 \Rightarrow S_1 S_2 B S_2 S_1 \Rightarrow S_1 S_2 S_2 B S_2 S_2 S_1 \Rightarrow S_1 S_2 S_2 S_1 S_2 S_2 S_1 \Rightarrow$$

$$
\begin{array}{ccc}
& & X\ X\ X\ X\ X\ X\ X \\
& X\ X\ X\ X\ X\ X\ X & X\ \bullet\ \bullet\ X\ \bullet\ \bullet\ X \\
X\ X\ X\ X\ X\ X\ X & X\ \bullet\ \bullet\ X\ \bullet\ \bullet\ X & X\ \bullet\ \bullet\ X\ \bullet\ \bullet\ X \\
S_1\ C\ C\ S_1\ C\ C\ S_1\ \Rightarrow & S_1\ D\ D\ S_1\ D\ D\ S_1\ \Rightarrow & S_1\ D\ D\ S_1\ D\ D\ S_1\ \Rightarrow \\
X\ X\ X\ X\ X\ X\ X & X\ \bullet\ \bullet\ X\ \bullet\ \bullet\ X & X\ \bullet\ \bullet\ X\ \bullet\ \bullet\ X \\
& X\ X\ X\ X\ X\ X\ X & X\ \bullet\ \bullet\ X\ \bullet\ \bullet\ X \\
& & X\ X\ X\ X\ X\ X\ X
\end{array}
$$

$$
\begin{array}{l}
X\ X\ X\ X\ X\ X\ X \\
X\ \bullet\ \bullet\ X\ \bullet\ \bullet\ X \\
X\ \bullet\ \bullet\ X\ \bullet\ \bullet\ X \\
X\ X\ X\ X\ X\ X\ X. \\
X\ \bullet\ \bullet\ X\ \bullet\ \bullet\ X \\
X\ \bullet\ \bullet\ X\ \bullet\ \bullet\ X \\
X\ X\ X\ X\ X\ X\ X
\end{array}
$$

3 Parallel Contextual Array Insertion Deletion Grammar

In this section we recall the definition of parallel contextual array insertion deletion grammar [6] and give an example.

Definition 8. *Let V be a finite alphabet. A column array context over V is of the form $c = \begin{bmatrix} u_1 \\ u_2 \end{bmatrix} \in V^{**}$, u_1, u_2 are of size $1 \times p$, $p \geq 1$.*

A row array context over V is of the form, $r = \begin{bmatrix} u_1 & u_2 \end{bmatrix} \in V^{**}$, u_1, u_2 are of size $p \times 1$, $p \geq 1$.

Definition 9. *The parallel column contextual insertion (deletion) operation is defined as follows: Let V be an alphabet, C be a finite subset of V^{**} whose elements are the column array contexts and $\varphi_c^i(\varphi_c^d) : V^{**} \times V^{**} \to 2^C$ be a choice mapping.*

We define $\varphi_c^i(\varphi_c^d) : V^{**} \times V^{**} \to 2^{V^{**}}$ such that, for arrays $A = \begin{bmatrix} a_{1j} & \cdots & a_{1(k-1)} \\ \vdots & \ddots & \vdots \\ a_{mj} & \cdots & a_{m(k-1)} \end{bmatrix}$, $B = \begin{bmatrix} a_{1k} & \cdots & a_{1(l-1)} \\ \vdots & \ddots & \vdots \\ a_{mk} & \cdots & a_{m(l-1)} \end{bmatrix}$, $j < k < l, a_{ij} \in V$, $\left(B = \begin{bmatrix} a_{1(k-p)} & \cdots & a_{1(l-1)} \\ \vdots & \ddots & \vdots \\ a_{m(k-p)} & \cdots & a_{m(l-1)} \end{bmatrix} \right)$, $I_c \in \varphi_c^i(A, B)(\varphi_c^d(A, B)), I_c(D_c) = \begin{bmatrix} u_1 \\ u_2 \\ \vdots \\ u_m \end{bmatrix}$, if $c_i = \begin{bmatrix} u_i \\ u_{i+1} \end{bmatrix} \in \varphi_c^i \left(\begin{matrix} a_{ij} & \cdots & a_{i(k-1)} & a_{ik} & \cdots & a_{i(l-1)} \\ a_{(i+1)j} & \cdots & a_{(i+1)(k-1)} & a_{(i+1)k} & \cdots & a_{(i+1)(l-1)} \end{matrix} \right)$ $\left(\varphi_c^d \left(\begin{matrix} a_{ij} & \cdots & a_{i(k-1)} & a_{i(k+p)} & \cdots & a_{i(l-1)} \\ a_{(i+1)j} & \cdots & a_{(i+1)(k-1)} & a_{(i+1)(k+p)} & \cdots & a_{(i+1)(l-1)} \end{matrix} \right) \right)$, $c_i \in C, 1 \leq i \leq m-1$, not all need to be distinct.

Given an array $X = [a_{ij}]_{m \times n}, a_{ij} \in V$, $X = X_1 \textcircled{$\parallel$} A \textcircled{$\parallel$} B \textcircled{$\parallel$} X_2$ ($X = X_1 \textcircled{$\parallel$} A \textcircled{$\parallel$} D_c \textcircled{$\parallel$} B \textcircled{$\parallel$} X_2$),

$$X_1 = \begin{bmatrix} a_{11} & \cdots & a_{1(j-1)} \\ \vdots & \ddots & \vdots \\ a_{m1} & \cdots & a_{m(j-1)} \end{bmatrix}, A = \begin{bmatrix} a_{1j} & \cdots & a_{1(k-1)} \\ \vdots & \ddots & \vdots \\ a_{mj} & \cdots & a_{m(k-1)} \end{bmatrix}, B = \begin{bmatrix} a_{1k} & \cdots & a_{1(l-1)} \\ \vdots & \ddots & \vdots \\ a_{mk} & \cdots & a_{m(l-1)} \end{bmatrix},$$

$$X_2 = \begin{bmatrix} a_{1l} & \cdots & a_{1n} \\ \vdots & \ddots & \vdots \\ a_{ml} & \cdots & a_{mn} \end{bmatrix}, 1 \le j \le k < l \le n+1 \ (or) \ 1 \le j < k \le l \le n+1, \text{ we write}$$

$X \Rightarrow^{col_i(col_d)} Y$ if $Y = X_1 \oplus A \oplus I_c \oplus B \oplus X_2$ $(Y = X_1 \oplus A \oplus B \oplus X_2)$, such that $I_c \in \varphi_c^i(A, B)$ $(D_c \in \varphi_c^d(A, B))$. $I_c(D_c)$ is called the inserted (deleted) column context. We say that Y is obtained from X by parallel column contextual insertion (deletion) operation. The following 4 special cases for $X = X_1 \oplus A \oplus B \oplus X_2$ are also considered,

1. For $j = 1$ we have $X_1 = \Lambda$.
2. For $j = k$, we have $A = \Lambda$. If $j = k \ne 1$, then $X_1 = \Lambda$ and $A = \Lambda$.
3. For $k = l$ (For $k + p = l$), we have $B = \Lambda$.
4. For $l = n + 1$, we have $X_2 = \Lambda$. If $k = l = n + 1$ (If $(k + p) = l = n + 1$), then $B = \Lambda$ and $X_2 = \Lambda$.

The case $j = k = l$ is not considered for parallel column contextual insertion (deletion) operation.

Similarly, we can define parallel row contextual insertion (deletion) operation by inserting (deleting) row context $\mathbf{I_r}(\mathbf{D_r})$ in between two sub-arrays A and B with the help of row operation $\ominus$ and set of row array contexts R. We have $X \Rightarrow^{row_i(row_d)} Y$ if $X = X_1 \ominus A \ominus B \ominus X_2 (X_1 \ominus A \ominus D_r \ominus B \ominus X_2)$ and $Y = X_1 \ominus A \ominus I_r \ominus B \ominus X_2$ $(X_1 \ominus A \ominus B \ominus X_2)$.

Definition 10. A parallel contextual array insertion deletion grammar is defined by $G = (V, T, M, C, R, \varphi_c^i, \varphi_r^i, \varphi_c^d, \varphi_r^d)$, where V is an alphabet, $T \subseteq V$ is a terminal alphabet, M is a finite subset of V^{**} called the base of G, C is a finite subset of V^{**} called column array contexts, R is a finite subset of V^{**} called row array contexts, $\varphi_c^i : V^{**} \times V^{**} \to 2^C$, $\varphi_r^i : V^{**} \times V^{**} \to 2^R$, $\varphi_c^d : V^{**} \times V^{**} \to 2^C$, $\varphi_r^d : V^{**} \times V^{**} \to 2^R$, are the choice mappings which perform the parallel column contextual insertion, row contextual insertion, column contextual deletion and row contextual deletion operations, respectively.

The insertion derivation with respect to G is a binary relation $\Rightarrow_i$ on V^{**} and is defined by $X \Rightarrow_i Y$, where $X, Y \in V^{**}$ if and only if $X = X_1 \oplus A \oplus B \oplus X_2$, $Y = X_1 \oplus A \oplus I_c \oplus B \oplus X_2$ or $X = X_3 \ominus A \ominus B \ominus X_4$, $Y = X_3 \ominus A \ominus I_r \ominus B \ominus X_4$ for some $X_1, X_2, X_3, X_4 \in V^{**}$ and I_c, I_r are inserted column and row contexts obtained by the parallel column or row contextual insertion operations according to the choice mappings.

The deletion derivation with respect to G is a binary relation $\Rightarrow_d$ on V^{**} and is defined by $X \Rightarrow_d Y$, where $X, Y \in V^{**}$ if and only if $X = X_1 \oplus A \oplus D_c \oplus B \oplus X_2$, $Y = X_1 \oplus A \oplus B \oplus X_2$ or $X = X_3 \ominus A \ominus D_r \ominus B \ominus X_4$, $Y = X_3 \ominus A \ominus B \ominus X_4$ for some $X_1, X_2, X_3, X_4 \in V^{**}$ and D_c, D_r are deleted column and row contexts with respect to the parallel column or row contextual deletion operations according to the choice mappings.

The direct derivation with respect to G is a binary relation $\Rightarrow_{i,d}$ on V^{**} which is either $\Rightarrow_i$ or $\Rightarrow_d$.

Definition 11. *Let* $G = (V, T, M, C, R, \varphi_c^i, \varphi_r^d, \varphi_c^d, \varphi_r^d)$ *be a parallel contextual array insertion deletion grammar. The language generated by* G, *denoted by* $L(G)$ *is defined by,*

$$L(G) = \{Y \in T^{**} | \exists X \in M \text{ with } X \Rightarrow_{i,d}^* Y\}.$$

The family of all array languages generated by parallel contextual array insertion deletion grammars (PCAIDGs) is denoted by $\mathfrak{L}(PCAIDG)$.

Example 2. We now give a PCAIDG $G_1 = (V, T, M, C, R, \varphi_c^i, \varphi_r^d, \varphi_c^d, \varphi_r^d)$ to generate $L \in$ (CF:CF)ML given in Example 1. Here,

$$V = \{X, Y, Z, \bullet\}, T = \{X, \bullet\}, M = \left\{ \begin{matrix} X\ X\ X\ X\ X \\ X\ \bullet\ X\ \bullet\ X \\ X\ X\ X\ X\ X \\ X\ \bullet\ X\ \bullet\ X \\ X\ X\ X\ X\ X \end{matrix} \right\},$$

$$C = \left\{ \begin{matrix} X\ Y\ X\ Z \\ \bullet\ Y\ X\ Z \end{matrix}, \begin{matrix} \bullet\ Y\ X\ Z \\ X\ Y\ X\ Z \end{matrix}, \begin{matrix} \bullet\ Y\ X\ Z \\ \bullet\ Y\ X\ Z \end{matrix}, \begin{matrix} X\ X\ X \\ X \end{matrix}, \begin{matrix} X\ \bullet \\ X\ \bullet \end{matrix}, \begin{matrix} X\ \bullet \\ X\ X \end{matrix}, \begin{matrix} X\ \bullet \\ X\ \bullet \end{matrix}, \begin{matrix} Y\ X\ Z \\ Y\ X\ Z \end{matrix} \right\},$$

$$R = \left\{ \begin{matrix} X\ \bullet \\ Y\ Y \\ X\ X \\ Z\ Z \end{matrix}, \begin{matrix} \bullet\ \bullet \\ Y\ Y \\ X\ X \\ Z\ Z \end{matrix}, \begin{matrix} \bullet\ X \\ Y\ Y \\ X\ X \\ Z\ Z \end{matrix}, \begin{matrix} X\ X \end{matrix}, \begin{matrix} X\ X \\ X\ \bullet \end{matrix}, \begin{matrix} X\ X \\ \bullet\ \bullet \end{matrix}, \begin{matrix} X\ X \\ \bullet\ X \end{matrix}, \begin{matrix} Y\ Y \\ X\ X \\ Z\ Z \end{matrix} \right\}.$$

The insertion rules are given by

(IR1) $\varphi_c^i \begin{bmatrix} X\ X \\ \bullet\ \end{bmatrix}, X \end{bmatrix} = \left\{ \begin{matrix} X\ Y\ X\ Z \\ \bullet\ Y\ X\ Z \end{matrix} \right\}$, $\varphi_c^i \begin{bmatrix} \bullet\ X \\ X \end{bmatrix}, X \end{bmatrix} = \left\{ \begin{matrix} \bullet\ Y\ X\ Z \\ X\ Y\ X\ Z \end{matrix} \right\}$,

$\varphi_c^i \begin{bmatrix} \bullet\ X \\ \bullet \end{bmatrix}, X \end{bmatrix} = \left\{ \begin{matrix} \bullet\ Y\ X\ Z \\ \bullet\ Y\ X\ Z \end{matrix} \right\}$.

(IR2) $\varphi_c^i \begin{bmatrix} Z\ X \\ Z \end{bmatrix}, \bullet \end{bmatrix} = \left\{ \begin{matrix} X\ X \\ X\ \bullet \end{matrix} \right\}$, $\varphi_c^i \begin{bmatrix} Z\ \bullet \\ Z \end{bmatrix}, X \end{bmatrix} = \left\{ \begin{matrix} X\ \bullet \\ X\ X \end{matrix} \right\}$, $\varphi_c^i \begin{bmatrix} Z\ \bullet \\ Z \end{bmatrix}, \bullet \end{bmatrix} = \left\{ \begin{matrix} X\ \bullet \\ X\ \bullet \end{matrix} \right\}$.

(IR3) $\varphi_r^i \begin{bmatrix} X\ \bullet, \begin{matrix} X\ X \\ X\ \bullet \end{matrix} \end{bmatrix} = \left\{ \begin{matrix} X\ \bullet \\ Y\ Y \\ X\ X \\ Z\ Z \end{matrix} \right\}$, $\varphi_r^i \begin{bmatrix} \bullet\ \bullet, \begin{matrix} X\ X \\ \bullet\ \bullet \end{matrix} \end{bmatrix} = \left\{ \begin{matrix} \bullet\ \bullet \\ Y\ Y \\ X\ X \\ Z\ Z \end{matrix} \right\}$,

$\varphi_r^i \begin{bmatrix} \bullet\ X, \begin{matrix} X\ X \\ \bullet\ X \end{matrix} \end{bmatrix} = \left\{ \begin{matrix} \bullet\ X \\ Y\ Y \\ X\ X \\ Z\ Z \end{matrix} \right\}$.

(IR4) $\varphi_r^i \begin{bmatrix} Z\ Z, X\ \bullet \end{bmatrix} = \left\{ \begin{matrix} X\ X \\ X\ \bullet \end{matrix} \right\}$, $\varphi_r^i \begin{bmatrix} Z\ Z, \bullet\ \bullet \end{bmatrix} = \left\{ \begin{matrix} X\ X \\ \bullet\ \bullet \end{matrix} \right\}$,

$\varphi_r^i \begin{bmatrix} Z\ Z, \bullet\ X \end{bmatrix} = \left\{ \begin{matrix} X\ X \\ \bullet\ X \end{matrix} \right\}$.

The deletion rules are given by

(DR1) $\varphi_c^d \begin{bmatrix} Z\ X \\ Z \end{bmatrix}, \bullet \end{bmatrix} = \left\{ \begin{matrix} X \\ X \end{matrix} \right\}$, $\varphi_c^d \begin{bmatrix} Z\ \bullet \\ Z \end{bmatrix}, X \end{bmatrix} = \left\{ \begin{matrix} X \\ X \end{matrix} \right\}$, $\varphi_c^d \begin{bmatrix} Z\ \bullet \\ Z \end{bmatrix}, \bullet \end{bmatrix} = \left\{ \begin{matrix} X \\ X \end{matrix} \right\}$.

(DR2) $\varphi_c^d \left[{X \atop \bullet}, {X \atop X} \right] = \left\{ {Y\,X\,Z \atop Y\,X\,Z} \right\}$, $\varphi_c^d \left[{\bullet \atop \bullet}, {X \atop X} \right] = \left\{ {Y\,X\,Z \atop Y\,X\,Z} \right\}$,

$\qquad \varphi_c^d \left[{\bullet \atop X}, {X \atop X} \right] = \left\{ {Y\,X\,Z \atop Y\,X\,Z} \right\}$.

(DR3) $\varphi_r^d \left[Z\,Z, X\,\bullet \right] = \left\{ X\,X \right\}$, $\varphi_r^d \left[Z\,Z, \bullet\,\bullet \right] = \left\{ X\,X \right\}$,

$\qquad \varphi_r^d \left[Z\,Z, \bullet\,X \right] = \left\{ X\,X \right\}$.

(DR4) $\varphi_r^d \left[X\,\bullet, X\,X \right] = \left\{ {Y\,Y \atop X\,X \atop Z\,Z} \right\}$, $\varphi_r^d \left[\bullet\,\bullet, X\,X \right] = \left\{ {Y\,Y \atop X\,X \atop Z\,Z} \right\}$,

$\qquad \varphi_r^d \left[\bullet\,X, X\,X \right] = \left\{ {Y\,Y \atop X\,X \atop Z\,Z} \right\}$.

A sample derivation to obtain a 7×7 picture of L is as follows:

$$
\begin{array}{l}
 \quad X\,X\,X\,X\,X\,X\,X \quad X\,X\,X\,X\,X\,X\,X \\
X\,X\,X\,X\,X\,X\,X \quad X \bullet \bullet X \bullet \bullet X \quad X \bullet \bullet X \bullet \bullet X \\
X \bullet \bullet X \bullet \bullet X \quad X \bullet \bullet X \bullet \bullet X \quad X \bullet \bullet X \bullet \bullet X \\
X \bullet \bullet X \bullet \bullet X \quad Y\,Y\,Y\,Y\,Y\,Y\,Y \quad Y\,Y\,Y\,Y\,Y\,Y\,Y \\
Y\,Y\,Y\,Y\,Y\,Y\,Y \Rightarrow_{IR4}^{\phi_i^r} X\,X\,X\,X\,X\,X\,X \quad = \quad X\,X\,X\,X\,X\,X\,X \Rightarrow_{DR4}^{\phi_d^r} \\
X\,X\,X\,X\,X\,X\,X \quad Z\,Z\,Z\,Z\,Z\,Z\,Z \quad Z\,Z\,Z\,Z\,Z\,Z\,Z \\
Z\,Z\,Z\,Z\,Z\,Z\,Z \quad X\,X\,X\,X\,X\,X\,X \quad X\,X\,X\,X\,X\,X\,X \\
X \bullet \bullet X \bullet \bullet X \quad X \bullet \bullet X \bullet \bullet X \quad X \bullet \bullet X \bullet \bullet X \\
X\,X\,X\,X\,X\,X\,X \quad X \bullet \bullet X \bullet \bullet X \quad X \bullet \bullet X \bullet \bullet X \\
 \quad X\,X\,X\,X\,X\,X\,X \quad X\,X\,X\,X\,X\,X\,X \\
X\,X\,X\,X\,X\,X\,X \\
X \bullet \bullet X \bullet \bullet X \\
X \bullet \bullet X \bullet \bullet X \\
X\,X\,X\,X\,X\,X\,X. \\
X \bullet \bullet X \bullet \bullet X \\
X \bullet \bullet X \bullet \bullet X \\
X\,X\,X\,X\,X\,X\,X
\end{array}
$$

Example 3. We now give another PCAIDG $G_2 = (V, T, M, C, R, \varphi_c^i, \varphi_r^d, \varphi_c^d, \varphi_r^d)$ to generate the following (CF:CF)ML:

$$
L_1 = \left\{
\begin{array}{cccc}
 & \bullet\,X\,\bullet & \bullet\,\bullet\,X\,\bullet\,\bullet \\
\bullet\,X\,\bullet & \bullet\,\bullet\,X\,\bullet\,\bullet & \bullet\,X\,\bullet & \bullet\,\bullet\,X\,\bullet\,\bullet \\
X\,X\,X, & X\,X\,X\,X\,X, & X\,X\,X, & X\,X\,X\,X\,X, \dots \\
\bullet\,X\,\bullet & \bullet\,\bullet\,X\,\bullet\,\bullet & \bullet\,X\,\bullet & \bullet\,\bullet\,X\,\bullet\,\bullet \\
 & & \bullet\,X\,\bullet & \bullet\,\bullet\,X\,\bullet\,\bullet
\end{array}
\right\} =
$$

$$
\left\{
\begin{array}{c}
(\ (\bullet)^n\ X\ (\bullet)^n\ )_m \\
(X)^n\ X\ (X)^n \\
(\ (\bullet)^n\ X\ (\bullet)^n\ )_m
\end{array}
\ \middle/\ n, m \geq 1 \right\}.
$$

Here $V = \{X, Y, \bullet\}$, $T = \{X, \bullet\}$, $M = \left\{\begin{array}{c} \bullet\,X\,\bullet \\ X\,X\,X \\ \bullet\,X\,\bullet \end{array}\right\}$,

$$
C = \left\{ \frac{\bullet\,Y}{\bullet\,Y}, \frac{\bullet\,Y}{X\,Y}, \frac{X\,Y}{\bullet\,Y}, \frac{Y\,\bullet}{Y\,\bullet}, \frac{Y\,\bullet}{Y\,X}, \frac{Y\,X}{Y\,\bullet}, \frac{X}{X}, \frac{Y\,X\,Y}{Y\,X\,Y} \right\},
$$

$$
R = \left\{ \frac{\bullet\,\bullet}{Y\,Y}, \frac{\bullet\,X}{Y\,Y}, \frac{X\,\bullet}{Y\,Y}, \frac{Y\,Y}{\bullet\,\bullet}, \frac{Y\,Y}{\bullet\,X}, \frac{Y\,Y}{X\,\bullet}, X\,X, \frac{Y\,Y}{X\,X} \right\}.
$$

The insertion rules are given by

(IR1) $\varphi_c^i \begin{bmatrix} \bullet & X \\ X, & X \end{bmatrix} = \left\{ \begin{array}{c} \bullet\,Y \\ X\,Y \end{array} \right\}$, $\varphi_c^i \begin{bmatrix} X & X \\ \bullet, & X \end{bmatrix} = \left\{ \begin{array}{c} X\,Y \\ \bullet\,Y \end{array} \right\}$, $\varphi_c^i \begin{bmatrix} \bullet & X \\ \bullet, & X \end{bmatrix} = \left\{ \begin{array}{c} \bullet\,Y \\ \bullet\,Y \end{array} \right\}$.

(IR2) $\varphi_c^i \begin{bmatrix} X & \bullet \\ X, & X \end{bmatrix} = \left\{ \begin{array}{c} Y\,\bullet \\ Y\,X \end{array} \right\}$, $\varphi_c^i \begin{bmatrix} X & X \\ X, & \bullet \end{bmatrix} = \left\{ \begin{array}{c} Y\,X \\ Y\,\bullet \end{array} \right\}$, $\varphi_c^i \begin{bmatrix} X & \bullet \\ X, & \bullet \end{bmatrix} = \left\{ \begin{array}{c} Y\,\bullet \\ Y\,\bullet \end{array} \right\}$.

(IR3) $\varphi_c^i \begin{bmatrix} \bullet & Y & X & Y \\ X' & Y & X & Y \end{bmatrix} = \left\{ \begin{matrix} X \\ X \end{matrix} \right\}$, $\varphi_c^i \begin{bmatrix} X & Y & X & Y \\ \bullet ' & Y & X & Y \end{bmatrix} = \left\{ \begin{matrix} X \\ X \end{matrix} \right\}$, $\varphi_c^i \begin{bmatrix} \bullet & Y & X & Y \\ \bullet ' & Y & X & Y \end{bmatrix} = \left\{ \begin{matrix} X \\ X \end{matrix} \right\}$.

(IR4) $\varphi_r^i \left[\bullet \; X, X \; X \right] = \left\{ \begin{matrix} \bullet & X \\ Y & Y \end{matrix} \right\}$, $\varphi_r^i \left[X \; \bullet, X \; X \right] = \left\{ \begin{matrix} X & \bullet \\ Y & Y \end{matrix} \right\}$, $\varphi_r^i \left[\bullet \; \bullet, X \; X \right] = \left\{ \begin{matrix} \bullet & \bullet \\ Y & Y \end{matrix} \right\}$.

(IR5) $\varphi_r^i \left[X \; X, \bullet \; X \right] = \left\{ \begin{matrix} Y & Y \\ \bullet & X \end{matrix} \right\}$, $\varphi_r^i \left[X \; X, X \; \bullet \right] = \left\{ \begin{matrix} Y & Y \\ X & \bullet \end{matrix} \right\}$, $\varphi_r^i \left[X \; X, \bullet \; \bullet \right] = \left\{ \begin{matrix} Y & Y \\ \bullet & \bullet \end{matrix} \right\}$.

(IR6) $\varphi_r^i \left[\begin{matrix} Y & Y \\ \bullet \; X, X \; X \\ Y & Y \end{matrix} \right] = \left\{ X \; X \right\}$, $\varphi_r^i \left[\begin{matrix} Y & Y \\ X \; \bullet, X \; X \\ Y & Y \end{matrix} \right] = \left\{ X \; X \right\}$,

$\varphi_r^i \left[\begin{matrix} Y & Y \\ \bullet \; \bullet, X \; X \\ Y & Y \end{matrix} \right] = \left\{ X \; X \right\}$.

(DR1) $\varphi_c^d \begin{bmatrix} X & \bullet \\ X' & X \end{bmatrix} = \left\{ \begin{matrix} Y & X & Y \\ Y & X & Y \end{matrix} \right\}$, $\varphi_c^d \begin{bmatrix} X & X \\ X' & \bullet \end{bmatrix} = \left\{ \begin{matrix} Y & X & Y \\ Y & X & Y \end{matrix} \right\}$, $\varphi_c^d \begin{bmatrix} X & \bullet \\ X' & \bullet \end{bmatrix} = \left\{ \begin{matrix} Y & X & Y \\ Y & X & Y \end{matrix} \right\}$.

(DR2) $\varphi_r^d \left[X \; X, \bullet \; X \right] = \left\{ \begin{matrix} Y & Y \\ X & X \\ Y & Y \end{matrix} \right\}$, $\varphi_r^d \left[X \; X, X \; \bullet \right] = \left\{ \begin{matrix} Y & Y \\ X & X \\ Y & Y \end{matrix} \right\}$. $\varphi_r^d \left[X \; X, \bullet \; \bullet \right] = \left\{ \begin{matrix} Y & Y \\ X & X \\ Y & Y \end{matrix} \right\}$.

A sample derivation to obtain a 5×5 picture of L_1 using G_2 is as follows:

$$
\begin{array}{c}
\bullet \ \bullet \ X \ \bullet \ \bullet \\
\bullet \ \bullet \ X \ \bullet \ \bullet \\
X \ X \ X \ X \ X \\
Y \ Y \ Y \ Y \ Y \\
X \ X \ X \ X \ X \\
Y \ Y \ Y \ Y \ Y \\
\bullet \ \bullet \ X \ \bullet \ \bullet \\
\bullet \ \bullet \ X \ \bullet \ \bullet
\end{array}
\Rightarrow_{DR2}^{\phi_r^d}
\begin{array}{c}
\bullet \ \bullet \ X \ \bullet \ \bullet \\
\bullet \ \bullet \ X \ \bullet \ \bullet \\
X \ X \ X \ X \ X. \\
\bullet \ \bullet \ X \ \bullet \ \bullet \\
\bullet \ \bullet \ X \ \bullet \ \bullet
\end{array}
$$

The corresponding $(CF : CF)MG$ generating L_1 is $M = (G, G')$ where $G = \langle N, T, P, S \rangle$ with

- $N = \{S, A\}$
- $T = \{S_1, S_2\}$
- $P = \{S \to S_1 A S_1, \ A \to S_1 A S_1, \ A \to S_2\}$

and

$G' = \{G_1, G_2\}$ with $G_1 = \langle N_1, T_1, P_1, S_1 \rangle$ and $G_2 = \langle N_2, T_2, P_2, S_2 \rangle$. Here

- $N_1 = \{S_1 \ , \ B\}$
- $T_1 = \{\bullet\}$
- $P_1 = \{S_1 \to \bullet B \bullet \ , \ B \to \bullet B \bullet \ , \ B \to X\}$

and

- $N_2 = \{S_2 \ , \ C\}$
- $T_2 = \{X\}$
- $P_1 = \{S_2 \to XCX \ , \ C \to XCX \ , \ C \to X\}$

A sample derivation of a picture given in L_1 using the above $(CF : CF)MG$ is given as follows:

$$
S \Rightarrow S_1 A S_1 \ \Rightarrow \ S_1 S_1 A S_1 S_1 \ \Rightarrow \ S_1 S_1 S_2 S_1 S_1 \ \Rightarrow \
\begin{array}{c}
\bullet \ \bullet \ X \ \bullet \ \bullet \\
B \ B \ C \ B \ B \\
\bullet \ \bullet \ X \ \bullet \ \bullet
\end{array}
\ \Rightarrow
$$

$$
\begin{array}{c}
\bullet \ \bullet \ X \ \bullet \ \bullet \\
X \ X \ X \ X \ X. \\
\bullet \ \bullet \ X \ \bullet \ \bullet
\end{array}
$$

4   Main Result

In this section we show our main result that $\mathfrak{L}(PCAIDG)$ includes $(CF : CF)ML$. For proving this, we make use of the Chomosky normal form (CNF) of a context-free grammar. This implies that any context free language without λ is generated by a grammar in which all productions are of the form $A \to BC$ or $A \to a$ where A, B and C are non-terminals and a is a terminal [1].

To prove $(CF : CF)ML \subsetneq \mathfrak{L}(PCAIDG)$ we write the production rules of both horizontal and vertical CFGs of $(CF : CF)MG$ in Chomsky Normal Form. For example, the rules of CFG G of $M = (G, G')$, of Example 1, in CNF are given by

$$S \to CD, \ C \to S_1, \ D \to AE, \ E \to S_1, \ A \to FG, \ F \to S_2, \ G \to BH,$$
$$H \to S_2, \ B \to IJ, \ I \to S_2, \ J \to BK, \ K \to S_2, \ B \to S_1.$$

Theorem 1. $(CF : CF)ML \subsetneq \mathfrak{L}(PCAIDG)$

Proof. The $(CF : CF)$ matrix grammar is given by $M_G = (G, G')$, where $G = \langle N, I, P, S \rangle$, is a context-free grammar in Chomsky normal form, $I = \{A_1, A_2, \ldots, A_k\}$, $G' = \{G_1, G_2, \ldots, G_k\}$, where each $G_j = \langle N_j, \Sigma, P_j, A_j \rangle$ where $1 \le j \le k$ is a context free grammar in Chomsky normal form.

The parallel contextual array insertion deletion grammar corresponding to the $(CF : CF)MG$ is given by

$G_{ID} = (V, T, M, C, R, \phi_r^i, \phi_c^i, \phi_r^d, \phi_c^d)$ where
$V = N \cup I \cup \Sigma \cup \{\#\}$.
$T = \Sigma$.
$M = \left\{ \begin{smallmatrix} \# & \# \\ \# & S \end{smallmatrix} \middle/ S \in N \text{ is the start symbol in } G \right\}$

$C = \left\{ \begin{smallmatrix} \# \\ A \end{smallmatrix}, \begin{smallmatrix} \# & \# \\ B & C \end{smallmatrix} \middle/ A \to BC \in P, A, B, C \in N \right\} \cup$

$\left\{ \begin{smallmatrix} \# \\ A \end{smallmatrix}, \begin{smallmatrix} \# \\ A_i \end{smallmatrix} \middle/ A \to A_i \in P, A \in N, A_i \in I \right\} \cup \left\{ \begin{smallmatrix} \# \\ \# \end{smallmatrix} \right\}$.

$R = \left\{ \# \ A_i, \begin{smallmatrix} \# & B_i \\ \# & C_i \end{smallmatrix} \middle/ A_i \to B_i C_i \in P_i, A_i, B_i, C_i \in N_i \right\} \cup$

$\left\{ \# \ A_i, \# \ a_i \middle/ A_i \to a_i \in P_i, A_i \in N_i, a_i \in \Sigma \right\} \cup$

$\left\{ A_i \ A_j, \begin{smallmatrix} B_i & B_j \\ C_i & C_j \end{smallmatrix} \middle/ A_i \to B_i C_i \in P_i, A_j \to B_j C_j \in P_j, A_i, B_i, C_i \in \right.$

$N_i, A_j, B_j, C_j \in N_j \Big\} \cup$

$\left\{ A_i \ A_j, \ a_i \ a_j \middle/ A_i \to a_i \in P_i, A_j \to a_j \in P_j, A_i \in N_i, A_j \in N_j, a_i, a_j \in \Sigma \right\} \cup$

$\left\{ \# \ \# \right\}$.

The column insertion deletion rules (horizontal productions) for horizontal growth: $\phi_c^i \left[\begin{smallmatrix} \# \\ S \end{smallmatrix}, \begin{smallmatrix} \lambda \\ \lambda \end{smallmatrix} \right] = \left\{ \begin{smallmatrix} \# & \# \\ A & B \end{smallmatrix} \middle/ S \to AB \in P \right\}, \phi_c^d \left[\begin{smallmatrix} \# \\ \# \end{smallmatrix}, \begin{smallmatrix} \# & \# \\ A & B \end{smallmatrix} \right] = \left\{ \begin{smallmatrix} \# \\ S \end{smallmatrix} \middle/ S \to \right.$

$AB \in P \Big\} \ \phi_c^i \left[\begin{smallmatrix} \# \\ A \end{smallmatrix}, \begin{smallmatrix} \lambda \\ \lambda \end{smallmatrix} \right] = \left\{ \begin{smallmatrix} \# & \# \\ B & C \end{smallmatrix} \middle/ A \to BC \in P \right\}$,

$$\phi_c^d \begin{bmatrix} \# & \# \ \# \\ D & B \ C \end{bmatrix} = \left\{ \frac{\#}{A} \middle/ A \to BC \in P, D \in N \cup I \right\},$$

$$\phi_c^i \begin{bmatrix} \# & \# \\ A & D \end{bmatrix} = \left\{ \frac{\# \ \#}{B \ C} \middle/ A \to BC \in P, D \in N \cup I \right\},$$

$$\phi_c^d \begin{bmatrix} \# & \# \ \# \\ \# & B \ C \end{bmatrix} = \left\{ \frac{\#}{A} \middle/ A \to BC \in P \right\},$$

$$\phi_c^i \begin{bmatrix} \# & \lambda \\ S & \lambda \end{bmatrix} = \left\{ \frac{\#}{A_i} \middle/ S \to A_i \in P \right\},$$

$$\phi_c^d \begin{bmatrix} \# & \# \\ \# & A_i \end{bmatrix} = \left\{ \frac{\#}{S} \middle/ S \to A_i \in P \right\},$$

$$\phi_c^i \begin{bmatrix} \# & \lambda \\ A & \lambda \end{bmatrix} = \left\{ \frac{\#}{A_i} \middle/ S \to A_i \in P \right\},$$

$$\phi_c^d \begin{bmatrix} \# & \# \\ D & A_i \end{bmatrix} = \left\{ \frac{\#}{A} \middle/ A \to A_i \in P, D \in N \cup I \right\},$$

$$\phi_c^i \begin{bmatrix} \# & \# \\ A & D \end{bmatrix} = \left\{ \frac{\#}{A_i} \middle/ A \to A_i \in P, D \in N \cup I \right\},$$

$$\phi_c^d \begin{bmatrix} \# & \# \\ \# & A_i \end{bmatrix} = \left\{ \frac{\#}{A} \middle/ A \to A_i \in P \right\}.$$

The row insertion deletion rules (vertical productions) for vertical growth:

$$\phi_r^i \begin{bmatrix} A_i \ A_j & , & \lambda \ \lambda \end{bmatrix} = \left\{ \frac{B_i \ B_j}{C_i \ C_j} \middle/ A_i \to B_i C_i \in P_i, A_j \to B_j C_j \in P_j \right\},$$

$$\phi_r^d \begin{bmatrix} \# \ \# & , & B_i \ B_j \\ & & C_i \ C_j \end{bmatrix} = \left\{ A_i \ A_j \middle/ A_i \to B_i C_i \in P_i, A_j \to B_j C_j \in P_j \right\}.$$

$$\phi_r^i \begin{bmatrix} A_i \ A_j & , & D_i \ D_j \end{bmatrix} = \left\{ \frac{B_i \ B_j}{C_i \ C_j} \middle/ A_i \to B_i C_i \in P_i, A_j \to B_j C_j \in P_j, \right.$$
$$D_i \in N_i \cup \Sigma, D_j \in N_j \cup \Sigma \Big\},$$

$$\phi_r^d \begin{bmatrix} D_i \ D_j & , & B_i \ B_j \\ & & C_i \ C_j \end{bmatrix} = \left\{ A_i \ A_j \middle/ A_i \to B_i C_i \in P_i, A_j \to B_j C_j \in P_j, \right.$$
$$D_i \in N_i \cup \Sigma, D_j \in N_j \cup \Sigma \Big\}.$$

$$\phi_r^i \begin{bmatrix} \# \ A_i & , & \lambda \ \lambda \end{bmatrix} = \left\{ \frac{\# \ B_i}{\# \ C_i} \middle/ A_i \to B_i C_i \in P_i \right\},$$

$$\phi_r^d \begin{bmatrix} \# \ \# & , & \# \ B_i \\ & & \# \ C_i \end{bmatrix} = \left\{ \# \ A_i \middle/ A_i \to B_i C_i \in P_i \right\}.$$

$$\phi_r^i \begin{bmatrix} \# \ A_i & , & \# \ D_i \end{bmatrix} = \left\{ \frac{\# \ B_i}{\# \ C_i} \middle/ A_i \to B_i C_i \in P_i, D_i \in N_i \cup \Sigma \right\},$$

$$\phi_r^d \begin{bmatrix} \# \ D_i & , & \# \ B_i \\ & & \# \ C_i \end{bmatrix} = \left\{ \# \ A_i \middle/ A_i \to B_i C_i \in P_i, D_i \in N_i \cup \Sigma \right\}.$$

$$\phi_r^i \begin{bmatrix} A_i \ A_j & , & \lambda \ \lambda \end{bmatrix} = \left\{ a_i \ a_j \middle/ A_i \to a_i \in P_i, A_j \to a_j \in P_j \right\},$$

$$\phi_r^d \begin{bmatrix} \# \ \# & , & a_i \ a_j \end{bmatrix} = \left\{ A_i \ A_j \middle/ A_i \to a_i \in P_i, A_j \to a_j \in P_j \right\}.$$

$$\phi_r^i \left[A_i \, A_j \;\;,\;\; D_i \, D_j \right] \;=\; \left\{ a_i \, a_j \Big/ A_i \;\rightarrow\; a_i \;\in\; P_i, A_j \;\rightarrow\; a_j \;\in\; P_j, \right.$$

$$\left. D_i \in N_i \cup \Sigma, D_j \in N_j \cup \Sigma \right\},$$

$$\phi_r^d \left[D_i \, D_j \;\;,\;\; a_i \, a_j \right] \;=\; \left\{ A_i \, A_j \Big/ A_i \;\rightarrow\; a_i \;\in\; P_i, A_j \;\rightarrow\; a_j \;\in\; P_j, \right.$$

$$\left. D_i \in N_i \cup \Sigma, D_j \in N_j \cup \Sigma \right\}.$$

$$\phi_r^i \left[\# \, A_i \;\;,\;\; \lambda \, \lambda \right] = \left\{ \# \, a_i \Big/ A_i \rightarrow a_i \in P_i \right\},$$

$$\phi_r^d \left[\# \, \# \;\;,\;\; \# \, a_i \right] = \left\{ \# \, A_i \Big/ A_i \rightarrow a_i \in P_i \right\},$$

$$\phi_r^i \left[\# \, A_i \;\;,\;\; \# \, D_i \right] = \left\{ \# \, a_i \Big/ A_i \rightarrow a_i \in P_i, D_i \in N_i \cup \Sigma \right\},$$

$$\phi_r^d \left[\# \, D_i \;\;,\;\; \# \, a_i \right] = \left\{ \# \, A_i \Big/ A_i \rightarrow a_i \in P_i, D_i \in N_i \cup \Sigma \right\},$$

$$\phi_r^d \left[\lambda \, \lambda \;\;,\;\; a_i \, a_j \right] = \left\{ \# \, \# \Big/ a_i, a_j \in \Sigma \right\},$$

$$\phi_r^d \left[\lambda \, \lambda \;\;,\;\; \# \, a_i \right] = \left\{ \# \, \# \Big/ a_i \in \Sigma \right\},$$

$$\phi_c^d \left[\begin{matrix} \lambda \\ \lambda \end{matrix} , \begin{matrix} a_i \\ b_i \end{matrix} \right] = \left\{ \begin{matrix} \# \\ \# \end{matrix} \Big/ a_i, b_i \in \Sigma \right\}.$$

The working of the PCAIDG is as follows: Depending upon the starting symbol S of the grammar $(CF:CF)MG$, the array of the axiom set M, $\begin{matrix} \# \, \# \\ \# \; S \end{matrix}$ is chosen accordingly. For the horizontal growth of pictures, appropriate parallel column contextual insertion deletion operations are applied alternatively according to the horizontal production rules of the grammar $(CF:CF)MG$. After the successful completion of application of horizontal production rules, parallel row contextual insertion deletion rules are applied alternatively corresponding to the vertical production rules of the grammar $(CF:CF)MG$. Finally, using parallel contextual row/column deletion operations, the #'s along the boundaries of the picture are deleted accordingly so that the resulting picture language belongs to the family of picture languages $\mathfrak{L}(PCAIDG)$.

We now prove the proper inclusion. James et al. [6] have proved that the language

$$L = \left\{ \begin{pmatrix} (\bullet)_n \\ (X)_n \\ (\bullet)_n \end{pmatrix}^n \begin{pmatrix} (X)_n \\ (X)_n \\ (\bullet)_n \end{pmatrix}^n \begin{pmatrix} (\bullet)_n \\ (\bullet)_n \\ (\bullet)_n \end{pmatrix}^n \Big/ n \geq 1 \right\} \text{ can be generated by a}$$

$PCAIDG$ but not by any $CSMG$ as can be seen in [16]. Hence it cannot be generated by any $(CF:CF)MG$ and hence $(CF:CF)ML$ is properly included in the family of $\mathfrak{L}(PCAIDG)$.

5   Applications

We now give few applications of the grammar $PCAIDG$:

1. In one of the pictures generated by $PCAIDG$ in Example 1,

$$\begin{array}{ccccc} X & X & X & X & X \\ X & \bullet & X & \bullet & X \\ X & X & X & X & X \\ X & \bullet & X & \bullet & X \\ X & X & X & X & X \end{array}$$

if we replace each X by $\square$, we get the following image

$$\begin{array}{ccccc} \square & \square & \square & \square & \square \\ \square & \bullet & \square & \bullet & \square \\ \square & \square & \square & \square & \square \\ \square & \bullet & \square & \bullet & \square \\ \square & \square & \square & \square & \square \end{array}.$$

2. Since James et al. [6] have proved that $\mathfrak{L}(CSML) \subsetneq \mathfrak{L}(PCAIDG)$, where $CSMGs$ generate wide varieties of pictures like kolam patterns, geometric patterns and other patterns [15–17], $PCAIDGs$ do the same and more.

6   Conclusion

In this paper, it is proved that the family of languages generated by the parallel contextual array insertion deletion grammars properly includes (CF:CF)ML. It can be shown that REC [3], $CSML$ [6], $(CF : RIR)$ [7] and $(CF : CF)ML$ [12] are incomparable but not disjoint. It is worth to investigate the position of $\mathfrak{L}(PCAIDG)$ in the hierarchy of families of two dimensional languages. An attempt has been made in this paper towards this direction. It should be noted that tabled matrix grammars [18] have higher generative capacity than Siromoney matrix models considered by us. It is interesting to compare the power of PCAIDG with tabled matrix grammars and many other models available in the literature and would be carried out as our future work.

Acknowledgment. We are grateful to the reviewers whose comments made to refine the paper significantly.

References

1. Meduna, A.: Automata and Languages: Theory and Applications. Springer, London (2000). https://doi.org/10.1007/978-1-4471-0501-5
2. Freund, R., Păun, Gh., Rozenberg, G.: Contextual array grammars. In: Subramanian, K.G., Rangarajan, K., Mukund, M. (eds.) Formal Models, Languages and Application Series in Machine Perception and Artificial Intelligence, World Scientific, vol. 66, 112–136 (2006)
3. Giammarresi, D., Restivo, A.: Two- dimensional languages. Handbook of Formal Languages **3**, 215–267 (1997)
4. Giammarresi, D., Restivo, A.: Recognizable picture languages. Int. J. Pattern Recogn. Artif. Intell. **6**, 241–256 (1992)

5. Helen Chandra, P., Subramanian, K.G., Thomas, D.G.: Parallel contextual array grammars and languages. Electron. Notes Discrete Math. **12**, 106–117 (2003)

6. James Immanuel, S., Thomas, D.G., Robinson, T.: Atulya K Nagar: parallel contextual array insertion deletion grammar. IWCIA, Lecture Notes in Computer Science **11255**, 28–42 (2018)

7. Jayasankar, S., Thomas D.G., James Immanuel, S., Meenakshi Paramasivan, T., Robinson, T., Atulya K. Nagar: Parallel contextual array insertion deletion P systems and siromoney matrix grammars. In: Pre-Proceedings of Asian Conference on Membrane Computing 2019 (ACMC 2019), 14–17 November Xiamen, China, pp. 134–151 (2019)

8. Kari, L.: On Insertion and deletion in formal languages. Ph.D. Thesis, University of Turku (1991)

9. King-Sun, F., Rosenfeld, A.: Pattern recognition and image processing. IEEE Trans. Comput. **C–25**(12), 1336–1346 (1976)

10. Krithivasan, K., Balan, M.S., Rama, R.: Array contextual grammars, In: Martin-Vide, C., Păun, Gh. (eds.) Recent Topics in Mathematical and Computational Linguistics. The Publishing house of the Romanian Academy, pp. 154–168 (2000)

11. Păun, G.: Marcus Contextual Grammars. Springer, Dordrecht (1997). https://doi.org/10.1007/978-94-015-8969-7

12. Radhakirshnan, V., Chakravathy, V.T., Krithivasan, K.: Some properties of matrix grammars, parallel image analysis. In: Sixth International Work shop on Parallel Image Processing and Analysis-Theory and Applications, 15–16 Jan 1999, pp. 213–225 (1999)

13. Rama, R., Smitha, T.A.: Some results on array contextual grammars. Int. J. Pattern Recogn. Artif. Intell. **14**, 537–550 (2000)

14. Rosenfeld, A.: Isotonic grammars, parallel grammars and picture grammars. In: Michie, D., Meltzer, D. (eds.)Machine Intelligence VI, pp. 281–294. University of Edinburgh Press, Scotland (1971)

15. Siromoney, G., Siromoney, R., Krithivasan, K.: Abstract families of matrices and picture languages. Comput. Graph. Image Process. **1**, 284–307 (1972)

16. Siromoney, G., Siromoney, R., Krithivasan, K.: Picture languages with array rewriting rules. Inf. Control **22**, 447–470 (1973)

17. Siromoney, G., Siromoney, R., Krithivasan, K.: Array grammars and kolam. Comput. Graph. Image Process. **3**, 63–82 (1974)

18. Siromoney, G., Subramanian, K.G., Rangarajan, K.: Parallel/sequential rectangular arrays with tables. Int. J. Comput. Math. Sect. A **6**, 143–158 (1977)

19. Subramanian, K.G., Van, D.L., Helen Chandra, P., Quyen, N.D.: Array grammars with contextual operations. Fund. Info. **83**, 411–428 (2008)

Digital Hyperplane Fitting

Phuc Ngo$^{(\boxtimes)}$

LORIA, Université de Lorraine, Nancy, France
hoai-diem-phuc.ngo@loria.fr

Abstract. This paper addresses the hyperplane fitting problem of discrete points in any dimension (*i.e.* in $\mathbb{Z}^d$). For that purpose, we consider a digital model of hyperplane, namely *digital hyperplane*, and present a combinatorial approach to find the optimal solution of the fitting problem. This method consists in computing all possible digital hyperplanes from a set **S** of n points, then an exhaustive search enables us to find the optimal hyperplane that *best fits* **S**. The method has, however, a high complexity of $O(n^d)$, and thus can not be applied for big datasets. To overcome this limitation, we propose another method relying on the Delaunay triangulation of **S**. By not generating and verifying all possible digital hyperplanes but only those from the elements of the triangulation, this leads to a lower complexity of $O(n^{\lceil \frac{d}{2} \rceil + 1})$. Experiments in 2D, 3D and 4D are shown to illustrate the efficiency of the proposed method.

Keywords: Optimal consensus · Exact computation · Discrete optimization · Optimal fitting · dD Delaunay triangulation

1 Introduction

Data fitting is the process of matching a set of data points with a model, possibly subject to constraints. This is an essential task in many applications of computer vision and image analysis; *e.g.* shape approximation [23,32], image registration [31,33], image segmentation [19,21]. In this context, the mostly considered models are the geometric ones such as a line, a circle in 2D or a plane, a surface in 3D. Among the models, the linear one has received greatest attention in theory and practice as many nonlinear models can be rearranged to a linear form by a suitable transformation of the model formulation [17,30]. In this paper, we are interested in the fitting problem of hyperplane – a linear model – for a set of discrete points **S** in any dimension; *i.e.* **S** $\subset \mathbb{Z}^d$ for d is the dimension of space. Then, this problem can be formulated as an optimization problem in which we find the parameters of the hyperplane that *best fits* **S**, namely *optimal solution*. It is clear that this problem depends on how we define the hyperplane model and the criteria for the best fitting, namely cost function of the optimization process. In practice, hyperplane fitting has a great interest in applications of classification for object detection and recognition [9].

Several works have been proposed in this context. We can mention, for instance, the methods based on regression [4,11,29,30]; *e.g.* least squares,

© Springer Nature Switzerland AG 2020
T. Lukić et al. (Eds.): IWCIA 2020, LNCS 12148, pp. 164–180, 2020.
https://doi.org/10.1007/978-3-030-51002-2_12

weighted least-squares, least-absolute-value regression and least median of squares (LMS). Generally, these approaches consider a hyperplane in the Euclidean space $\mathbb{R}^d$ and find the model that minimizes the sum of the geometric distance from all given points to the model. However, the provided solution is known to be unstable and sensitive to large deviant data points, namely *outliers* [24]. Other well-known approaches for fitting hyperplane using voting scheme include the Hough Transform (HT) [15,18], the RANdom SAmple Consensus (RANSAC) [16] and associated variations [12]. More precisely, HT considers a dual space, namely *Hough parameter space* of the input set **S**. This space is discretized into cells and used as an accumulator of votes from points of **S**. Typically, for each point $x \in \mathbf{S}$, a vote is added to cell in the accumulator that could generate from x. Then, the optimal solution is computed from the cell having the maximal of votes. RANSAC method chooses d points from **S** at random to form a candidate hyperplane passing through these points. Then, it calculates the distance of every points of **S** to the candidate hyperplane, and the points within a threshold distance are considered to be in the *consensus set* of the hyperplane. A score associated to the hyperplane is computed based on its consensus set; *e.g.* the size of the consensus set. This process is iterated a certain times and reports the hyperplane of maximal score as the optimal solution; *i.e.* the best-fitting hyperplane for **S**. Both HT and RANSAC are simple, efficient and robustness to outliers. They, however, have a computational complexity growing with the number of model parameters [12,15,16,18]. In addition, the above approaches were originally designed for points in $\mathbb{R}^d$ using the Euclidean hyperplane model. Of course, the discrete space of $\mathbb{Z}^d$ is a subspace of $\mathbb{R}^d$, then the fitting problem can be solved using these approaches. However, because of their continuous consideration, applying them for points of $\mathbb{Z}^d$ requires the use of floating point numbers which may induce the numerical error. Furthermore, due to the discrete nature of points in Z^d, computing an actually optimal solution of discrete points is practically impossible in a continuous space as there is an infinity of solutions.

Under the assumption of input points in discrete space of $\mathbb{Z}^d$, this paper addresses the problem of hyperplane fitting in a fully discrete context. More specifically, we consider the digital model of hyperplane of $\mathbb{Z}^d$, namely *digital hyperplane*, and propose methods for digital hyperplane fitting using exact computation. For that purpose, we first present a combinatorial approach for solving this problem. The method consists in generating all the possible digital hyperplanes associated to a set of n point $\mathbf{S} \subset \mathbb{Z}^d$. Contrarily to $\mathbb{R}^d$, this set –despite a potentially high complexity– remains finite and thus allows an explicit exploration to find the optimal solution via a discrete optimization scheme. The method guarantees the global optimality, it has however a computational complexity of $O(n^d)$. In practice, it is unsolvable for $n = 10^6$ in 3D[1]. This high complexity practically forbids its use for big dataset.

---

[1] Supposing it needs 1 µs for generating and testing one hyperplane, then it takes about $3 * 10^7$ years to find the optimal solution.

In order to solve this fitting problem in a practical context, we propose a new method to find locally optimal solution. The method is based on a heuristic involving the Delaunay triangulation [13,27]. Basically, it consists in computing the Delaunay triangulation of the input points, then using the triangulated elements to generate hyperplanes and find an optimal solution for the fitting problem stated above. It should be mentioned that Delaunay triangulations are used in numerous applications of computational geometry [8], geometric modelling [6] and computer graphics [3,14]. It is not only well-known for its optimal properties [2,22,27] but also for its advantage to be incrementally computed in $O(n^{\lceil \frac{d}{2} \rceil + 1})$ complexity [7,10].

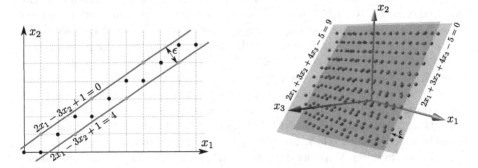

Fig. 1. Digital hyperplanes in 2D and 3D. Left: a digital line and right: a digital plane.

2   Preliminaries

2.1   Digital Hyperplane

An affine hyperplane in Euclidean space $\mathbb{R}^d$ of dimension $d \geq 2$ is defined by the set of points $\mathrm{x} = (x_1, x_2, \ldots, x_d) \in \mathbb{R}^d$ satisfying the following equation:

$$\mathcal{H} = \{\mathrm{x} \in \mathbb{R}^d : \sum_{i=1}^{d} a_i x_i + a_{d+1} = 0\}, \tag{1}$$

with $a_i \in \mathbb{R}$ are coefficients of the hyperplane, w.l.o.g. $\mathcal{H}$ can be unambiguously represented by its parameters and denoted by $\mathcal{H} = (a_i)_{i=1}^{d+1}$. In other words, a hyperplane is the solution of a single linear equation (Eq. 1). Lines and planes are respectively hyperplanes in 2 and 3 dimensions.

The digitization of hyperplane in the discrete space of $\mathbb{Z}^d$ is called *digital hyperplane*, and defined as follows. Note that this definition is similar to the one in [28] with a slight difference at the double less than or equal to ($\leq$) in Eq. 2.

Definition 1. *A digital hyperplane in $\mathbb{Z}^d$, $d \geq 2$, is defined by the set of discrete points* $\mathrm{x} = (x_1, x_2, \ldots, x_d) \in \mathbb{Z}^d$ *satisfying the inequalities:*

$$\mathcal{DH}_\omega = \{\mathrm{x} \in \mathbb{Z}^d : 0 \leq \sum_{i=1}^{d} a_i x_i + a_{d+1} \leq \omega\}, \tag{2}$$

with $a_i \in \mathbb{Z}$ are coefficients of the digital hyperplane, and $\omega \in \mathbb{Z}$ a given constant.

Such a digital plane $\mathcal{DH}_\omega$ can be represented by its parameters and denoted by $\mathcal{DH}_\omega = (a_i)_{i=1}^{d+1}$. Geometrically, $\mathcal{DH}_\omega$ is a set of discrete points lying in between two parallel hyperplanes $\sum_{i=1}^{d} a_i x_i + a_{d+1} = 0$ and $\sum_{i=1}^{d} a_i x_i + a_{d+1} = \omega$; these two parallel hyperplanes are called the *support hyperplanes*, and the points that are on the support hyperplanes are called *support points* of $\mathcal{DH}_\omega$. The distance between the two hyperplanes is $\epsilon = \frac{w}{\sqrt{\sum_{i=1}^{d} a_i^2}}$ which refers to the *euclidean thickness* of $\mathcal{DH}_\omega$, while w refers to *arithmetical thickness* of $\mathcal{DH}_\omega$ [28]. An example in 2D and 3D is given in Fig. 1.

From linear algebra, a hyperplane in dimension d is a $(d-1)$-dimensional subspace; *i.e.* it is defined by $(d-1)$ linearly independent vectors. These $(d-1)$ vectors can be created from d distinct points of $\mathcal{H}$. In other words, given any d points in the hyperplane in general linear position, *i.e.* they are all $(d-1)$ linearly independent, there is a unique hyperplane of $\mathbb{R}^d$ passing through them. In case of $\mathbb{Z}^d$, from Eq. 2, we also need d linearly independent support points to determine a digital hyperplane. However, contrarily to the Euclidean space, for a given set of d support points and a value ω, the digital hyperplane $\mathcal{DH}_\omega$ passing though these points is not unique. This is illustrated in Fig. 2.

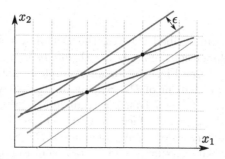

Fig. 2. Different digital lines of thickness ω passing through two points (in black).

2.2 Delaunay Triangulation

The *Delaunay triangulation* was introduced by Boris Delaunay in [13]. It is initially defined for a given set of n points $\mathbf{S} = \{\mathrm{x}_i \in \mathbb{R}^2 \mid i = 1..n\}$ as a triangulation $\mathcal{DT}(\mathbf{S})$ such that the circumcircle associated to any triangle in

$\mathcal{DT}(\mathbf{S})$ does not contain any other points of $\mathbf{S}$ in its interior. In other words, a Delaunay triangulation fulfills the *empty circle property* (also called *Delaunay property*): the circumscribing circle of any triangle of the triangulation encloses no other data point. Such a triangulation can be seen as a partition of the convex hull of $\mathbf{S}$ into triangles whose vertices are the points of $\mathbf{S}$, and it maximizes the minimum angle of all the angles of the triangles in $\mathcal{DT}(\mathbf{S})$. An illustration is given in Fig. 3(a). It should be mentioned that in some degenerate cases, the Delaunay triangulation is not guaranteed to exist or be unique; *e.g.* for a set of linear points there is no Delaunay triangulation, for four or more points on the same circle the Delaunay triangulation is not unique.

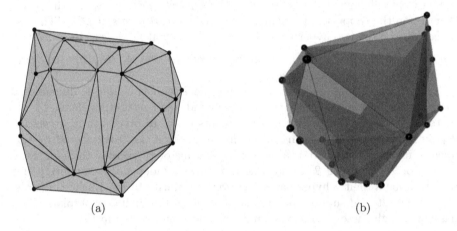

(a)   (b)

Fig. 3. Illustration of Delaunay triangulation in (a) 2D and (b) 3D.

By considering circumscribed spheres, the Delaunay triangulation extends to three and higher dimensions, namely *dD Delaunay triangulation* [27] for d is the dimension of space (see Fig. 3(b) for an example in 3D). Then, we call an *i-face* for $i \in [0, d]$ is an element of $\mathcal{DT}(\mathbf{S})$ containing $i + 1$ vertices of $\mathbf{S}$. Then, a vertex is a 0-face, an edge is a 1-face, a triangle is 2-face, a tetrahedron is a 3-face, a ridge is a $(d-2)$-face, a facet is a $(d-1)$-face and a full cell is a d-face. In [22,27], discussions on optimal properties of $\mathcal{DT}(\mathbf{S})$ in d dimension have been presented such as the maximum min-containment radius, uniformity of size and shape. In particular, as mentioned in [27] the dD Delaunay triangulation can be transformed into a convex-hull problem in dimension $d + 1$. Henceforth, convex-hull algorithms can be used to obtain the Delaunay triangulation. In this context, there exist efficient and incremental algorithms [5,10,27] to construct the Delaunay triangulation. Furthermore, it is shown in [26] that the dD Delaunay triangulation of n points in $\mathbb{R}^d$ contains $O(n^{\lceil \frac{d}{2} \rceil})$ faces.

In this paper, we consider the dD Delaunay triangulation for the set of discrete points $\mathbf{S} = \{x_i \in \mathbb{Z}^d \mid i = 1..n\}$, and use the implementation of Delaunay triangulation proposed in CGAL [10] because of its robustness, efficiency, ease of use and flexibility. It is shown in [7] that the worst case complexity, without spatial sort of input points, of the method is $O(n^{\lceil \frac{d}{2} \rceil + 1})$. With spatial sort and random points, one can expect a much better complexity of $O(n \log n)$.

3 Digital Hyperplane Fitting

From Definition 1, we can describe our fitting problem for a set of discrete points $\mathbf{S}$ as finding a digital hyperplane $\mathcal{DH}_\omega = (a_i)_{i=1}^{d+1}$ of given ω that encloses the most number of points of $\mathbf{S}$. Such a hyperplane is called *optimal hyperplane*, the points of $\mathbf{S}$ belonging to $\mathcal{DH}_\omega$, *i.e.* $\mathbf{x}_i \in \mathbf{S} \cap \mathcal{DH}_\omega$, are called *inliers*, the other points of $\mathbf{S}$ are called *outliers*. In other words, the digital hyperplane fitting aims to solve an optimization problem being expressed as maximizing the number of inliers.

Definition 2. *Given* $\mathbf{S} = \{\mathbf{x}_i \in \mathbb{Z}^d \mid i = 1..n,$ *for* $n \geq d\}$ *and a constant* $\omega \in \mathbb{Z}$. *The best fitting hyperplane of* $\mathbf{S}$ *is defined as*

$$\mathcal{DH}_\omega^* = \underset{\mathcal{DH}_\omega \in \mathbb{F}(\mathbf{S})}{\arg \max} \; \{\mathbf{S} \cap \mathcal{DH}_\omega\}$$

where $\mathbb{F}(\mathbf{S})$ *is the search space and it contains the set of all hyperplanes of given* ω *generated from* $\mathbf{S}$.

Due to the discrete nature of the problem, it should be mentioned that the search space $\mathbb{F}(\mathbf{S})$ can be huge but finite as $\mathbf{S}$ is finite and the points $\mathbf{x}_i \in \mathbf{S}$ have finite coordinates. Roughly speaking, a brute-force search within $\mathbb{F}(\mathbf{S})$ would lead to a globally optimal solution for the digital hyperplane fitting of $\mathbf{S}$ in Definition 2. Finding $\mathbb{F}(\mathbf{S})$ in 2D (resp. 3D) case is solved in [34]. More specifically, using rotation and translation techniques, it is proved that a digital lines (resp. planes) can be determined with at least 2 (resp. 3) support points. Therefore, the whole search space $\mathbb{F}(\mathbf{S})$ can be constructed from all possible pairs (resp. triplets) of points in $\mathbf{S}$ for digital line (resp. plane) fitting.

Property 1 ([34]). Given a set of points $\mathbf{S} \subset \mathbb{Z}^2$ (resp. $\mathbb{Z}^3$), and a set of inliers, namely *consensus set*, for a given digital line (resp. plane). It is possible to find a new digital line (resp. plane) with the same consensus set, such that it has at least 2 (resp. 3) inliers as support points.

This is clearly understandable as a digital line (resp. plane) can be computed from two (resp. three) support points. This result can be extended to higher dimension thanks to the very definition of digital hyperplane (see Definition 1).

Property 2. Given a set of points $\mathbf{S} \subset \mathbb{Z}^d$, and a set of inliers for a given digital hyperplane. There exists an other hyperplane with the same inlier set such that it has at least d inliers as support points.

From *Property 2*, by taking all possible d-uplet of points in $\mathbf{S}$ as support points, one can construct the whole search space $\mathbb{F}(\mathbf{S})$ of $\mathbf{S}$ for digital hyperplane fitting.

Proposition 1. *Given* $\mathbf{S} = \{x_i \in \mathbb{Z}^d \mid i = 1..n\}$ *and a value* ω, *the number of digital hyperplanes* $\mathcal{DH}_\omega$ *generated from* $\mathbf{S}$ *is* $O(n^d)$.

Proof. From Definition 1 and *Property* 2, we need d linearly independent points as support points to determine a digital hyperplane $\mathcal{DH}_\omega$. In order to select d points in $\mathbf{S}$, we need a complexity of $O(n^d)$ since there are n points in $\mathbf{S}$ and the linearly independent test of these points is $O(1)$.

Let consider the two support hyperplanes of $\mathcal{DH}_\omega$:

$$\sum_{i=1}^{d} a_i x_i + a_{d+1} = 0 \tag{3}$$

$$\sum_{i=1}^{d} a_i x_i + a_{d+1} = \omega \tag{4}$$

It should be recalled that, for a given set of d support points, the digital hyperplane $\mathcal{DH}_\omega$ passing though them is not unique. Different cases may appear to the d selected support points of $\mathcal{DH}_\omega$. Typically, there are i points on the support hyperplane in Eq. 3, and $d - i$ points on the one in Eq. 4, for $i = 0, .., d$. In particular, $i = 0$ or $i = d$ mean all points belong respectively to Eq. 3 or Eq. 4. Due to the symmetry of selecting points, *e.g.* having i points on Eq. 3 and $d - i$ points on Eq. 4 is equivalent to $d - i$ points on Eq. 3 and i points on Eq. 4, the total number of possible hyperplanes $\mathcal{DH}_\omega$ passing though these d support points is $\binom{d}{0} + \frac{1}{2} \sum_{i=1}^{d-1} \binom{d}{i} + \binom{d}{d} = 1 + 2^{d-1}$. This leads to the final complexity of $O(2^{d-1} n^d)$ for generating all digital hyperplanes of a given set $\mathbf{S}$. For a fixed dimension space d, this complexity becomes $O(n^d)$. $\qquad\square$

From *Proposition* 1, one can generate the whole search space $\mathbb{F}(\mathbf{S})$ from $\mathbf{S}$. Then, by verifying the inliers of each hyperplane in $\mathbb{F}(\mathbf{S})$, we can find the optimal hyperplane as the one that maximizes the number inliers. In this context, some solutions have been proposed for the specific cases of 2D and 3D, for instance in [34] a combinatorial approach using dual space is presented for digital line and plane fitting which a time complexity of $O(n^d \log n)$ for $d = 2, 3$, and an improved algorithm for 2D cases with $O(n^2)$ time-complexity using a topological sweep method in [20]. In [1], a study for efficient digital hyperplane fitting in $\mathbb{Z}^d$ with bounded error is investigated. Still in [1], a conjecture of optimal computational complexity for this problem in any dimension is provided, and it is $O(n^d)$.

4   Hyperplane Fitting Using dD Delaunay Triangulation

The combinatorial approach in the previous section, by generating all possible digital hyperplanes from $\mathbf{S}$, allows to find the optimal solution for digital hyperplane fitting. It has, however, a high complexity of $O(n^d)$ with n is the number of discrete points in $\mathbf{S}$ and d is the dimension space of points in $\mathbf{S}$. This forbids the use of the method in many applications with big datasets.

Faced with this dilemma, a new approach of digital hyperplane fitting is proposed in this section. The approach is based on a heuristic involving the dD Delaunay triangulation. Roughly speaking, the method uses the triangulated elements to filter the *admissible* combinations of discrete points and to generate digital hyperplanes for the considered fitting problem.

One of the interesting aspect of the Delaunay triangulation is that it implicitly presents an information of distribution/density of points in the space; *i.e.* the points being close to each other form small and thin cells, while those being far create excessively large and long cells (see Fig. 4 for an illustration in 2D). This enables us to relate and to recognize points belonging to the same hyperplane; *i.e.* points appearing to lie reasonably on a hyperplane are close and arranged in a linear form. Roughly speaking, the fitting problem can be solved using the dD Delaunay triangulation, according to two criteria: (1) the candidate hyperplanes should be on d-faces whose *width* is smaller than ω and (2) the best fitted hyperplane is the one containing the most number of inliers.

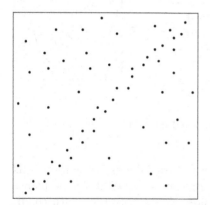 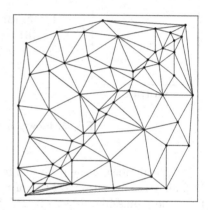

Fig. 4. A set of points (left) and its Delaunay triangulation (right). Points appearing to lie reasonably on a line are close and arranged in a linear form and their corresponding triangles by the Delaunay triangulation are smaller than the others.

Let $\mathbf{S} = \{x_i \in \mathbb{Z}^d \mid i = 1..n\}$ be the input set of points and $\mathcal{DT}(\mathbf{S})$ be the dD Delaunay triangulation of $\mathbf{S}$. Let t be a d-face in $\mathcal{DT}(\mathbf{S})$ such that t is formed by the vertices $y_0, ..., y_d \in \mathbf{S}$, w.l.o.g. t can be denoted by $t = (y_0, ..., y_d)$. We define the width –or height– of t as the minimal euclidean distance of a vertex of t to the hyperplane passing through all other vertices in t.

$$w(t) = \min \ \{||y_j, \mathcal{H}(f_j)||_2^2 \text{ for } j = 0, ..., d\}$$

where $f_j = (y_0, ..., y_{j-1}, y_{j+1}, ..., y_d)$ is the hyperplane opposite to y_j in t. We call a d-face $t \in \mathcal{DT}(\mathbf{S})$ is *admissible* for the digital hyperplane fitting of $\mathbf{S}$ with a given ω if $w(t) \leq \omega$. This is called the *width condition* and will be used in the fitting method to filter the points and to generate the digital hyperplane for examination.

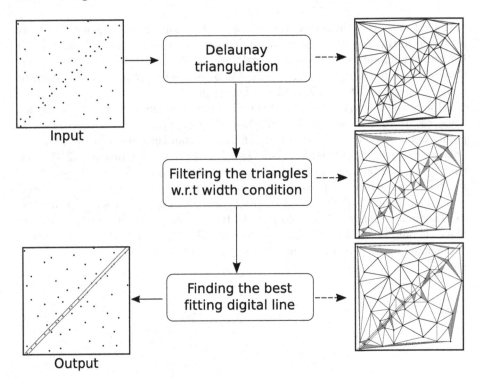

Fig. 5. Flowchart of the proposed method.

The main idea of the proposed method is as follows. We first compute the dD Delaunay triangulation $\mathcal{DT}(\mathbf{S})$ of $\mathbf{S}$, and then generate the digital hyperplanes from the admissible d-faces in $\mathcal{DT}(\mathbf{S})$; *i.e.* those that satisfy the width condition. For each computed digital hyperplane, we verify the inliers and report the optimal solution for the digital hyperplane fitting of $\mathbf{S}$ as the one maximizing the number of inliers. The algorithm is summarized in Algorithm 1, and Fig. 5 illustrates the method in 2D (the idea is exactly the same in any dimension). It is stated in [26] the total number of faces in $\mathcal{DT}(\mathbf{S})$ is $O(n^{\lceil \frac{d}{2} \rceil})$. In other words, the d-faces in $\mathcal{DT}(\mathbf{S})$ being finite, this process has a guaranteed termination. Furthermore, the computations in Algorithm 1 can be performed using only integer/rational numbers since all inputs are given in integer numbers.

By not generating and verifying all digital hyperplanes from $\mathbf{S}$ but only from the admissible d-faces in $\mathcal{DT}(\mathbf{S})$, this method leads to a much lower algorithmic complexity. More precisely, the worst-case complexity to compute the dD Delaunay triangulation of $\mathbf{S}$ is $O(n^{\lceil \frac{d}{2} \rceil+1})$, finding admissible d-faces and generating the corresponding digital hyperplanes cost $O(n^{\lceil \frac{d}{2} \rceil})$, and the inliers verification of each hyperplane is computed in $O(n)$. Therefore, the final computing complexity of Algorithm 1 is $O(n^{\lceil \frac{d}{2} \rceil+1})$ for n is the number of points in $\mathbf{S}$ and d is the dimension space of the points.

Algorithm 1: Digital hyperplane fitting with fixed thickness

---

1 **Algorithm** HyperplaneFitting

   Input: A set $\mathbf{S} = \{x_i \in \mathbb{Z}^d \mid i = 1..n\}$ of n points and a value ω

   Output: The best fitted digital hyperplane

   Variables : $\mathcal{DT}(\mathbf{S})$: the set of d-faces of the Delaunay Triangulation of $\mathbf{S}$

   C: the set of d-faces satisfying width condition

2 | $\mathcal{DT}(\mathbf{S}) = \{t_i = (y_j)_{j=0}^d \text{ for } i = 1..m \mid y_j \in \mathbf{S}\}$

   | /* Finding d-faces t_i satisfying the width condition   */

3 | $C = \emptyset$

4 | **foreach** $t_i \in \mathcal{DT}(\mathbf{S})$ **do**

5 | | **foreach** $y_j \in t_i$ **do**

6 | | | $f_j = (y_0, ..., y_{j-1}, y_{j+1}, ..., y_d)$ // the $(d-1)$-face opposite to vertex y_j in t_i

7 | | | $\mathcal{H}(f_j) = (a_i)_{i=0}^{d+1}$ // the hyperplane passing through d points of f_j (see Eq. 1)

8 | | | $d = distance(y_j, \mathcal{H}(f_j))$ // the distance of y_j to $\mathcal{H}(f_j)$

9 | | | **if** $d^2 \leq \frac{\omega^2}{\sum_{i=1}^d a_i^2}$ **then**

10 | | | | $\mathcal{DH}_\omega(t_i) = (a_i)_{i=0}^{d+1}$ // the digital hyperplane associated to t_i (see Eq. 2)

11 | | | | $C = C \cup \mathcal{DH}_\omega(t_i)$

   | /* Computing the best fitted digital hyperplane   */

12 | $max = 0$

13 | $\mathcal{DH}_\omega^* = (0, ..., 0)$

14 | **foreach** $\mathcal{DH}_\omega^i \in C$ **do**

15 | | $n_i = \text{CountInliers}(\mathbf{S}, \omega, \mathcal{DH}_\omega^i)$ // (see the below Procedure)

16 | | **if** $n_i > max$ **then**

17 | | | $max = n_i$

18 | | | $\mathcal{DH}_\omega^* = \mathcal{DH}_\omega^i$

19 | **return** $\mathcal{DH}_\omega^*$

---

1 **Procedure** CountInliers

   Input: A set $\mathbf{S} = \{x_i \in \mathbb{Z}^d \mid i = 1..n\}$ of n points, a value ω and a digital hyperplane $\mathcal{DH}_\omega = (a_i)_{i=0}^{d+1}$

   Output: The number of inliers of $\mathbf{S}$ w.r.t $\mathcal{DH}_\omega$

2 | $count = 0$

3 | **foreach** $x = (x_0, x_1, ..., x_d) \in \mathbf{S}$ **do**

4 | | $v = \sum_{i=1}^d a_i x_i + a_{d+1}$

5 | | **if** $v \geq 0$ *and* $v \leq \omega$ **then**

6 | | | $count = count + 1$

7 | **return** $count$

5   Experimental Results

We have implemented in C++ the proposed method of fitting hyperplanes described in Sect. 4 using the Triangulations and Delaunay Triangulations package in CGAL [10]. This package provides functions to compute Delaunay triangulation of points in dimension 2, 3 and d. In particular, the dD Delaunay triangulation is computed by constructing convex hull in $d+1$ dimensions. This makes the method flexible and can handle any dimension. It is, however, much slower than the libraries specifically designed for 2D and 3D Delaunay Triangulation. As mentioned, the computation of dD Delaunay triangulation for n input points is at most $O(n^{\lceil \frac{d}{2} \rceil + 1})$, and it is $O(n)$ and $O(n^2)$ in 2D and 3D, respectively. In other words, the proposed algorithm is computationally more efficient in 2D and 3D with the specific implementations in CGAL. The source code is also available for testing at https://github.com/ngophuc/HyperplaneFitting.

In the following, we present the some experimental results in dimension 2, 3 and 4 to demonstrate the validity and efficiency of our method. It should be noticed that the proposed method is general and could work in any dimension, as well as its implementation remains conceptually unchanged in any dimension. Furthermore, the fitting problem in 3D and 4D is relatively expensive to solve as the runtime complexity is $O(n^3)$. All experiments are performed on a standard PC usinng Intel Core i5 processor.

5.1   2D Case: Digital Line Fitting

At first, the experiments are carried out on 2D data points generated with the digital lines of equation

$$0 \leq 2x_1 + 3x_2 - 12 \leq \omega \tag{5}$$

with $\omega = 1, 2$ and 3. For each line, we randomly generate, according to Eq. 5, 100 inliers and $100k$ outliers with $k = 1, ..., 10$; *i.e.* 30 test datasets with ground-truth (see Fig. 6 for some examples). All data points are generated in a window of $[-100, 100]^2$. We report the inliers of the fitted line for each experiment and compare with the ground-truth by Eq. 5. Results are given in Tables 1 and 2.

Table 1. Results of fitted digital lines by the proposed method in Fig. 6.

Figure	Thickness	#Points	#Inliers	Runtime
Figure 6(a)	$\omega = 1$	1100	100	62.175 ms
Figure 6(b)	$\omega = 2$	200	103	20.531 ms
Figure 6(c)	$\omega = 3$	800	108	165.905 ms

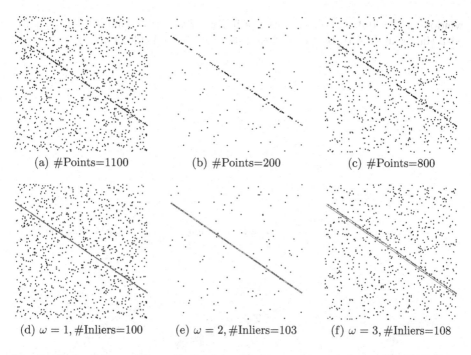

(a) #Points=1100   (b) #Points=200   (c) #Points=800

(d) $\omega = 1$, #Inliers=100   (e) $\omega = 2$, #Inliers=103   (f) $\omega = 3$, #Inliers=108

Fig. 6. Evaluation on 2D synthetic data. First row: input points with 100 inliers, different numbers of outliers and different thickness ω. Second row: results of fitting digital lines obtained by the proposed method.

Table 2. Measured performance of the proposed method on 2D synthetic data. S is the set of all ground-truth inliers, D the set of all inliers detected by the proposed method.

Measures	Results
Runtime (on average)	71.69 ms
Precision (%): $P = \#(D \cap S)/\#D$	96.99 ± 4.1
Recall (%): $R = \#(D \cap S)/\#S$	100
F-measure (%): $F = 2 \times P \times R/(P + R)$	98.42 ± 2.2

5.2   3D Case: Digital Plane Fitting

Next experiments are on volume data points which are generated as follows. We consider digital planes of equation

$$0 \leq 2x_1 + 3x_2 + x_3 - 9 \leq \omega \qquad (6)$$

with $\omega = 1, 3$ and 5. Similarly to 2D, we randomly generate, for each plane, the 100 inliers and $100k$ outliers with $k = 1, ..., 10$. That makes 30 test datasets with ground-truth (see Fig. 7 for some examples). All data points are generated in a window of $[-100, 100]^3$. We report the inliers of the fitted planes and compare with the ground-truth by Eq. 6. Results are shown in Tables 3 and 4.

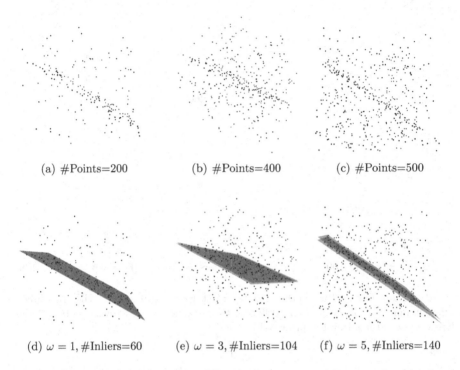

(a) #Points=200   (b) #Points=400   (c) #Points=500

(d) $\omega = 1$, #Inliers=60   (e) $\omega = 3$, #Inliers=104   (f) $\omega = 5$, #Inliers=140

Fig. 7. Evaluation on 3D synthetic data. First row: input points with 100 inliers, different numbers of outliers and different thickness ω. Second row: results of fitting digital planes obtained by the proposed method.

Furthermore, the proposed method allows to work with large datasets in an efficient way. As illustrated in Fig. 8 and Table 3, the algorithm takes around 23 s to deal with 2989 input points.

Table 3. Results of fitted digital planes by the proposed method on Fig. 7 and Fig. 8

Figure	Thickness	#Points	#Inliers	Runtime
Figure 7(a)	$\omega = 1$	200	60	35.291 ms
Figure 7(b)	$\omega = 3$	400	104	172.838 ms
Figure 7(c)	$\omega = 5$	800	140	1063.31 ms
Figure 8(a)	$\omega = 1$	2989	578	21718.1 ms
Figure 8(a)	$\omega = 3$	2989	1040	23328.6 ms

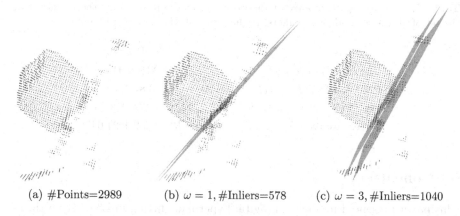

(a) #Points=2989   (b) $\omega = 1$, #Inliers=578   (c) $\omega = 3$, #Inliers=1040

Fig. 8. Evaluation on 3D data: (a) input points, (b) and (c) are fitting digital planes of (a) obtained by the proposed method for $\omega = 1$ and 3, respectively.

Table 4. Measured performance of the proposed method on 3D synthetic data. S is the set of all ground-truth inliers, D the set of all inliers detected by the proposed method.

Measures	Results
Runtime (on average)	533.68 ms
Precision (%): $P = \#(D \cap S)/\#D$	81.28 ± 15.92
Recall (%): $R = \#(D \cap S)/\#S$	77.7 ± 22.75
F-measure (%): $F = 2 \times P \times R/(P + R)$	76.16 ± 16.26

5.3   4D Case: Digital Hyperplane Fitting

Next experiments are on 4D data points generated with the following digital hyperplanes:

$$0 \leq 2x_1 + 3x_2 + x_3 + 7x_4 - 9 \leq \omega \tag{7}$$

with $\omega = 1, 3$ and 5. Then, we randomly generate, for each hyperplane, the 100 inliers and $100k$ outliers with $k = 1, ..., 10$. That makes 30 test datasets with ground-truth. All data points are generated in a window of $[-100, 100]^4$. We report the inliers of the fitted hyperplane for each experiment and compare with the ground-truth by Eq. 7. Results are shown in Table 5.

Overall, the experiments demonstrate the efficiency and effectiveness of the proposed method. It is robust to the number of outliers and has a good performance in term of runtime. In particular, the proposed method can be applied to large datasets and in high dimensions, which are difficult with traditional methods. However, the experiment was conducted mostly with synthetic data. This allows us to evaluate the behaviour of the proposed method. In practice, the sets of input points – particularly, in 2D and 3D – can be obtained by feature extraction or segmentation algorithm.

Table 5. Measured performance of the proposed method on 4D synthetic data. S is the set of all the ground-truth inliers, D the set of all the detected inliers.

Measures	Results
Runtime (on average)	11582.41 ms
Precision (%): $P = \#(D \cap S)/\#D$	73.84 ± 20.54
Recall (%): $R = \#(D \cap S)/\#S$	69.58 ± 31.3
F-measure (%): $F = 2 \times P \times R/(P + R)$	67.62 ± 24.04

6   Conclusion

This paper presented methods of digital hyperplane fitting in $\mathbb{Z}^d$ of given thickness ω. Two strategies have been proposed. The first one consists of generating all possible digital hyperplanes from a set $\mathbf{S}$ of n points. Then, performing an exhaustive search overall generated hyperplanes allows to find the global optimum of the fitting problem. However, this approach costs $O(n^d)$ which is a polynomial complexity of degree equal to the dimension of the problem. This limits its use in practical contexts. To overcome this issue, we proposed another method with a heuristic based on Delaunay triangulation to find a *local optimum* of digital hyperplane fitting problem. More precisely, instead of examining all digital hyperplanes generated from $\mathbf{S}$, we verify only the hyperplanes generated from the d-cells of the Delaunay triangulation of $\mathbf{S}$ whose width is smaller than ω. This method leads to a much lower algorithmic complexity of $O(n^{\lceil \frac{d}{2} \rceil + 1})$ and it is efficient in dealing with large datasets. Furthermore, the presented method can be applied to points in $\mathbb{R}^d$ with no special change.

Experiments have been conducted to validate the feasibility of the proposed method. Nonetheless, it is mostly with synthetic data. In future works, we would like to test the proposed method on real data and to provide comparisons with other methods in the literature such as [1,16,18,34]. Another perspective is the application of the proposed method for shape fitting problem. As it is shown in [25], by a transformation of the model formulation, the digital plane fitting can be used to solve digital annulus fitting.

References

1. Aiger, D., Kenmochi, Y., Talbot, H., Buzer, L.: Efficient robust digital hyperplane fitting with bounded error. In: Debled-Rennesson, I., Domenjoud, E., Kerautret, B., Even, P. (eds.) DGCI 2011. LNCS, vol. 6607, pp. 223–234. Springer, Heidelberg (2011). https://doi.org/10.1007/978-3-642-19867-0_19
2. Amenta, N., Attali, D., Devillers, O.: A tight bound for the delaunay triangulation of points on a polyhedron. Discrete Comput. Geom. **48**, 19–38 (2012)
3. Amenta, N., Choi, S., Dey, T.K., Leekha, N.: A simple algorithm for homeomorphic surface reconstruction. Int. J. Comput. Geom. Appl. **2**(01n02), 213–222 (2000)
4. Arlinghaus, S.L.: Practical Handbook of Curve Fitting. CRC Press, Boca Raton (1994)

5. Barber, C.B., Dobkin, D.P., Dobkin, D.P., Huhdanpaa, H.: The quickhull algorithm for convex hulls. ACM Trans. Math. Softw. **22**(4), 469–483 (1996)
6. Bern, M., Eppstein, D.: Mesh generation and optimal triangulation. In: Lecture Notes Series on Computing. Computing in Euclidean Geometry, pp. 47–123 (1995)
7. Boissonnat, J.D., Devillers, O., Hornus, S.: Incremental construction of the Delaunay graph in medium dimension. In: Annual Symposium on Computational Geometry, pp. 208–216 (2009)
8. Boissonnat, J.D., Yvinec, M.: Algorithmic Geometry. Cambridge University Press, Cambridge (1998)
9. Cevikalp, H.: Best fitting hyperplanes for classification. IEEE Trans. Pattern Anal. Mach. Intell. **39**(6), 1076–1088 (2016)
10. CGAL-Team: CGAL: the computational geometry algorithms library (2019). http://www.cgal.org
11. Chernov, N.: Circular and Linear Regression: Fitting Circles and Lines by Least Squares. CRC Press, Boca Raton (2010)
12. Chum, O.: Two-view geometry estimation by random sample and consensus. PhD thesis, Czech Technical University, Prague, Czech Republic (2005)
13. Delaunay, B.: Sur la sphère vide. Bulletin de l'Académie des Sciences de l'URSS, Classe des Sciences Mathématiques et Naturelles (1934)
14. Dey, T.K.: Curve and Surface Reconstruction: Algorithms with Mathematical Analysis. Cambridge University Press, New York (2006)
15. Duda, R., Hart, P.: Use of the hough transformation to detect lines and curves in pictures. Commun. ACM **15**, 11–15 (1972)
16. Fischler, M., Bolles, R.: Random sample consensus: a paradigm for model fitting with applications to image analysis and automated cartography. Commun. ACM **24**(6), 381–395 (1981)
17. Gajic, Z.: Linear Dynamic Systems and Signals. Prentice Hall, Upper Saddle River (2003)
18. Hough, P.V.C.: Machine analysis of bubble chamber pictures. In: International Conference on High Energy Accelerators and Instrumentation, pp. 554–556 (1959)
19. Kenmochi, Y., Buzer, L., Sugimoto, A., Shimizu, I.: Discrete plane segmentation and estimation from a point cloud using local geometric patterns. J. Autom. Comput. **5**(3), 246–256 (2008)
20. Kenmochi, Y., Talbot, H., Buzer, L.: Efficiently computing optimal consensus of digital line fitting. In: Proceedings of ICPR, pp. 1064–1067 (2010)
21. Köster, K., Spann, M.: An approach to robust clustering - application to range image segmentation. IEEE Trans. Pattern Anal. Mach. Intell. **25**(2), 430–444 (2000)
22. Musin, O.R.: Properties of the delaunay triangulation. In: Proceedings of the Thirteenth Annual Symposium on Computational Geometry (1997)
23. Ngo, P., Nasser, H., Debled-Rennesson, I.: A discrete approach for decomposing noisy digital contours into arcs and segments. In: Chen, C.-S., Lu, J., Ma, K.-K. (eds.) ACCV 2016. LNCS, vol. 10117, pp. 493–505. Springer, Cham (2017). https://doi.org/10.1007/978-3-319-54427-4_36
24. NIST/SEMATECH: e-Handbook of statistical methods (2012). http://www.itl.nist.gov/div898/handbook/
25. Phan, M.S., Kenmochi, Y., Sugimoto, A., Talbot, H., Andres, E., Zrour, R.: Efficient robust digital annulus fitting with bounded error. In: Gonzalez-Diaz, R., Jimenez, M.-J., Medrano, B. (eds.) DGCI 2013. LNCS, vol. 7749, pp. 253–264. Springer, Heidelberg (2013). https://doi.org/10.1007/978-3-642-37067-0_22

26. Raimund, S.: The upper bound theorem for polytopes: an easy proof of its asymptotic version. Comput. Geom. **5**, 115–116 (1995)
27. Rajan, V.T.: Optimality of the delaunay triangulation in R^d. Discrete Comput. Geom. **12**(1), 189–202 (1994)
28. Reveillès, J.P.: Géométrie discrète, calculs en nombre entiers et algorithmique. Thèse d'état. Université Louis Pasteur, Strasbourg (1991)
29. Rousseeuw, P.J.: Least median of squares regression. J. Am. Stat. Assoc. **79**(388), 871–880 (1984)
30. Searle, S.R., Gruber, M.H.: Linear Models, 2nd edn. Wiley, Hoboken (2016)
31. Shum, H., Ikeuchi, K., Reddy, R.: Principal component analysis with missing data and its application to polyhedral object modeling. IEEE Trans. Pattern Anal. Mach. Intell. **17**(9), 854–867 (1995)
32. Sivignon, I., Dupont, F., Chassery, J.M.: Decomposition of a three-dimensional discrete object surface into discrete plane pieces. Algorithmica **38**(1), 25–43 (2004)
33. Zitová, B., Flusser, J.: Image registration methods: a survey. Image Vis. Comput. **21**(11), 977–1000 (2003)
34. Zrour, R., et al.: Optimal consensus set for digital line and plane fitting. Int. J. Imaging Syst. Technol. **21**(1), 45–57 (2011)

Methods and Applications

Repairing Binary Images Through the 2D Diamond Grid

Lidija Čomić[1(✉)] and Paola Magillo[2]

[1] Faculty of Technical Sciences, University of Novi Sad, Novi Sad, Serbia
comic@uns.ac.rs
[2] DIBRIS, University of Genova, Genova, Italy
magillo@dibris.unige.it

Abstract. A 2D binary image is well-composed if it does not contain a 2×2 configuration of two diagonal black and two diagonal white squares. We propose a simple repairing algorithm to construct two well-composed images I_4 and I_8 starting from an image I, and we prove that I_4 and I_8 are homotopy equivalent to I with 4- and 8-adjacency, respectively. This is achieved by passing from the original square grid to another one, rotated by $\pi/4$, whose pixels correspond to the original pixels and to their vertices. The images I_4 and I_8 are double in size with respect to the image I. Experimental comparisons and applications are also shown.

Keywords: Digital topology · Well-composed images · Repairing 2D digital binary images

1 Introduction

In 2D, an image I in the square grid is well-composed if it does not contain blocks of 2×2 squares with alternating colors in chessboard configuration [2,3,18,19]. The process of transforming a given image into a well-composed one, that is in some sense similar to the original, is called *repairing*. Many image processing algorithms are simpler and faster when applied on well-composed images, making image repairing and study of different types of well-composedness a vivid research area [4–6,9,10].

Here, we address 2D image repairing by passing to another square grid, rotated by $\pi/4$ with respect to the original one and scaled by factor $1/\sqrt{2}$, in which each diamond (rotated square) corresponds either to a square or to a vertex in the original grid. In the rotated grid, we construct two well-composed images I_4 and I_8, homotopy equivalent to I with 4- and 8-adjacency, respectively.

The advantages of our approach are that the two repaired images are still in the square grid, so they can be processed with classical methods, in a simplified way thanks to well-composedness; we can choose between two types of adjacency in repairing; and the size of the resulting image in the diamond (rotated square) grid is just double that of the original one.

© Springer Nature Switzerland AG 2020
T. Lukić et al. (Eds.): IWCIA 2020, LNCS 12148, pp. 183–198, 2020.
https://doi.org/10.1007/978-3-030-51002-2_13

The contributions of this paper are:

- A simple repairing procedure, which results in two well-composed images I_4 and I_8, each of which is twice as large as the initial image I.
- A proof that the two repaired images I_4 and I_8 are well-composed and homotopy equivalent to the image I with 4- and 8-adjacency, respectively.
- Comparison with the state-of-the-art, showing the usefulness or our approach.

2   Background Notions

The square grid is the tessellation of the plane into Voronoi regions associated with points with integer coordinates. Each region (pixel) is a unit square, with sides parallel to the coordinate x and y axes [16,17]. (For a set S of points in the plane, the Voronoi region associated with a point p in S contains all points in the plane that are closer to p than to any other point in S [1,22].) Given a square P, the four squares sharing an edge with P are said to be *4-adjacent* to P; the eight squares sharing a vertex with P are said to be *8-adjacent* to P, and the four of them which are not 4-adjacent are called *strictly 8-adjacent*.

A 2D *binary digital image* I is a finite set of squares in the square grid. The squares in I are called *black* (object) squares. The squares in the complement I^c of I are called *white* (background). Connectedness is an equivalence relation obtained as the reflexive and transitive closure of adjacency. The classes of the image I with respect to α-connectedness are called connected α-components, for $\alpha \in \{4, 8\}$. Finite connected components of the complement of I are called *holes*. To maintain some similarity between the digital and continuous topology, the components and holes of I are defined with opposite types of adjacency.

A 2D image I has a *gap* at a vertex v if v is incident to two strictly 8-adjacent white squares (and two strictly 8-adjacent black ones) [7,8,11]. A gap-free image is called *well-composed* [18].

Depending on the chosen adjacency relation, two cell complexes can be naturally associated with a given image I. Recall that a cell complex Q is a collection of cells (homeomorphic images of the unit ball, that fit nicely together: the boundary of each cell and the intersection of any two cells (if nonempty) is composed of cells of lower dimension). In the plane, the underlying space $|Q|$ of a complex Q is the set of points in $\mathbb{R}^2$ that belong to some cell in Q.

For an image with 8-adjacency, the associated complex Q_8 is cubical and consists of the squares in I and all their edges and vertices. For an image with 4-adjacency, the associated polygonal complex Q_4 can be obtained from Q_8 by inserting a vertex at the center of each edge incident to a critical vertex, duplicating the critical vertices, and moving the two copies slightly towards the interior of the two incident black squares. The difference between the two complexes around a critical vertex is illustrated in Fig. 1.

A formal definition of homotopy [15] is out of the scope of this paper. Intuitively, if one shape can be continuously deformed into the other then the two shapes are homotopy equivalent. Two 2D homotopy equivalent shapes have the same number of connected components and holes, with the same containment relations among them.

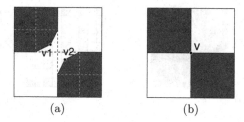

(a)   (b)

Fig. 1. Complexes Q_4 (a) and Q_8 (b) in the neighborhood of a critical vertex v. The transformation in (a) affects only one quarter of each pixel incident to v.

3   Related Work

Several repairing algorithms have been proposed in the literature, each with its features (regarding size, topology preservation, choice of adjacency and type of the output complex), as well as its benefits and drawbacks. We review briefly the repairing algorithm proposed for 2D images [24], and the 2D versions of the algorithms proposed for 3D images [12,25,26].

The method by Rosenfeld et al. [24] for 2D binary images uses a rescaling by factor 3 in both x and y directions of the square grid. In the rescaled grid, changing all black squares involved in a critical configuration to white squares, or vice versa, removes all critical configurations, without creating other ones (see Fig. 2). The repaired image is homotopy equivalent to the initial image with the appropriate adjacency relation.

The randomized algorithm by Siqueira et al. [25] iteratively changes white squares to black ones, until a well-composed image is obtained. The grid resolution of the repaired image is the same as that of the input image, but there is no guarantee that the topology (homotopy) of the input image is preserved.

The algorithm by Gonzalez-Diaz et al. [12] creates a polygonal well-composed complex homotopy equivalent to the input image with 8-adjacency by increasing the grid resolution four times in each coordinate direction, thickening the neighborhood of critical vertices and subdividing it into polygons. The idea of the method is illustrated in Fig. 3. Later [13], the rescaling factor was reduced to 2.

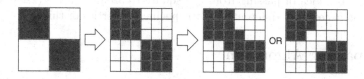

Fig. 2. Repairing process according to Rosenfeld et al. [24]. Each square becomes a block of 3×3 squares, which become black or white according to 8-adjacency or 4-adjacency (last two images).

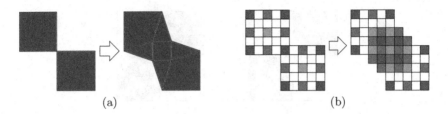

Fig. 3. Repairing process according to González-Díaz et al. [12]. The square complex becomes a polygonal complex, original squares become 2-cells of different shapes, depending on how many of their vertices are critical. (a) Geometry, (b) matrix representation, where blue, red and green squares correspond to 0-, 1- and 2-dimensional cells, as in [12]. (Color figure online)

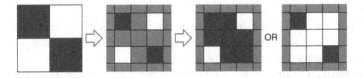

Fig. 4. Repairing process according to Stelldinger et al. [26]. Each square, edge and vertex becomes a square, which becomes black or white according to 8-adjacency or 4-adjacency (last two images).

The algorithm by Stelldinger et al. [26] increases the grid resolution twice in both coordinate directions, by creating an additional square for each edge and each vertex in the grid. If 4-adjacency is considered for I, the squares corresponding to the edges and vertices in Q are black only if all the incident squares in I are black. If 8-adjacency is considered, then the squares corresponding to the edges and vertices in the associated complex Q (edges and vertices incident to some black square in I) are also black (see Fig. 4).

The algorithms by Čomić and Magillo [9,10] repair 3D images by passing from the cubic to the body centered cubic (BCC) or the face centered cubic (FCC) grid, respectively. These grids are defined by the Voronoi tessellation of $\mathbb{R}^3$ associated with the centers and the vertices, or the centers and the midpoints of the edges of each unit cube. The 2D repairing algorithm proposed here uses the same basic idea of passing from the square to an alternative (2D diamond) grid, but is necessarily different, due to specific properties of this grid.

4   Repairing Algorithm

We propose to pass from the square to the 2D diamond grid, by extending the set $\mathbb{Z}^2$ of square centers with the set $(\mathbb{Z}+1/2)^2$ of square vertices and tessellating the plane into Voronoi regions associated with the extended set. Each region is a rotated square (a diamond) with sides parallel to the lines $x \pm y = 0$. We call even diamonds the ones associated with square centers, and odd diamonds the ones associated with square vertices.

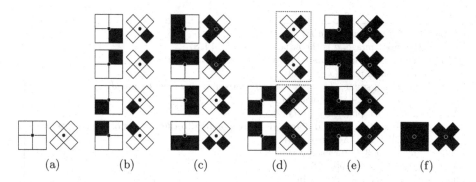

Fig. 5. All possible configurations at a vertex in the square grid, and corresponding repaired configurations in the diamond grid. The four incident squares of the vertex can be: (a) all white, (b) one black and three white, (c) and (d) two black and two white, (e) three black and one white, (f) all black. The diamond corresponding to the vertex is added in the two upper cases of (c), in (e), in (f) and, for the second version of the algorithm, in (d). Note that y-axis points downwards, as usual for images.

We create two well-composed images I_4 and I_8 in the diamond grid, homotopy equivalent to the image I with 4- and 8-adjacency, respectively. We include in the repaired images I_4 and I_8 all even diamonds corresponding to the black squares. We include also all odd diamonds corresponding to

1. the vertices incident to four black squares;
2. the vertices incident to three black and one white square;
3. the vertices incident to two vertically (horizontally) 4-adjacent black squares and two white ones, if the vertex is in the direction of the positive x axis (y axis) of the shared edge.

In the repaired image I_8, we include also the odd diamonds corresponding to

4. the critical vertices,

while we do not include those diamonds in the image I_4.

There are 16 different types of vertices in the square grid, depending on the configuration of the black and white incident squares. We show these configurations and the corresponding black diamonds in Fig. 5. Figure 6 presents the pseudocode of the repairing algorithm. Figure 7 shows the effect of the two versions of the repairing algorithm on a sample image.

5 Well-Composedness and Homotopy Equivalence

We show that our repairing algorithm produces well-composed images homotopy equivalent to the given image I with the chosen adjacency relation.

```
Repair(Image Q, int a, Image D)
// Q is the input image, a is the adjacency type ∈ {4, 8},
// D is the output image (rotated 90 degrees), given as all white
1    for each black pixel (x, y) in Q
2    // (x, y) is a square, (x − y, x + y) the corresponding even diamond
3        set (x − y, x + y) as black in D
4    for each black pixel (x, y) in Q
5    // v will contain the odd diamonds corresponding to the vertices of the square (x, y)
6        v[0..3] = Get4Adj(x − y, x + y)
7        for i = 0..3 // for each vertex
8            if MustFill(v[i], D, a)
9                set v[i] as black in D


MustFill(int x, int y, Image D, int a)
// (x, y) is an odd diamond, a is the adjacency type ∈ {4, 8},
// D is the output image where black even pixels have been set,
// the function checks if (x, y) must be filled
1    p[0..3] = Get4Adj(x, y) // even diamonds
2    c = the number of black diamonds among p[0..3]
3    if (c ≤ 1) return false
4    if (c ≥ 3) return true
5    // c = 2, check configuration
6    if (a = 4) // 4-adjacency
7        return (v[0] is black) and (v[2] is white)
8    // 8-adjacency
9    return (v[0] is black) or ((v[1] is black) and (v[3] is black))


Get4Adj(int x, int y)
// (x, y) is a diamond, the function returns its four 4-adjacent diamonds
1    return [(x, y − 1), (x + 1, y), (x, y + 1), (x − 1, y)]
```

Fig. 6. Pseudocode of the repairing algorithm and of auxiliary functions used in it.

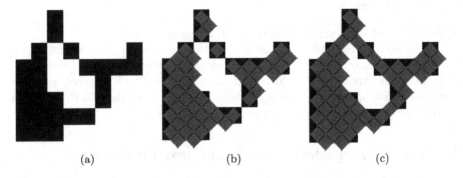

(a)   (b)   (c)

Fig. 7. (a) An image I in the square grid and the two repaired images I_4 (b) and I_8 (c) in the rotated grid (red diamonds) with the two versions of the algorithm. (Color figure online)

5.1   Well-Composedness

Proposition 1. *The two images I_4 and I_8 produced by our repairing algorithm are well-composed.*

Proof: A critical vertex in the diamond grid is incident to exactly two black diamonds that are both even or both odd. Our rules prevent the creation of either type of criticality:

- For each pair of 8-adjacent black even diamonds, corresponding to a pair of 4-adjacent squares in I, the odd diamond (4-adjacent to both) in the conventional direction, corresponding to a vertex incident to both squares in I, is in I_4 and I_8, thus preventing the creation of critical vertices incident to two even black diamonds.
- For each pair of 8-adjacent black odd diamonds, corresponding to two adjacent vertices (incident to the same edge e) in the square grid, the two even diamonds, corresponding to the two squares incident to e, cannot be both white. Since a filled vertex has always at least two incident black squares, the only possible configuration would be an array of 2×3 or 3×2 squares, where the two central ones are white, and the remaining four are black. In this case, thanks to the choice of the conventional direction, rule 3 of our algorithm would not fill both vertices.   □

5.2   Homotopy Equivalence

Proposition 2: *The spaces $|I_4|$ and $|I_8|$ are homotopy equivalent to the spaces $|Q_4|$ and $|Q_8|$, respectively.*

Proof: For $i \in \{4, 8\}$, we construct simplicial complexes ΣQ_i and ΣI_i that triangulate Q_i and I_i, respectively. Recall that a k-simplex is the convex hull of $k + 1$ affinely independent points. In 2D, simplexes are vertices, edges and triangles. A simplicial complex Σ is a finite set of simplexes, such that

- for each simplex in Σ, all its faces are in Σ and
- the intersection of two simplexes in Σ is either empty or composed of their common faces.

We show that $|Q_i|$ and $|I_i|$, $i \in \{4, 8\}$, are homotopy equivalent by constructing a sequence of collapses and expansions [27] that transform ΣQ_i to ΣI_i. In 2D, an elementary collapse removes from a simplicial complex Σ

- a triangle t and a (free) edge e, if t is the only triangle incident to e, or
- an edge e and a (free) vertex v, if e is the only edge incident to v.

Expansion is inverse to collapse, it introduces into Σ a pair of simplexes such that one is a free face of another. Both operations preserve the homotopy type of $|\Sigma|$ [27].

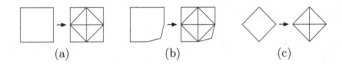

Fig. 8. Triangulation of (a) squares and (b) polygonal cells (for $i = 4$) in Q_i, and (c) of diamonds in I_i.

We triangulate Q_i by introducing a vertex at the midpoint of each edge, inscribing the corresponding even diamond in each square C in Q_i (for $i = 4$, in the corresponding polygonal cell if some vertex of C is critical), and connecting the center c of each diamond to the four diamond vertices. We triangulate I_i by introducing the center c of each diamond and connecting c to the four diamond vertices. The triangulations ΣQ_i and ΣI_i are illustrated in Fig. 8.

The two triangulations ΣQ_i and ΣI_i coincide on $|Q_i| \cap |I_i|$. We will transform ΣQ_i into ΣI_i through a process that first collapses extra triangles of ΣQ_i, and then creates the missing triangles through expansion.

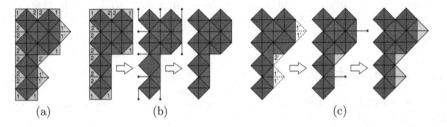

Fig. 9. (a) Superposition of the triangulations ΣQ_i and ΣI_i for an example without critical configurations (there is no difference between $i = 4$ and $i = 8$). Red triangles belong to both triangulations, yellow triangles belong to ΣQ_i only, and cyan triangles belong to ΣI_i only. Yellow and cyan triangles are labeled with their type. (b) Triangles of ΣQ_i, which are not in ΣI_i, are removed through collapse. (c) Triangles of ΣI_i, which are not in ΣQ_i, are created through expansion. (Color figure online)

We collapse the triangles in ΣQ_i that are outside ΣI_i. They are contained in odd diamonds centered at the vertices in I incident to

1. exactly one black square,
2. exactly two edge-adjacent black squares and are not in the conventional direction, or
3. (for $i = 4$) exactly one of the two polygonal 2-cells corresponding to the two strictly vertex-adjacent black squares (these are the two copies of the critical vertices).

For each triangle t of type 1 or 3, we collapse t with one of its two free edges, and we collapse the remaining edge with its unique free vertex. Triangles of type 2 come in pairs. We collapse the two triangles, each with its unique free edge, and we collapse their shared edge with its unique free vertex. This stage is illustrated in Figs. 9 (b) and 10 (b), (d).

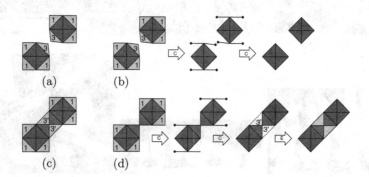

Fig. 10. (a) Triangle types of ΣQ_4 and ΣI_4 for a critical configuration of 2×2 pixels. (b) Transformation of ΣQ_4 into ΣI_4. (c) Triangle types of ΣQ_8 and ΣI_8 for the same configuration. (d) Transformation of ΣQ_8 into ΣI_8. In (b) and (d), arrows with letters C and E denote collapsing of extra triangles and expansion to create missing triangles. In (b) expansion is not needed.

We expand the triangles in ΣI_i outside of ΣQ_i. They are contained in odd diamonds centered at the vertices in I incident to

1'. exactly two edge-adjacent black squares and are in the conventional direction,
2'. exactly three black squares, or
3'. (for $i = 8$) exactly two strictly vertex-adjacent black squares (critical vertices).

We expand each triangle t of type 2' or 3' by adding t together with its unique free edge. Triangles of type 1' come in pairs. We expand first their shared edge together with its unique free vertex, and we expand the two triangles, each with its unique free edge. This stage is illustrated in Figs. 9 (c) and 10 (d).

6 Experimental Comparisons and Results

We implemented our image repairing algorithm and, for comparison purposes, the algorithm in [26]. We have chosen this algorithm as our competitor because it is the one using the least additional memory among those which are able to preserve image homotopy and which produce an image in the square grid. Both algorithms have been implemented in the two versions, i.e., producing a repaired image homotopy equivalent to the given one with both 4- and 8-adjacency.

We tested the algorithms on several gray-scale images from the pixabay repository (https://pixabay.com/en/photos/grayscale/) after converting them

acid (960 × 625)   birch (960 × 639)   car (960 × 602)

chess (960 × 560)   fog (960 × 353)   hands (889 × 720)

Fig. 11. Our binary versions of original images, with the size of bounding box.

to binary images by applying a threshold equal to 128 (where gray values are from 0 to 255). The used images are shown in Fig. 11. In order to analyze a possible dependency of the results from image resolution, for each image we produced two lower resolutions by resizing it to 1/2 and 1/4 of the original size (in both x and y direction).

All algorithms are implemented in C language and executed on a PC equipped with an Intel CPU i7-2600K CPU at 3.4 GHz with 32 Gbyte of RAM.

Table 1 shows the sizes of the input images, the number of critical configurations in them, the sizes of the repaired images, and the execution times. As expected, our repaired image has half the size of the one produced by our competitor [26]. The time taken by our algorithm is from 67% to 75% that of the competitor algorithm in the case of repairing with 8-adjacency, and from 55% to 62% with 4-adjacency.

Figure 12 shows a repaired image. From the point of quality, the result produced by our competitor [26] is lighter than the original with 4-adjacency and darker with 8-adjacency. We analyze this behavior in detail on toy inputs in Fig. 13. The input images have black and white lines of the same width. This property is maintained in the images repaired by our algorithm. The competitor algorithm [26], instead, shrinks black lines and expands white lines with 4-adjacency, and does the opposite with 8-adjacency. This may be a problem when preserving the area of black zones is important.

Getting back to Fig. 12, the original image has 72964 black pixels. Our repaired images have 142434 and 150106 black pixels, and the pixel size is 1/2 of the original one. Thus, the areas of black zones in our repaired images are 71217 (97.6% of the original) and 75053 (102.9% of the original), respectively.

The difference in areas with respect to the original image is below 3%. Images repaired with our competitor [26] have 198640 and 384452 black pixels, and the size of each pixel is 1/4 of the original one. Thus, the areas of black zones in the repaired images are 49660 (68.0% of the original) and 96113 (131.7% of the original), respectively. Here, the difference with respect to the original is more than 30%.

On the other side, our algorithm tends to smooth right angles, as we see in Fig. 13 (b), and is not completely symmetric in the four cardinal directions. Another evident feature of our results is rotation. This is a problem when the image is seen by human users (where well-composedness is not relevant), while it does not affect computations made on images (where well-composedness may be important).

As the diamond grid is a (rotated) square grid, all classical image processing algorithms can be applied to the two repaired images. Moreover, some algorithms are simpler on well-composed images. Of course, execution time increases with image size. In the following, we analyze the impact of the size of the repaired image on the performance of image processing algorithms, since our repaired images are half in size, with respect to the ones repaired by our best competitor [26] (as we have just shown). As meaningful examples, we consider contour extraction and shrinking, i.e., a very simple and a rather complex task.

Table 1. The table shows: number of black squares (size) of input images and of output repaired images obtained by our method (our) and by the competitor one in [26] (comp.); number of critical vertices in the input images; execution times (in milliseconds); ratio between running times of our method and of the competitor one [26]. Images have been repaired according to 4- and 8-adjacency.

Image	Input image		4-adjacency					8-adjacency				
		Critical	Our		Comp. [26]		Time	Our		Comp. [26]		Time
	Size	Vertices	Size	Time	Size	Time	%	Size	Time	Size	Time	%
Acid 1	234K	1308	466K	122	870K	204	60	467K	122	999K	167	73
Acid 1/2	58K	387	115K	31	208K	51	61	115K	31	253K	42	74
Acid 1/4	14K	218	28K	7	48K	13	54	28K	7	65K	10	70
Birch 1	340K	22K	678K	206	1038K	330	62	700K	203	1665K	274	74
Birch 1/2	88K	4485	176K	51	288K	83	60	181K	50	414K	68	74
Birch 1/4	23K	574	46K	12	79K	21	57	47K	12	105K	17	71
Car 1	448K	2239	896K	230	1697K	386	59	898K	229	1880K	316	72
Car 1/2	112K	540	225K	57	424K	97	59	225K	57	473K	79	72
Car 1/4	28K	103	56K	14	106K	24	58	56K	14	119K	19	74
Chess 1	213K	180	425K	105	834K	179	59	426K	105	867K	146	72
Chess 1/2	53K	29	106K	26	205K	45	58	106K	26	219K	36	72
Chess 1/4	13K	6	26K	6	50K	11	55	26K	6	56K	9	67
Fog 1	196K	700	391K	98	758K	166	58	392K	97	805K	135	72
Fog 1/2	49K	146	98K	24	190K	41	59	98K	24	202K	34	71
Fog 1/4	12K	22	25K	6	47K	10	60	25K	6	51K	8	67
Hands 1	422K	1030	845K	209	1659K	356	59	846K	209	1720K	290	72
Hands 1/2	106K	280	212K	52	414K	89	58	212K	52	433K	73	71
Hands 1/4	27K	33	53K	13	103K	22	59	53K	13	110K	18	72

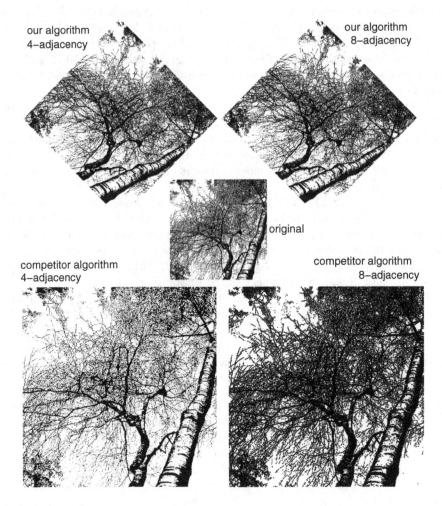

Fig. 12. A portion of test image "birch" with its repaired versions.

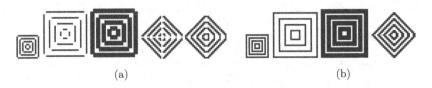

Fig. 13. Two toy images and their repaired versions. From left to right: input image, output images by competitor algorithm [26] with 4- and 8-adjacency, output images by our algorithm with 4- and 8-adjacency. Image (b) is well-composed and our algorithm gives the same output with both adjacencies.

Finding the contours of an image means finding sequences of black border squares bounding each connected component and each hole of the image I. Connected components and holes can be considered with either 4- or 8-adjacency. For a well-composed image, the adjacency type makes no difference, and 4-connected contours can be extracted: each contour is a circular list of black squares where each square is 8-adjacent to at least one white square, and 4-adjacent to the previous square in the list. We implemented the extraction of 4-connected contours on well-composed images by means of the classical contour following approach [14,21]. We have a current border square at each step, and we decide how to extend the contour based on the configuration (black or white) of the eight squares adjacent to it. We adopt a compressed representation of the contour, from [20], that is, we do not record all squares of the contour, but just the ones corresponding to the corners. This saves space and allows easy rescaling of the contour when the image is enlarged or shrunk.

Table 2 (a) compares the times for contour extraction from images repaired with our algorithms and with the competitor one [26]. As image sizes are halved, execution times on our repaired images are almost halved, with more gain for 8-adjacency. The ratio is from 0.44 to 0.55 for 8-adjacency and from 0.5 to 0.75 for 4-adjacency, and there is no specific trend with image resolution.

The aim of shrinking is to reduce each connected component C of an image I to a single pixel if C is homotopy equivalent to a disk, to a simple closed chain of pixels if C is homotopy equivalent to a circle, etc. This is commonly achieved by iteratively removing squares from the image, i.e., changing their status from black to white. Squares can only be removed if their removal does not change either the number of components or the number of holes according to the chosen (4- or 8-) adjacency. The process is iterated until no more squares may be removed. We consider shrinking of a well-composed image, with 4-adjacency. The decision whether a square is removable or not only depends on the status of the eight squares in its neighborhood (see [16,23] for details and algorithms). Squares with disjoint neighborhoods do not affect the removability of each other. In the method we implemented, black squares are classified into four subsets according to the parity of their (x, y) coordinates. The algorithm performs a cycle on the four subsets, examining one of them at a time. Removable squares of the current subset do not interfere and are removed together. Table 2 (b) compares the times for shrinking images repaired with our algorithms and with the competitor one [26]. Here running time also depends heavily on image characteristics, besides input size, but we can see that execution times are roughly halved also in this case. Ratios are a bit smaller than for contour extraction, ranging from 0.34 to 0.5 with 8-adjacency and from 0.32 to 0.6 with 4-adjacency, and present no specific trend with resolution.

Table 2. Running times for (a) contour extraction and (b) shrinking, applied on images repaired by our approach and by the competitor one [26] (in milliseconds). From top to bottom: full resolution, scaled 1/2, and scaled 1/4.

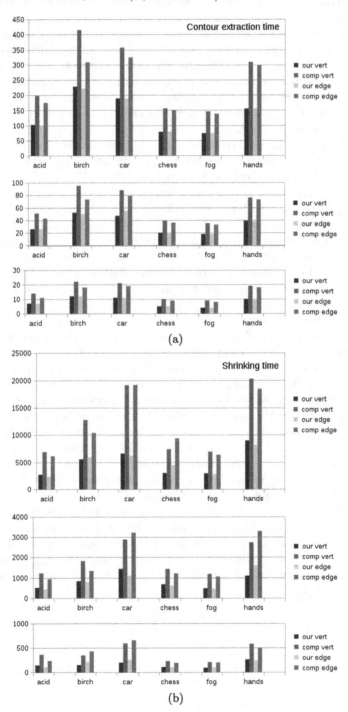

7   Summary

We have proposed a simple method to transform a given 2D binary image into two homotopy equivalent well-composed images (depending on the chosen adjacency) by using the 2D diamond grid. The resulting images are just double in size with respect to the original one, and this improves the best existing homotopy preserving method [26], where the image size is doubled in both coordinate directions (leading to a factor 4). The diamond grid is a rotated square grid, thus all known image processing algorithms can be applied to our repaired images, and their smaller size, with respect to the state-of-the-art of homotopy equivalent image repairing, saves processing time.

Acknowledgments. This work has been partially supported by the Ministry of Education and Science of the Republic of Serbia within the Project No. 34014.

References

1. de Berg, M., Cheong, O., van Kreveld, M.J., Overmars, M.H.: Computational Geometry: Algorithms and Applications, 3rd edn. Springer, Heidelberg (2008). https://doi.org/10.1007/978-3-540-77974-2
2. Boutry, N., Géraud, T., Najman, L.: How to make nD images well-composed without interpolation. In: 2015 IEEE International Conference on Image Processing, ICIP 2015, pp. 2149–2153 (2015)
3. Boutry, N., Géraud, T., Najman, L.: A tutorial on well-composedness. J. Math. Imaging Vis. **60**(3), 443–478 (2017). https://doi.org/10.1007/s10851-017-0769-6
4. Boutry, N., Géraud, T., Najman, L.: How to make n-D plain maps defined on discrete surfaces Alexandrov-well-composed in a self-dual way. J. Math. Imaging Vis. **61**, 849–873 (2019). https://doi.org/10.1007/s10851-019-00873-4
5. Boutry, N., González-Díaz, R., Jiménez, M.-J.: One more step towards well-composedness of cell complexes over nD pictures. In: Couprie, M., Cousty, J., Kenmochi, Y., Mustafa, N. (eds.) DGCI 2019. LNCS, vol. 11414, pp. 101–114. Springer, Cham (2019). https://doi.org/10.1007/978-3-030-14085-4_9
6. Boutry, N., González-Díaz, R., Jiménez, M.-J.: Weakly well-composed cell complexes over nD pictures. Inf. Sci. **499**, 62–83 (2019)
7. Brimkov, V.E., Maimone, A., Nordo, G., Barneva, R.P., Klette, R.: The number of gaps in binary pictures. In: Bebis, G., Boyle, R., Koracin, D., Parvin, B. (eds.) ISVC 2005. LNCS, vol. 3804, pp. 35–42. Springer, Heidelberg (2005). https://doi.org/10.1007/11595755_5
8. Brimkov, V.E., Moroni, D., Barneva, R.: Combinatorial relations for digital pictures. In: Kuba, A., Nyúl, L.G., Palágyi, K. (eds.) DGCI 2006. LNCS, vol. 4245, pp. 189–198. Springer, Heidelberg (2006). https://doi.org/10.1007/11907350_16
9. Čomić, L., Magillo, P.: Repairing 3D binary images using the BCC grid with a 4-valued combinatorial coordinate system. Inf. Sci. **499**, 47–61 (2019)
10. Čomić, L., Magillo, P.: Repairing 3D binary images using the FCC grid. J. Math. Imaging Vis. **61**(9), 1301–1321 (2019). https://doi.org/10.1007/s10851-019-00904-0

11. Françon, J., Schramm, J.-M., Tajine, M.: Recognizing arithmetic straight lines and planes. In: Miguet, S., Montanvert, A., Ubéda, S. (eds.) DGCI 1996. LNCS, vol. 1176, pp. 139–150. Springer, Heidelberg (1996). https://doi.org/10.1007/3-540-62005-2_12

12. González-Díaz, R., Jiménez, M.-J., Medrano, B.: 3D well-composed polyhedral complexes. Discrete Appl. Math. **183**, 59–77 (2015)

13. González-Díaz, R., Jiménez, M.-J., Medrano, B.: Efficiently storing well-composed polyhedral complexes computed over 3D binary images. J. Math. Imaging Vis. **59**(1), 106–122 (2017). https://doi.org/10.1007/s10851-017-0722-8

14. Gose, S.J.E., Johnsonbaugh, R.: Pattern Recognition and Image Analysis. Prentice-Hall Inc., Upper Saddle River (1996)

15. Hatcher, A.: Algebraic Topology. Cambridge University Press, Cambridge (2001)

16. Klette, R., Rosenfeld, A.: Digital Geometry. Geometric Methods for Digital Picture Analysis. Morgan Kaufmann Publishers, San Francisco, Amsterdam (2004)

17. Kong, T.Y., Rosenfeld, A.: Digital topology: introduction and survey. Comput. Vis. Graph.- Image Process. **48**(3), 357–393 (1989)

18. Latecki, L.J.: 3D well-composed pictures. CVGIP: Graph. Model Image Process. **59**(3), 164–172 (1997)

19. Latecki, L.J., Eckhardt, U., Rosenfeld, A.: Well-composed sets. Comput. Vis. Image Underst. **61**(1), 70–83 (1995)

20. Miyatake, T., Matsushima, H., Ejiri, M.: Contour representation of binary images using run-type direction codes. Mach. Vis. Appl. **9**, 193–200 (1997)

21. Pavlidis, T.: Algorithms for Graphics and Image Processing. Computer Science Press, Boca Raton (1982)

22. Preparata, F.P., Shamos, M.I.: Computational Geometry - An Introduction. MCS. Springer, New York (1985). https://doi.org/10.1007/978-1-4612-1098-6

23. Preston Jr., K., Duff, M.J.B.: Modern Cellular Automata. AAPR. Springer, Boston (1984). https://doi.org/10.1007/978-1-4899-0393-8

24. Rosenfeld, A., Kong, T.Y., Nakamura, A.: Topology-preserving deformations of two-valued digital pictures. Graph. Models Image Process. **60**(1), 24–34 (1998)

25. Siqueira, M., Latecki, L.J., Tustison, N.J., Gallier, J.H., Gee, J.C.: Topological repairing of 3D digital images. J. Math. Imaging Vis. **30**(3), 249–274 (2008)

26. Stelldinger, P., Latecki, L.J., Siqueira, M.: Topological equivalence between a 3D object and the reconstruction of its digital image. IEEE Trans. Pattern Anal. Mach. Intell. **29**(1), 126–140 (2007)

27. Whitehead, J.H.C.: Simplicial spaces, nuclei and m-groups. Proc. London Math. Soc. **45**, 243–327 (1938)

Transmission Based Adaptive Automatic Tube Voltage Selection for Computed Tomography

Gábor Lékó[(✉)] and Péter Balázs

Department of Image Processing and Computer Graphics, University of Szeged,
Árpád tér 2, Szeged 6720, Hungary
{leko,pbalazs}@inf.u-szeged.hu

Abstract. Computed Tomography (CT) is a widely used x-ray based imaging modality in radiodiagnostics and non-destructive testing. For medical considerations as well as for economical and environmental reasons, it is desirable to reduce the dose of radiation and to optimize energy used for acquiring x-ray projection images. Especially, in case of elongated objects, using a constant energy spectrum radiation may not provide realistic information about the interior of the examined object. In this paper we provide an adaptive tube voltage selection method, which determines the proper amount of radiation energy on-the-fly, during the acquisition, based on projection information. By experiments on software phantom images we show that this adaptive approach can produce CT images of better quality, and in the same time, by consuming less energy.

Keywords: Tube voltage selection · Online method · Non-destructive testing · Radiation dose optimization

1 Introduction

In Computed Tomography (CT) [3], x-ray radiation is used to produce the projections of an object. Gathering these projections from different angles one can reconstruct the interior of the subject of investigation. CT is one of the most frequently used modalities in radiodiagnostics and non-destructive testing (NDT). However, undergoing multiple high radiation dose examinations may cause adverse health effects, even lethal diseases in human body [10]. Besides, high radiation costs high energy. In both human studies and NDT, it is a reasonable goal to use as low energy and radiation as possible for the acquisition of projections, while still preserving the required reconstruction quality. When using a constant energy spectrum, the investigation of elongated objects may be problematic. Using too low energy, the photons cannot penetrate the object in the direction of elongation. On the other hand, too high energy can cause the photons to traverse the object without significant attenuation, in the direction perpendicular to the elongation of the object. A solution could be to use different energy levels when producing projections from different directions.

© Springer Nature Switzerland AG 2020
T. Lukić et al. (Eds.): IWCIA 2020, LNCS 12148, pp. 199–208, 2020.
https://doi.org/10.1007/978-3-030-51002-2_14

Automatic Tube Current Modulation (ATCM) and Automatic Tube Voltage Selection (ATVS) are well-researched areas. In most of the cases, the CT scanners contain intelligent scanning technologies, which can automatically recommend tube current and voltage settings to provide the lowest radiation dose and a high image quality. For this purpose, widely used scanners are CARE kV and CARE Dose4D from Siemens Medical Solutions, Forcheim, Germany [7], and the scanners of GE Healthcare, with the previously mentioned similar embedded technologies. Papers [1,2,8,19] show examples for using ATCM and/or ATVS combining with novel iterative reconstruction techniques to achieve good image quality with low dose.

In [9], a detailed list is given about the CT parameters that influence the radiation dose. There, the author claims that variations in tube potential should not be considered for pure dose reduction purposes except in the case of CT angiography. However, numerous articles investigate the effects of using only low tube voltage level or ATVS for different radiation tasks (see, e.g., [16–18,20,21]). They prove that optimizing tube voltage itself can reduce the radiation dose and it can increase the image contrast for structures with a high effective atomic number, such as calcium and iodine. Moreover, with low voltage level more energy efficient acquisitions could be performed. From the viewpoint of lifetime of the x-ray tube, it is also beneficial to use lower energies.

Several studies focus on CT parameter optimization using different quality measures like transmission, linearity, signal-to-noise ratio, contrast-to-noise ratio, sharpness, etc [6,14,15]. Using the proper combination of these measures could lead to even better optimized CT parameters.

The aim of this paper is to provide an adaptive tube voltage selection method. In this context "adaptive" means that the tube voltage will be re-selected during the acquisition, projection-by-projection. To our knowledge, this is the first approach that attempts to select tube voltage on-the-fly based on projection (sinogram) information. This approach could be effectively used, especially in case of elongated objects.

The structure of the paper is the following. In Sect. 2 we describe our proposed voltage selection method. In Sect. 3 we give details about the experimental frameset, while in Sect. 4 we present the experimental results. Finally, Sect. 5 is for the conclusions.

2   Proposed Method

First, we need a figure of merit to quantitatively evaluate the radiation used, more precisely, the projection images produced. For that aim we use *transmission*, which describes the ratio between the minimal and the maximal x-ray intensity at the detector. CT standards, which are still in development, suggest transmission values of 14% (ISO 15708) or 10% (prEN 16016) at the path of highest absorption for optimal scan quality. However, depending on the scanning and reconstruction problem, optimal transmission can be even lower or higher. In our experiments, 20% proved to be a good choice.

Our proposed voltage selection method is described in Algorithm 1. In the beginning, we define a couple of parameters. The optimal transmission is 0.2 (20%) as we mentioned before. Furthermore, a tolerance value is associated with this optimum. By empirical investigations we set this value to 0.25, meaning that the transmission is still acceptable if it is not lower than $0.2 \cdot 0.75$ and not greater than $0.2 \cdot 1.25$. We also need to define tube voltage levels and an angle set for the acquisitions (see more details in the experimental section). As a first part (Steps 2–7 of Algorithm 1), we acquire the first projection on all the different energy levels. Then the energy level with the closest transmission value to the optimum will be chosen for further acquisitions. In the second part (Steps 9–19 of Algorithm 1), we acquire a projection on the previously chosen energy level. If its transmission value is in a tolerance distance to the optimum then we stay on the current energy level. Otherwise, if it is out of the acceptable range then we step to a higher/lower energy level for the next projection. This second part is repeated until the predefined number of projections is reached. This way, based on the shape of the object, the resulting sinogram may contain projections from different energy levels. Thus, one can collect the most informative projections (energy levels) belonging to the given angles. We will refer to the sinogram obtained this way as the *adaptive sinogram*.

Algorithm 1: Voltage selection

Require: $\mathbf{tr_{opt}}$ - optimal transmission; $\mathbf{tr_{tol}}$ - tolerance for the transmission level;
  TVL - set of predefined tube voltage levels; **A** - set of predefined angles

1: $\mathbf{tr_{opt}} \leftarrow 0.2$, $\mathbf{tr_{tol}} \leftarrow 0.25$
2: **for all** $lvl \in$ **TVL do**
3:    acquire projection from starting angle on lvl
4:    $closest_{lvl} \leftarrow lvl$ with the closest transmission value to $\mathbf{tr_{opt}}$
5:    $closest_{pr} \leftarrow$ projection acquired on $closest_{lvl}$
6:    $closest_{tr} \leftarrow$ transmission value of $closest_{pr}$
7: **end for**
8: $sinogram_{opt} \leftarrow sinogram_{opt} \cup closest_{pr}$
9: **for all** angle $\theta \in$ **A do**
10:    $actual_{pr} \leftarrow$ acquire projection from angle θ on $closest_{lvl}$
11:    $sinogram_{opt} \leftarrow sinogram_{opt} \cup actual_{pr}$
12:    **if** $|\mathbf{tr_{opt}} - closest_{tr}| \geq \mathbf{tr_{opt}} \cdot \mathbf{tr_{tol}}$ **then**
13:       **if** $\mathbf{tr_{opt}} - closest_{tr} < 0$ **then**
14:          $closest_{lvl} \leftarrow$ decrease $closest_{lvl}$ to the closest lower level in **TVL**
15:       **else**
16:          $closest_{lvl} \leftarrow$ increase $closest_{lvl}$ to the closest higher level in **TVL**
17:       **end if**
18:    **end if**
19: **end for**

It would be ideal if the sum of intensities at the detectors were equal for all individual projections. However, since we are using different energy levels for different projections, this is not always the case. This can be an issue, during the reconstruction, since reconstruction algorithms typically assume a constant energy spectrum for acquiring the projections. To resolve this issue, in Algorithm 2, we propose a sinogram equalization method. In the pseudo code, $\odot$ and $\oslash$ stands for the pointwise product and division, respectively. First, we create a mask, which covers the region(s) of the sinogram where the shadow of the object is present (Step 1). Then, we take the sum of the pixels covered by the mask and determine its maximum value (Steps 2–3). A scaling vector is created by taking the ratio between the maximum value and the previous pixel sum (Step 4). Finally, the scaling vector is used to re-scale the adaptive sinogram (Step 5). After applying Algorithm 2, the sum of the intensities will be close to each other, in all projections, and thus, with this approach one can approximate the ideal case. We will refer to the sinogram created by Algorithm 2 as the *corrected sinogram*.

Algorithm 2: Sinogram equalization

Require: sinogram$_{adapt}$
1: $sinogram_{bw} \leftarrow$ thresholded **sinogram$_{adapt}$**
2: $vector_{obj} \leftarrow$ column sum of $(sinogram_{opt} \odot sinogram_{bw})$
3: $vector_{max} \leftarrow$ column maximums of $vector_{obj}$
4: $vector_{scale} \leftarrow vector_{max} \oslash vector_{obj}$
5: $sinogram_{corr} \leftarrow$ **sinogram$_{adapt}$** $\odot$ $vector_{scale}$

---

3 Test Data and Experimental Settings

To conduct experiments, we used the GATE simulation software [4,5], an advanced open source software developed by the international OpenGATE collaboration and dedicated to numerical simulations in medical imaging and radiotherapy. During the simulations, we used polychromatic source. To simulate a perfect energy spectrum, we used Spekcalc [11–13] which is an executable for calculating the x-ray emission spectra from tungsten anodes such as those used in diagnostic radiology and kV radiotherapy x-ray tubes. In our experiments, the difference between the peak and minimum energy was set to 30 keV (the default of the software) for each energy level, except on the ones, where the peak energy was too low to set the minimum to be 30 keV smaller.

We studied 6 different phantom models shown in Fig. 1. P1 is a perfect sphere on which we show how the transmission behaves if an object is not elongated, P2–P5 are elongated binary objects and P6 is a self-made approximation of the well-known Forbild hip phantom[1]. For the reconstructions, we took the middle slices of these objects (see Fig. 4).

During the simulations in GATE, parallel geometry was assumed. The length of the simulated detector line was 72 mm and its height was 0.1 mm. The width

---

[1] http://www.imp.uni-erlangen.de/phantoms/index.htm.

of one single detector was set to 0.4 mm. This ensures a precision of 0.4 mm on the x-axis and 0.1 mm on the y-axis. Equivalently, it means 180 projection rays in each direction which suitably covers images of size 128 × 128, in each direction. We collected 285 projections on the 0°–179° interval, based on the Nyquist–Shannon sampling theorem. The rotation time was 1 s/projection. In GATE, the tube current cannot be controlled, only the number of photons to emit, which was set to 200000 in our case. A simulation could take a long time depending on the number of the simulated photons. We chose the value 200000 based on empirical tests, it provided an acceptable consensus between running time and quality. After acquiring the projections, the reconstructions were performed using the Filtered Backprojection (FBP) technique, implemented in C++, using the CUDA sdk[2] with GPU acceleration.

The quality of the reconstructions was evaluated by SNR (signal-to-noise ratio). The SNR is given as the ratio of the mean gray value inside the object μ_o, and its standard deviation σ_o:

$$SNR = \frac{\mu_o}{\sigma_o}. \tag{1}$$

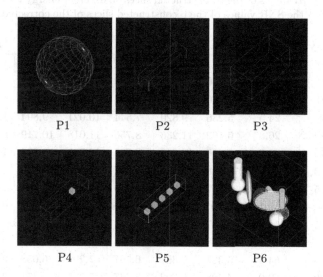

Fig. 1. The 3D models of the used phantoms. P1: a sphere, P2: an elongated cuboid, P3: a wider cuboid, P4: a P2 sized cuboid with one small hole in it, P5: a P2 sized cuboid with five small (equal sized) holes in it, P6: an approximation of the Forbild hip phantom.

---

[2] https://www.developer.nvidia.com/cuda-zone.

4   Results

We predefined 15 different energy levels. Based on their polychromatic spectra, their mean energies could also be determined. In Table 1 we present the SNR values of the reconstructed images belonging to sinograms of the 15 different energy levels. To ensure a fair and realistic comparison with the corrected sinogram later, Algorithm 2 was executed also on these 15 sinograms. The highest (best) SNR values are highlighted in every row to indicate which energy levels provided the best quality. We will refer to the sinograms of these reconstructions as the *original best sinograms*. The last row shows the SNR values of the reconstructed images belonging to the corrected sinograms (created using Algorithms 1 and 2). In the case of P1, one can see that the corrected and the original best SNRs are the same, since the sphere is the perfect opposite of an elongated object. It means that tube voltage selection is unnecessary in this case. The corrected SNRs of P2–P5 are better than that of the optimal best reconstructions. Taking P6 under examination, the SNR of the corrected reconstruction is almost as good as that the original best sinogram can provide.

Table 1. The SNR values of the reconstructed slices on different energy levels [keV] and (in the last row) the SNR values of the reconstructed slices of the corrected sinograms.

Energy	Mean E.	P1	P2	P3	P4	P5	P6
2–20	15.7	1.874	3.204	2.604	3.309	3.593	1.074
3–25	19.3	3.156	4.900	4.080	5.033	5.373	1.095
3–30	22.1	4.550	8.880	6.782	8.697	9.392	1.110
5–35	24.5	5.236	9.820	7.544	10.021	9.894	1.119
10–40	26.8	6.162	11.236	8.756	11.018	10.719	**1.124**
15–45	28.8	6.310	10.648	8.796	10.335	9.924	1.119
20–50	31.7	6.548	11.149	9.242	10.889	10.689	1.042
25–55	35.6	7.357	12.728	10.290	12.376	**11.412**	0.957
30–60	40.1	**7.797**	**13.237**	**10.598**	**12.875**	11.051	0.883
35–65	44.9	7.157	10.621	9.634	10.298	9.417	0.803
40–70	49.9	6.178	8.896	7.994	8.835	7.872	0.722
45–75	54.9	5.321	7.387	6.567	7.291	6.658	0.632
50–80	59.9	4.408	5.992	5.337	5.987	5.451	0.540
55–85	64.4	3.742	5.125	4.414	5.073	4.618	0.462
60–90	60.3	3.049	4.082	3.481	4.023	3.640	0.378
Corrected	Corrected	7.797	14.175	10.729	13.757	11.610	1.111

In Fig. 2, we present the transmission values of P2 on the different energy levels. The *Opt. trans.* line shows where the transmission should be for an optimal sinogram. In this case, Algorithm 1 chose the 9th energy level (30–60 keV) first, then it decreased to the 5th energy level (10–40 keV), and then increased back to the 9th. In the case of P1, this diagram would contain almost straight lines. Figure 3 shows the difference between the sinograms of P2. Figure 3a is the original best of P2 acquired on the energy level highlighted in Table 1, Fig. 3b is the adaptive sinogram created with Algorithm 1, and Fig. 3c is the corrected sinogram after applying Algorithm 2 on Fig. 3b. The reconstructed phantoms can be seen in Fig. 4, using the corrected sinograms.

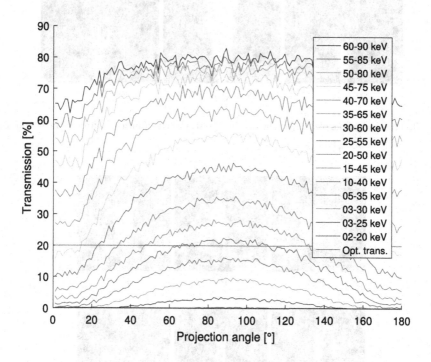

Fig. 2. The transmission values of the scan of P2.

Beside the slightly improved SNR values, another advance of the adaptive voltage selection strategy is lower energy consumption, based on the fact that from certain angles even a lower energy level could ensure optimal transmission. In the first and second row of Table 2, one can see the cumulated energies for creating the original best and the adaptive (and corrected) sinogram, respectively. Note that sinogram correction is simply achieved by executing Algorithm 2, i.e., it does not affect cumulated energy. As we mentioned before, P1 is the perfect opposite of an elongated object, therefore, the cumulated energy is the same in both cases. In all the other cases, the used energy for acquiring the adaptive (corrected) sinograms is lower than it is needed for the original best sinograms. The saved energy in percentage is shown in the third row. A significant energy reduction can be observed.

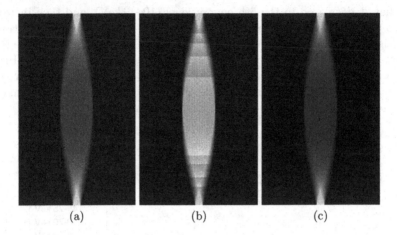

(a)   (b)   (c)

Fig. 3. The sinograms of P2. (a) the original best, (b) the adaptive and (c) the corrected one.

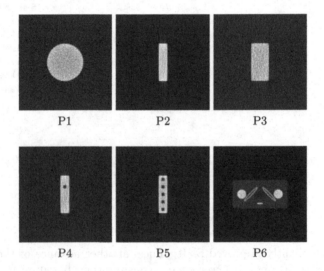

Fig. 4. The reconstructed middle slices of the phantoms.

Table 2. The cumulated energies needed for the original best and the corrected sinograms, and the percentage of the saved energies, respectively.

	P1	**P2**	**P3**	**P4**	**P5**	**P6**
Orig. Best [keV]	11428.5	11428.5	11428.5	11428.5	10146	7638
Corrected [keV]	11428.5	8930.4	10395	8930.4	8776.7	6906.1
Saved energy [%]	0	21.86	9.04	21.86	13.5	9.58

5   Conclusions

In this paper, we proposed an adaptive Automatic Tube Voltage Selection method complemented with a sinogram equalization algorithm for Computed Tomography. The method adjusts the tube voltage on-the-fly during the acquisition based on the transmission value of the previous projection. This method relies only on the projection data and does not require any further information. After producing the adaptive sinogram, an equalization is performed. By experimental tests on software simulated phantoms using the GATE toolbox, we found that our method is able to produce optimized sinograms and by that ensuring lower energy consumption and slightly better reconstruction quality. The presented method could be utilized in industrial non-destructive testing and it has potential even in medical cases.

Acknowledgements. Gábor Lékó was supported by the UNKP-19-3 New National Excellence Program of the Ministry of Human Capacities. This research was supported by the project "Integrated program for training new generation of scientists in the fields of computer science", no. EFOP-3.6.3-VEKOP16-2017-00002. This research was supported by grant TUDFO/47138-1/2019-ITM of the Ministry for Innovation and Technology, Hungary. The authors would also like to thank Csaba Olasz for his help in using GATE Simulation Software, László G. Varga for providing the reconstruction toolbox for the experimental tests and Gábor Petrovszki for conducting preliminary experiments on hardware phantoms to investigate the effect of tube voltage selection.

References

1. Chen, J.H., Jin, E.H., He, W., Zhao, L.Q.: Combining automatic tube current modulation with adaptive statistical iterative reconstruction for low-dose chest CT screening. PloS one **9**, e92414 (2014)
2. Gervaise, A., Naulet, P., Beuret, F., Henry, C., Pernin, M., Portron, Y., Lapierre-Combes, M.: Low-dose CT with automatic tube current modulation, adaptive statistical iterative reconstruction, and low tube voltage for the diagnosis of renal colic: Impact of body mass index. Am. J. Roentgenol. (AJR) **202**, 553–560 (2014)
3. Herman, G.T.: Fundamentals of Computerized Tomography: Image Reconstruction from Projections, 2nd edn. Springer Publishing Company, Incorporated, London (2009)
4. Jan, S., et al.: GATE: A simulation toolkit for PET and SPECT. Phys. Med. Biol. **49**(19), 4543–4561 (2004)
5. Jan, S., et al.: GATE v6: A major enhancement of the GATE simulation platform enabling modelling of CT and radiotherapy. Phys. Med. Biol. **56**(4), 881–901 (2011)
6. Krämer, A., Kovacheva, E., Lanza, G.: Projection based evaluation of CT image quality in dimensional metrology (2015)
7. Lee, K.H., et al.: Attenuation-based automatic tube voltage selection and tube current modulation for dose reduction at contrast-enhanced liver CT. Radiology **265**(2), 437–447 (2012)
8. Lv, P., Liu, J., Zhang, R., Jia, Y., Gao, J.: Combined use of automatic tube voltage selection and current modulation with iterative reconstruction for CT evaluation of small hypervascular hepatocellular carcinomas: Effect on lesion conspicuity and image quality. Korean J. Radiol. **16**(3), 531–540 (2015)

9. Nagel, H.D.: CT Parameters that Influence the Radiation Dose, pp. 51–79. Springer, Berlin, Heidelberg (2007)
10. Ngaile, J., Msaki, P.: Estimation of patient organ doses from CT examinations in Tanzania. J. Appl. Clin. Med. physics/Am. Coll. Med. Phys. **7**, 80–94 (2006)
11. Poludniowski, G., Landry, G., DeBlois, F., Evans, P.M., Verhaegen, F.: SpekCalc: A program to calculate photon spectra from tungsten anode x-ray tubes. Phys. Med. Biol. **54**(19), N433–N438 (2009)
12. Poludniowski, G.: Calculation of x-ray spectra emerging from an x-ray tube. part II. X-ray production and filtration in x-ray targets. Med. Phys. **34**, 2175–2186 (2007)
13. Poludniowski, G., Evans, P.M.: Calculation of x-ray spectra emerging from an x-ray tube. Part I. Electron penetration characteristics in x-ray targets. Med. Phys. **34**, 2164–2174 (2007)
14. Reiter, M., Harrer, B., Heinzl, C., Salaberger, D., Gusenbauer, C., Kuhn, C., Kastner, J.: Simulation Aided Study for Optimising Industrial X-ray CT Scan Parameters for Non-Destructive Testing and Materials Characterisation, vol. 6, pp. 20–22 (2011)
15. Reiter, M., Heinzl, C., Salaberger, D., Weiss, D., Kastner, J.: Study on parameter variation of an industrial computed tomography simulation tool concerning dimensional measurement deviations (2010)
16. Rhee, D., Kim, S.w., Moon, Y., Kim, J., Jeong, D.: Effects of the difference in tube voltage of the CT scanner on dose calculation. J. Korean Phys. Soc. **67**, 123–128 (2015)
17. Schindera, S., Winklehner, A., Alkadhi, H., Goetti, R., Fischer, M., Gnannt, R., Szucs-Farkas, Z.: Effect of automatic tube voltage selection on image quality and radiation dose in abdominal CT angiography of various body sizes: A phantom study. Clin. Radiol. **68**, e79–e86 (2012)
18. Seyal, A., Arslanoglu, A., Abboud, S., Sahin, A., Horowitz, J., Yaghmai, V.: CT of the Abdomen with Reduced Tube Voltage in Adults: A Practical Approach. Radiographics **35**, 1922–1939 (2015). A review publication of the Radiological Society of North America, Inc
19. Shin, H.J., et al.: Radiation dose reduction via sinogram affirmed iterative reconstruction and automatic tube voltage modulation (CARE kV) in abdominal CT. Korean J. Radiol. Off. J. Korean Radiol. Soc. **14**, 886–893 (2013)
20. Tang, K., Wang, L., Li, R., Lin, J., Zheng, X., Cao, G.: Effect of low tube voltage on image quality, radiation dose, and low-contrast detectability at abdominal multidetector CT: phantom study. J. Biomed. Biotechnol. **2012**(130169), 1–6 (2012)
21. Yu, L., Li, H., Fletcher, J., McCollough, C.: Automatic selection of tube potential for radiation dose reduction in CT: A general strategy. Med. Phys. **37**, 234–243 (2010)

MicrAnt: Towards Regression Task Oriented Annotation Tool for Microscopic Images

Miroslav Jirik[1,2(✉)] ⓘ, Vladimira Moulisova[2] ⓘ, Claudia Schindler[3],
Lenka Cervenkova[2] ⓘ, Richard Palek[2,4] ⓘ, Jachym Rosendorf[2,4] ⓘ,
Janine Arlt[3] ⓘ, Lukas Bolek[2] ⓘ, Jiri Dejmek[2] ⓘ, Uta Dahmen[3] ⓘ,
Kamila Jirikova[5] ⓘ, Ivan Gruber[1] ⓘ, Vaclav Liska[2,4] ⓘ, and Milos Zelezny[1] ⓘ

[1] NTIS - New Technologies for the Information Society, Faculty of Applied Sciences,
University of West Bohemia, Pilsen, Czech Republic
mjirik@ntis.zcu.cz
[2] Biomedical Center, Faculty of Medicine in Pilsen, Charles University,
Pilsen, Czech Republic
[3] Experimental Transplantation Surgery Department,
Universitätsklinikum Jena, Jena, Germany
[4] Department of Surgery, University Hospital and Faculty of Medicine in Pilsen,
Charles University, Pilsen, Czech Republic
[5] Department of Ethology and Companion Animal Science, Faculty of Agrobiology,
Food and Natural Resources, Czech University of Life Sciences,
Prague, Czech Republic

Abstract. Annotating a dataset for training a Supervised Machine Learning algorithm is time and annotator's attention intensive. Our goal was to create a tool that would enable us to create annotations of the dataset with minimal demands on expert's time. Inspired by applications such as Tinder, we have created an annotation tool for describing microscopic images. A graphical user interface is used to select from a couple of images the one with the higher value of the examined parameter. Two experiments were performed. The first compares the speed of annotation of our application with the commonly used tool for processing microscopic images. In the second experiment, the texture description was compared with the annotations from MicrAnt application and commonly used application. The results showed that the processing time using our application is 3 times lower and the Spearman coefficient increases by 0.05 than using a commonly used application. In an experiment, we have shown that the annotations processed using our application increase the correlation of the studied parameter and texture descriptors compared with manual annotations.

Keywords: Microscopy · Annotation · Scaffold · Liver ·
Decellularization

ⓒ Springer Nature Switzerland AG 2020
T. Lukić et al. (Eds.): IWCIA 2020, LNCS 12148, pp. 209–218, 2020.
https://doi.org/10.1007/978-3-030-51002-2_15

1   Introduction

Microscopy is a widespread method without which it is impossible to imagine the operation of any laboratory in the world [11]. In the last decade, some operations previously performed by human operators have begun to be performed by advanced algorithms based on machine learning methods. Training of artificial intelligence algorithms requires enormous size datasets that contain not only the data itself but also the relevant annotations. The rapid development of computer vision, which has been brought about by progress in the area of convolutional neural networks in recent years, has highlighted this problem.

Creating data annotations is time-consuming and requires the full attention of the operator. Whole Slide Imaging brings the ability to distribute digitized glass slides and increase reproducibility [3], but also the need to process large amounts of image data. There are tools to address such tasks, such as Amazon Mechanical Turk which allows you to distribute an extensive task cheaply. Paolacci et al. [13], addressed potential concerns by presenting demographic data about subject population and offered advice on how to manage a common subject pool. However, this type of service is suitable for simple tasks that do not require extensive knowledge and training. Describing microscopy images is beyond such limitations. Extensive expertise is required from the operator performing such a description. The time of such an expert is very expensive and often not even available. Therefore, we decided to design a tool that makes the annotation of microscopic images as efficient as possible. Such a suitable tool allows creating datasets for training machine learning algorithms in a short time with minimal costs.

2   Methods

Supervised Machine Learning methods with a teacher require different types of datasets. To solve the classification task, it is necessary to assign discrete values to the image or part of the image to indicate the target class. Our main goal, however, was to create a tool to support the creation of annotations for the task of regression. In such cases, the part of the microscopic image or the entire microscopic image is assigned a numerical value that represents a qualitative description of the parameter under investigation.

2.1   Region Selection

The first step called "Region Selection" is to prepare the image data by selecting an image area. It is done by using free Hamamatsu NDP.view 2 Application [6] and its Free Hand Region Annotation tool. Outputs are stored in proprietary file format and it is separated into two files. The image `.ndpi` is used to store image pyramid bitmap data. It is similar to `TIFF` format with some custom TIFF tags which makes it incompatible with standard tiff-reader. The file format is discussed in [4, 10].

The annotation is stored in .ndpa file. There are four important properties of the region stored in .ndpa file and used in our algorithm:

- shape
- color
- title
- detail

Application MicrAnt is open-source software written in Python [15] and the repository [8] is hosted on the GitHub server. The packages Scipy [9], Numpy [12] used for general purpose and the Scikit-image [16] is used for image processing. Image data are read by the Open Slide package described in [5].

MicrAnt is dependent on particular software in the "Region Selection" step. The .ndpi image format is widely used in microscopy. The use of open-source packages allows us to easily use another format by implementing an image reader for each specific image format.

2.2 Image Data and Metadata

When the annotation with specified color 1 is loaded from .ndpa file the image area is read from .ndpi file according to annotation position, width, height and selected level of details. When the annotator's decision should be based on a wider image area the additional black color region can be added to input annotation data. In that case, the image view size is given by the surrounding black annotation region.

2.3 Annotation

In the Initial Annotation step, the goal is to provide the first numerical annotation to all microscopy image regions. The annotation parameter name is set by the operator. Then all the images are randomly organized and the operator is asked to set the value of the parameter for every region. Alternatively, this can be done also by using details property in NDP.view 2. All collected data are stored in the Open XLS file and can be used for regression tasks by using floating-point number values and classification tasks by using discrete values.

In the next step, the annotation parameter is fine-tuned for the regression task. The goal of this step is to increase the precision of annotation. The image regions are sorted by the annotation parameter collected in the Initial Annotation step. A couple of images with a small difference in the application parameter is shown to the operator. By visual examination of this couple, the operator selects the image with a higher value of the examined parameter. Screenshot of MicrAnt application can be seen on Fig. 2. The fine-tune procedure is inspired by the bubble-sort algorithm. It changes just the order of the elements. To allow the evolution of the values of the parameter we designed the update procedure. The order and value of the examined parameters are not changed if the value of the examined parameter on the left image is lower than the value on the right image. The update procedure is applied if the

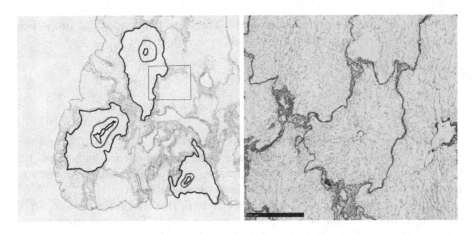

Fig. 1. H&E stained image of the porcine liver. On the left image is region selection. Magenta and red closed curves are used as a region selection. The black curve can be used for defining a larger view around the colored selection. The right image shows the lobule detail (Color figure online).

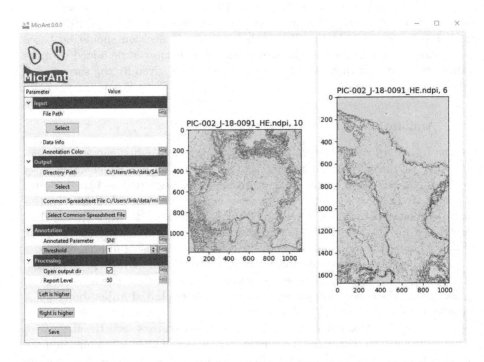

Fig. 2. Screenshot of MicrAnt application. The user selects the image with higher value of the examined parameter.

value on the right image is lower. Then the new value for both images is calculated
according to the Eqs. 1, 2 and 3.

$$P \in (p_0, p_1, \ldots) \tag{1}$$

$$p_i = p_i + \mathrm{rnd}(0, 0.5) \cdot (p_j - p_i) \tag{2}$$

$$p_j = p_j + \mathrm{rnd}(0, 0.5) \cdot (p_i - p_j) \tag{3}$$

where p_i and p_j are the annotation parameters and P is the family of all annotations. Equations 2 and 3 represent the updates. The output is stored in the
open XLS file.

2.4   Intensity Rescale

The intensity of the color channels of each pixel is dependent on the specific
staining performance. Sometimes the contrast may be below the optimal level.
To increase readability, a method of normalization of intensities was introduced.

$$k = \frac{i_0 - i_1}{q_5 - q_{95}} \tag{4}$$

$$q = y_i - k q_{95} \tag{5}$$

$$i_o = \mathrm{sgn}\left(k i_i + q\right) \tag{6}$$

The input intensities (i_i) are transformed to output intensities (i_o) so that the
5th and 95th percentiles (q_5 and q_{95}) of the intensity values are mapped between
-0.9 and 0.9 (i_0 and i_1). The range of target values is then limited by the
sigmoidal function. This limits the outflow of intensities outside the range. See
Fig. 3.

3   Results and Discussion

Two experiments were prepared to prove two different aspects of our approach.
We used H&E (Haemotoxylin and Eosin) stained images of decellularized porcine
liver scaffolds (Fig. 1). This type of tissue consisting of extracellular matrix pro-
teins after cell removal has great potential in tissue engineering because of the
possibility of artifical organ preparation [1]. Scaffolds are produced by our decellularization method.

After thawing the frozen livers slowly at $4\,^{\circ}\mathrm{C}$ and rinsing with saline solution,
scaffolds were generated by circulating perfusion with two types of detergent
solutions (1% Triton X-100 followed by 1% sodium dodecyl-sulphate (SDS)).
Decellularization procedure was stopped when the livers turned homogenously
white. Scaffolds were washed with saline to remove residual detergent solution.
Tissue samples were fixed in 10% neutral-buffered formalin, embedded in paraf-
fin, sectioned and stained in H&E.

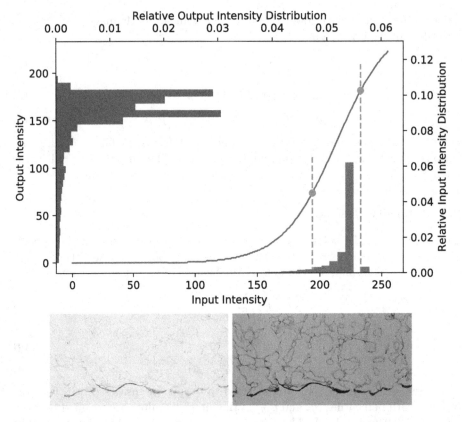

Fig. 3. Intensity Normalization. The top image shows the intensity transformation function. The input image is the bottom left; the bottom right image shows the image after intensity transformation.

The quality estimation of the scaffold is based on fragmented criteria and concentrate mainly on bulk properties. Morphological evaluation is mostly qualitative and superficial. In the future we would like to prepare the scaffold quality measurement based on texture properties. The data annotation is the first step.

The intralobular network represents the network of sinusoidal vessels; thus we introduced the Sinusoidal Integrity parameter (SNI) to quantify the level of preservation of these fine ECM structures after decellularization. SNI is describing the quality of preserved liver sinusoid. This parameter is related to the quality of the decellularization procedure. SNI is zero when no structure is preserved in liver lobule after decellularization and it is two for perfectly preserved sinusoid on the H&E image.

Table 1. Time consumed for annotation.

Step	Lobule selection	Annotation	Annotation
Application	NDP.view 2	NDP.view 2	MicrAnt
Time	6:14	11:03	10:11
Number of samples	30	30	100
Time [s]	374	663	611
Time per sample [s]	12.46	22.10	6.11

In the first experiment, the annotation time was measured. There are three steps in the annotation procedure: Lobuli selection, initial annotation, and fine annotation. Our work is focused on the second and third steps. The lobule selection is done by the NDP. view 2. The second step can be done with both applications (NDP. view 2 and MicrAnt) while the refinement can be done only with MicrAnt.

We used H&E stained images of liver scaffolds. Time consumed by annotation of 30 regions in NDP.view 2 was measured. The operator used standard procedure to select lobules in the image by the Freehand Region drawing tool. Then the initial annotation was done by inserting SNI values into the Annotation Details window. Both times were measured. In MicrAnt the operator annotated 100 regions with lobules. The time required for this operation was measured. Refinement and the initial annotation in MicrAnt are comparable from the time requirements point of view. Data can be found in Table 1.

The second experiment is focused on the calculation of the correlation between parameter SNI and texture features extracted from the region. We used 21 H&E stained whole slide scans of liver scaffolds with 293 selected lobules.

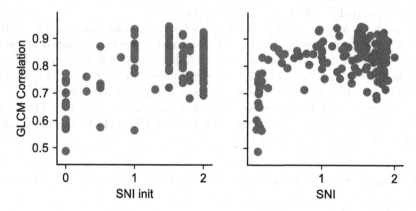

Fig. 4. GLCM correlation texture feature and SNI initial annotation. Manual annotation is shown on the left image and the annotation after MicrAnt refinement step is shown on the right image.

We compared the correlation of texture features with the initial annotation and with the refined annotation. We used the Correlation (Fig. 4) and Correlation Variance (Fig. 5) feature from a group of Gray Level Co-occurrence Matrix texture features described by Haralick in [7] and Tuceryan in [14]. The Spearman correlation coefficient was measured after the Initial Annotation step and then after refinement with MicrAnt. The calcuated data can be found in Table 2.

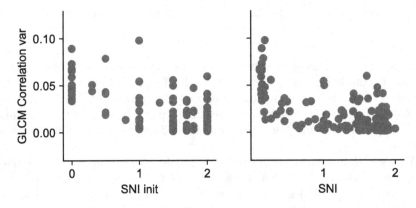

Fig. 5. GLCM correlation variance texture feature and SNI parameter annotated manually and by MicrAnt. Left and right images show the manual annotation and MicrAnt refined annotation respectively.

In the second part of the second experiment, the regression based on the texture features was prepared. We compared the regression trained with init annotation and finetuned annotation. The output can be found in Fig. 6. The mean absolute error was measured and it is 0.2638 and 0.2402 for the initial annotation and the refined annotation respectively.

An example of MicrAnt practical use can be a significant simplification of the whole scan analysis done by ScaffAn tool which was recently developed for quality assessment of decellularized scaffolds suitable for in vitro tissue engineering [17]. The MicrAnt application's code is open-source and the repository is hosted on GitHub [8]. In the last decade, Continuous Integration has become a global phenomenon and the best practice of modern software engineering. The automatic tests were implemented and we use Travis CI [2] and GitHub integration to trigger the build.

Table 2. Spearman Correlation Coeficient between GLCM features and SNI annotation.

Texture Feature	NDP.view 2	MicrAnt
GLCM Correlation	0.259785	0.325264
GLCM Correlation var	0.409871	0.432594

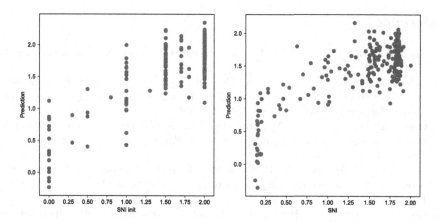

Fig. 6. Prediction of SNI parameter based on GLCM texture features. The left image shows the prediction based on init annotation while the right image shows the prediction trained with finetuned annotation.

4   Conclusion

We introduced a new open-source application MicrAnt for fast annotation of microscopy image data. Annotation is a three-step procedure. The first step is region selection and it is done by the external free application. In the following step, the initial annotation is done in MicrAnt by setting a numeric value to choose the annotation parameter. Most important is the third step. Here the annotation is made more precisely by comparing a couple of regions.

By the first experiment we showed that the annotation refinement with MicrAnt is faster than the reannotation of all data in the external application. The annotation refinement had also positive consequences for machine learning algorithms trained by this data, as it was shown in the second experiment.

In the future, we would like to prepare import plugin for Zeiss .czi image format. We also plan to use a quicksort algorithm scheme in the initial annotation step to make the procedure faster.

Acknowledgments. This work was supported by Charles University Research Centre program UNCE/MED/006 "University Center of Clinical and Experimental Liver Surgery" and Ministry of Education project ITI CZ.02.1.01/0.0/0.0/17_048/0007280: Application of modern technologies in medicine and industry. The research was also supported by the project LO 1506 of the Czech Ministry of Education, Youth and Sports. The authors appreciate the access to computing and storage facilities owned by parties and projects contributing to the National Grid Infrastructure MetaCentrum provided under the program "Projects of Large Research, Development, and Innovations Infrastructures" (CESNET LM2015042).

References

1. Atala, A., Kurtis Kasper, F., Mikos, A.G.: Engineering complex tissues. Sci. Transl. Med. **4**(160), 1–11 (2012)
2. Beller, M., Gousios, G., Zaidman, A.: Oops, my tests broke the build: an explorative analysis of travis CI with GitHub. In: IEEE International Working Conference on Mining Software Repositories, pp. 356–367 (2017)
3. Campbell, W.S., Foster, K.W., Hinrichs, S.H.: Application of whole slide image markup and annotation for pathologist knowledge capture. J. Pathol. Inform. **4**, 2–8 (2013)
4. Deroulers, C., Ameisen, D., Badoual, M., Gerin, C., Granier, A., Lartaud, M.: Analyzing huge pathology images with open source software. Diagn. Pathol. **8**(1), 1–8 (2013)
5. Goode, A., Gilbert, B., Harkes, J., Jukic, D., Satyanarayanan, M.: OpenSlide: a vendor-neutral software foundation for digital pathology. J. Pathol. Inform. **4**, 27 (2013). https://www.ncbi.nlm.nih.gov/pubmed/24244884, www.ncbi.nlm.nih.gov/pmc/articles/PMC3815078/
6. Hamamatsu: NDP.view2 (2019). https://www.hamamatsu.com/
7. Haralick, R.M.: Statistical and structural approaches to texture. Proc. IEEE **67**(5), 786–804 (1979)
8. Jirik, M.: MicrAnt - Github repository (2020). https://github.com/mjirik/micrant
9. Jones, E., Oliphant, T., Peterson, P., et al.: SciPy: Open source scientific tools for Python. http://www.scipy.org/
10. Khushi, M., Edwards, G., de Marcos, D.A., Carpenter, J.E., Graham, J.D., Clarke, C.L.: Open source tools for management and archiving of digital microscopy data to allow integration with patient pathology and treatment information. Diagn. Pathol. **8**(1), 1–7 (2013)
11. Majno, P., Mentha, G., Toso, C., Morel, P., Peitgen, H.O., Fasel, J.H.: Anatomy of the liver: an outline with three levels of complexity - a further step towards tailored territorial liver resections. J. Hepatol. **60**(3), 654–662 (2014). https://doi.org/10.1016/j.jhep.2013.10.026
12. Oliphant, T.: Guide to NumPy. CreateSpace Independent Publishing Platform (2006). http://www.numpy.org/
13. Paolacci, G., Chandler, J., Ipeirotis, P.G.: Running experiments on Amazon mechanical turk. Judgment Decis. Making **5**(5), 411–419 (2010)
14. Tuceryan, M., Jain, A.K.: Texture analysis. In: The Handbook of Pattern Recognition and Computer Vision, 2nd edn. (1998). http://citeseerx.ist.psu.edu/viewdoc/summary?doi=10.1.1.38.5980
15. Van Rossum, G., Drake Jr., F.L.: Python tutorial. Centrum voor Wiskunde en Informatica Amsterdam (1995)
16. van der Walt, S., Schönberger, J.L., Nunez-Iglesias, J., Boulogne, F., Warner, J.D., Yager, N., Gouillart, E., Yu, T.: scikit-image: image processing in Python. PeerJ **2**, e453 (2014)
17. Moulisov á, V., et al.: Novel morphological multi-scale evaluation system for quality assessment of decellularized liver scaffolds. J. Tissue Eng. **11**, 2041731420921121 (2020)

Graph Cuts Based Tomography Enhanced by Shape Orientation

Marina Marčeta[✉] and Tibor Lukić

Faculty of Technical Sciences, University of Novi Sad, Novi Sad, Serbia
marceta.marina@gmail.com, tibor@uns.ac.rs

Abstract. The topic of this paper includes graph cuts based tomography reconstruction methods in binary and multi-gray level cases. A energy-minimization based reconstruction method for binary tomography is introduced. This approach combines the graph cuts and a gradient based method, and applies a shape orientation as an a priori information. The new method is capable for reconstructions in cases of limited projection view availability. Results of experimental evaluation of the considered graph cuts type reconstruction methods for both binary and multi level tomography are presented.

Keywords: Discrete tomography · Shape orientation · Energy minimization methods

1 Introduction

The word tomography comes from Greek words *tomos* which means slice and *graphein* which means to write and it denotes an area in image processing that deals with reconstructing images from the given projection data. Its main focus usually are the objects which are not easily accessible or visible.

A wave penetrates through an unknown object and collects the projection data from the object. In order to gather enough data for a successful reconstruction, waves usually need to penetrate the object from large number of directions. The unknown object that tomography pursues to restore is identified as a function with a domain that can be discrete or continuous and a range that is a given set of (usually) real numbers. Therefore, in order to obtain the image of the unknown object, it is needed to reconstruct this function based on the known data (integrals or sums over subsets of its domain). In *Discrete Tomography* (DT) [11,12] the range of the function is a finite set. DT that deals with the problem of reconstruction of binary images is named binary tomography (BT). If DT deals with the reconstruction of digital images which consist of numerous gray levels, it is referenced as multi-level discrete tomography.

Although problems of multi-level discrete tomography image reconstruction can occur frequently in the application, to the best of our knowledge, there are only few reconstruction algorithms that deal with such problems. Discrete Algebraic Reconstruction Technique (DART) [1], Multi-Well Potential based method

© Springer Nature Switzerland AG 2020
T. Lukić et al. (Eds.): IWCIA 2020, LNCS 12148, pp. 219–235, 2020.
https://doi.org/10.1007/978-3-030-51002-2_16

(MWPDT) [17], a combination of non-local projection constraints with a continuous convex relaxation of the multilabeling problem [25] and the Non-Linear Discretization function based reconstruction algorithm (NLD) [24] are among them. A recently introduced method (GCDT) [20], which combines a gradient based algorithm with a graph cuts type optimization method, showed good performance for this type of problem. This paper gives an overview and experimental evaluation of most often used algorithms for multi-level tomography reconstruction problem.

The good performance of the GCDT algorithm for multi-level case motivate us to make a step further and apply and adjust this approach to an other interesting problem: binary tomography for limited projection availability. We propose a new method which incorporates an a prior knowledge about the solution into the reconstruction process. The smooth solution is determined by the regularized gradient based SPG algorithm [4] which is subsequently binarized applying a max-flow type graph cut algorithm, introduced in [9] and further analyzed in [7,10,15]. The added prior information is the shape orientation descriptor [26]. Our motivation for the selection of this type of prior information is lies in the fact that it shows excellent performance in combination with convex-concave and gradient based approaches, see the reconstruction method (BTO) [18].

The paper has the following structure. Section 2 gives the description of the basic reconstruction problem. In Sect. 3 the new reconstruction method is presented. Experimental results are provided in Sect. 4. Finally, the conclusion is given in the Sect. 5.

2   Reconstruction Problem

In this paper we consider the DT reconstruction problem, represented by a linear system of equations

$$Au = b, \quad A \in \mathbb{R}^{M \times N}, \quad u \in \Lambda^N, \quad b \in \mathbb{R}^M, \quad \Lambda = \{\lambda_1, \lambda_2, ..., \lambda_k\}, \quad k \geq 2 \quad (1)$$

where k is the number of different gray level values. The set Λ is given by the user. The matrix A is a so-called *projection matrix*, whose each row corresponds to one projection ray. The corresponding components of the vector b contain the detected projection values, while the vector u represents the unknown image to be reconstructed. The i-th row entries $a_{i,\cdot}$ of A represent the length of the intersection of the pixels and the i-th projection ray passing through them. The projection value measured by a projection ray is calculated as a sum of products of the pixel's intensity and the corresponding length of the projection ray through that pixel. Projections are taken from different directions. For each projection direction a number of parallel projection rays are taken (parallel beam projection). The projection direction is determined by the angle β. The distance between two adjacent parallel projection rays can vary depending on the reconstruction problem, we set this distance to be equal to the side length of pixels. The reconstruction problem means finding the solution image u of the linear system of Eq. (1), where the projection matrix A and the projection vector b are given. This system is often undetermined ($N > M$).

3 Graph Cuts Reconstruction Method Assisted by Shape Orientation

A directed, weighted graph $G = (X, \rho)$, consists of a set of nodes X and a set of directed edges ρ that connect them. The nodes,in image processing interpretations, mostly correspond to pixels or voxels in 3D. All edges of graph are assigned some weight or cost.

Let $G = (X, \rho)$ be a directed graph with non-negative edge weights that has two special nodes or terminals, the source A and the sink B. An $a-b$-cut (which is referred informally as a cut) $C = A, B$ is a partition of the terminals in X into two disjoint sets A and B so that $a \in A$ and $b \in B$. The cost of the cut is the sum of costs of all edges that go from A to B:

$$c(A, B) = \sum_{x \in A, y \in B, (x,y) \in \rho} c(x, y).$$

The minimum $a-b$-cut problem is to find a cut C with the minimum cost among all cuts. Algorithms to solve this problem can be found in [7].

The approach that uses graph cuts for energy minimization has, as basic technique, construction of a specialized graph for the energy function to be minimized such that the minimum cut on the graph also minimizes the energy. The form of the graph depends on the exact form of X and on the number of labels. The minimum cut, in turn, can be computed very efficiently by max flow algorithms.

These methods have been successfully used in the last 20 years for a wide variety of problems, naming image restoration [8,9], stereo and motion [2,14], image synthesis [16], image segmentation [6] and medical imaging [5,13].

The Potts model in graph cuts theory is based on the minimization of the following energy

$$E(d) = \sum_{p \in \mathcal{P}} D(p, d_p) + \sum_{(p,q) \in \mathcal{N}} K_{(p,q)} \cdot (1 - \delta_{d_p, d_q}), \tag{2}$$

where $d = \{d_p | p \in \mathcal{P}\}$ represents the labelling of the image pixels $p \in \mathcal{P}$. By $D(p, d_p)$ we denote the data cost term, where $D(p, d_p)$ is a penalty or cost for assigning a label d_p to a pixel p. $K_{(p,q)}$ is an interaction potential between neighboring pairs p and q, $\mathcal{N}$ is a set of neighboring pairs. Function δ_{d_p, d_q} is Kronecker delta function.

3.1 Shape Orientation

In this section we give a short description and a calculation method of the shape orientation.

The geometrical moment of a digitize image u is defined by

$$m_{p,q}(u) = \sum_{(i,j) \in \Omega} u(i,j) i^p j^q,$$

where $\Omega \subseteq \mathbb{R}^2$ denotes the image domain.

The center of gravity (or centroid) of an image (or a shape) u is defined by

$$(C_x(u), C_y(u)) = \left(\frac{m_{1,0}(u)}{m_{0,0}(u)}, \frac{m_{0,1}(u)}{m_{0,0}(u)} \right).$$

The centroid enables the definition of the centralized moment which is translation invariant. The centralized moment of an image u of order $p + q$ is given by

$$\overline{m}_{p,q}(u) = \sum_{(i,j) \in \Omega} u(i,j)(i - C_x(u))^p (j - C_y(u))^q.$$

The shape orientation is an often used and well known shape descriptor [26]. The orientation is determined by the angle α, which represents the slope of the axis of the second moment of inertia (orientation axis) of the considered shape [23]. The orientation (angle α) for the the given image u can be calculated by the following equation:

$$\frac{\sin(2\alpha)}{\cos(2\alpha)} = \frac{2 \cdot \overline{m}_{1,1}(u)}{\overline{m}_{2,0}(u) - \overline{m}_{0,2}(u)}. \tag{3}$$

Moments in (3) are translation invariant, making the orientation invariant to translation transformations, for more details see [18, 26].

3.2   The Proposed Method

Our tomography reconstruction approach is a combination of the graph cuts method and a gradient based minimization method. In the first step, we determine the data cost values for each image pixels. The data cost values are determined as intensity values of the continuous or smooth approximation (solution) of the final reconstruction image. The smooth solution is obtained by the following energy-minimization problem

$$\min_{u \in [0,1]^N} E_Q(u) := w_P \|Au - b\|_2^2 + w_H \sum_{i=1}^N \sum_{j \in \Upsilon(i)} (u_i - u_j)^2 \tag{4}$$
$$+ w_O \left(\phi(u) - \alpha^* \right)^2 + \mu \langle u, \tau - u \rangle,$$

where $\tau = [1, 1, \ldots, 1]^T$ is a vector of size N. Parameter w_P regulates the influence of the data fitting term, w_H controls the second term, whose role is to enhance the homogeneity or compactness of the solution. By $\Upsilon(i)$ two neighbor pixel indexes (in x and y axis directions) of pixel i is denoted. The orientation of the solution u is denoted by $\phi(u)$, while α^* is given true orientation. Parameter w_O determines the impact of the orientation regularization. The task of the concave regularization term $\langle u, \tau - u \rangle$ (inner product of vectors u and $1 - u$) is to push pixel intensities toward binary values. Parameter μ regulates the influence of this binarization term and its value is gradually increased during the reconstruction process. The problem (4), for each fixed μ, is solved by the Spectral

Projected Gradient (SPG) iterative optimization algorithm, originally proposed by Birgin et al. [3]. Motivation for application of this algorithm is supported by its successful application in similar problems, see [19, 21, 22]. The reconstruction process (4) is terminated before the fully binary solution is achieved. The termination criterion is given by

$$\langle u, \tau - u \rangle < E_{bin},$$

where E_{bin} regulates the degree of binarization of the solution u. In our experiments its value is set as 100. This partially binarization of the smooth solution is applied in order to improve the determination of data cost terms for graph cuts method.

In the next step we have to fully binarize the smooth solution of the problem (4), obtained by the SPG algorithm. For this task we apply the graph cuts method based on the Potts model, described in Sect. 3. The data cost term D in (2) is determined using information provided by the smooth solution u. More precisely, we define it in the following way

$$D(p, 0) = u(p),$$
$$D(p, 1) = 1 - u(p),$$

where $u(p)$ represents the intensity of a pixel p. The idea is to make data cost small or cheap in the vicinity of the given gray values. The neighbor pairs are defined based on 1-neighboring system, i.e., $(p, q) \in \mathcal{N}$ if the image coordinates of p and q differ for one value only. The interaction potential $K_{(p,q)}$ (see (2)) in our experiments is set as a constant and its value is 1. Now, the energy function in (2) is determined and ready to be minimized. For this task we use the GCO graph cuts based optimization algorithm, introduced in [9] and further analyzed in [7, 10, 15]. The GCO algorithm determines the label values d_p for each pixel p. Each label value is assigned to one predefined gray level in the following way: $d_p = 0 \rightarrow 0$ and $d_p = 1 \rightarrow 1$. Therefore, the obtained label values also determine intensities of pixels in the final (binary) solution, therefore the reconstruction process is terminated. We denote this method by Graph Cuts Tomography Assisted by the Orientation prior (GCORIENTBT) reconstruction method. If in the Eq. 4 we set values of parameters w_o, w_H and μ as 0, and solve it using the combination of SPG and GCO algorithms, we are getting algorithm introduced in [20] and denoted by GCDT.

4   Experimental Results

In this section we experimentally evaluate the proposed graph cuts based reconstruction methods, denoted by GCDT and GCORIENTBT. In the experiments for multi gray level tomography we use 6 test images (phantoms), as originals in reconstructions, presented in Fig. 1. Phantoms PH1, PH2 and PH3 contain 3 gray levels, while phantoms PH4, PH5 and PH6 contain 6 gray levels. In addition, GCDT was also tested on binary images using phantoms PH7, PH8, PH9 as well as on two medical images: binary segmented CT image of a bone implant, inserted in a leg of a rabbit (PH10) and a Binary segmented florescence image of Calcein stained Chinese hamster ovary cell (PH11). We consider reconstructions of these images obtained from different projection directions. A total of 128 parallel rays are taken for each projection direction for multi gray level images and 64 projection rays for binary images. In all cases, the projection directions are uniformly selected between 0 and 180°. The obtained results are compared with two reconstruction methods suggested for multi level discrete tomography, so far: 1) Multi Well Potential based method (MWPDT) proposed in [17] (this method is developed and used only for phantoms with 3 gray levels); and 2) DART method, proposed in [1]. We also include into the evaluation process a simple method based on classical threshold, denoted by TRDT. Additionally, binary images are compared with similar reconstruction method for binary tomography introduced in [18] and denoted by BTO. All reconstruction methods (BTO, DART, GCORIENTBT, MWPDT, GCDT and TRDT) are implemented completely in Matlab.

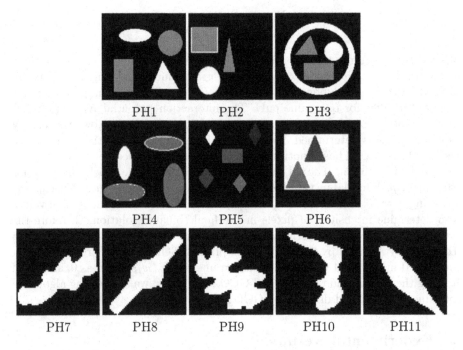

Fig. 1. Original test images (128x128). Phantoms PH1, PH2 and PH3 contain 3 different gray levels (0,0.5,1), PH4, PH5 and PH6 contain 6, while PH7, PH8, PH9, PH10, PH11 present binary images.

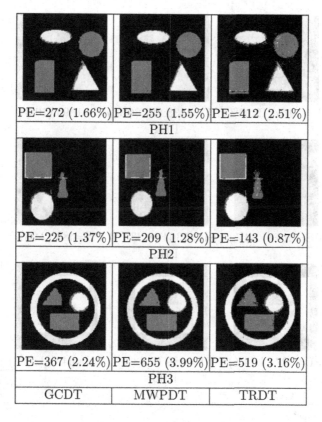

| GCDT | MWPDT | TRDT |

Fig. 2. Reconstructions of the test images using data from 6 projection directions.

In the evaluation process, we analyze the quality of the reconstructions and required running times. The quality of the reconstructions are expressed by the pixel error (PE), i.e., the absolute number of the misclassified pixels, and by the misclassification rate $(m.r.)$, i.e., the pixel error measure relative to the total number of image pixels. Also, as a qualitative error measure, we consider the projection error, defined by $PRE = \|Au^r - b\|$, where u^r represents the reconstructed image. This error express the accordance of the reconstruction with the given projection data.

In Table 1 we present pixel errors for reconstructions of three phantom images (PH1, PH2 and PH3) obtained from different number of projections by three different methods (MWPDT, GCDT and TRDT). From total of 12 different reconstruction problems, GCDT method provided best results in 10 cases, while in 2 cases the dominant was the MWP method. Regrading the PRE values, see Table 2, the proposed GCDT method dominated in 8 cases, while MWP in 4 cases.

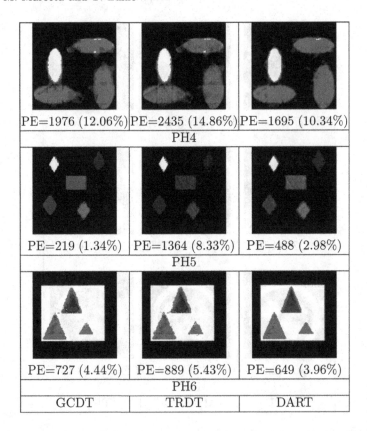

Fig. 3. Reconstructions of the test images using data from 6 projection directions.

Best running times in all experiments were achieved by the MWP method. GCDT and TRDT methods use the smooth solution/reconstruction as a first step, before the "discretization process" starts.

The smooth solution is achieved as a final termination, with high precision, which requires significantly higher number of iterations compared to MWPDT method in total, thus resulting in greater consumption of time. In Figs. 2 and 4 reconstructions from 6 and 15 projection directions of images containing 3 gray level are presented.

Reconstruction results of phantoms with 6 different gray levels (Table 3) show that, compared to TRDT and DART, GCDT method prevails in 10 out of 12 cases, whilst DART performes the best in 2 cases. In Figs. 3 and 5 reconstructions from 6 and 15 projection directions respectively are presented.

In addition to multi level discrete tomography, we have tested our algorithm on binary images. It can be noticed on the Fig. 6 that GCDT method gives poor results in the cases of the reconstruction from two projections. On the other hand, already from higher number of projections, GCDT shows competitive performance. We have tried to avoid this drawback by adding orientation as

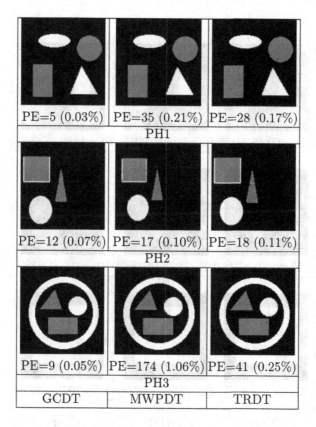

Fig. 4. Reconstructions of the test images using data from 15 projection directions.

a priori information to GCDT method, thus building the GCORIENTBT algorithm. Later, we have compared the GCORIENTBT algorithm with three other reconstruction methods (BTO, DART, GCDT) (Tables 4 and 5). In 12 out of 15 cases, GCORIENTBT gives the best reconstruction (smallest PE/m.r.). It can be noticed that, as expected, by adding the orientation prior to GCDT method, significantly better results for BT are obtained. The advantage of GCORIENTBT is in running time as well. Execution time of GCORIENTBT is in most of the cases even more than 50% shorter compared to its best competitor BTO.

Summarizing the results obtained by the total of 24 multi- gray level analyzed reconstruction tasks, see Tables 1 and 3, the quality of the reconstruction, indicated by m.r.for the proposed GCDT method was the best in 20 cases, i.e., in 83% of the analyzed cases. We emphasize that the results of the GCDT method, in cases when they are the best, are significantly better (in the most of the cases more than 50% better). Farther, GCORIENTBT performed better in 80% of the cases.

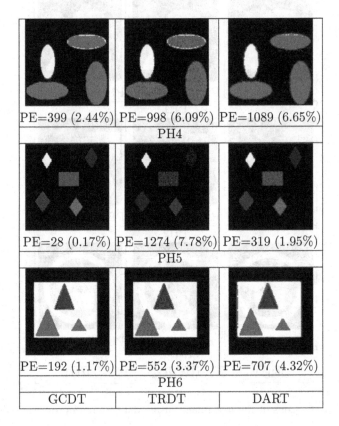

Fig. 5. Reconstructions of the test images using data from 15 projection directions.

Table 1. Experimental results for PH1, PH2 and PH3 images, using three different reconstruction methods. The abbreviation d indicates the number of projections. The best performance is bold fonted.

d		PH1				PH2				PH3			
		6	9	12	15	6	9	12	15	6	9	12	15
MWP	(PE)	**255**	159	59	35	**143**	138	20	18	655	456	275	174
	(m.r. %)	**1.55**	0.97	0.36	0.21	**0.87**	0.84	0.12	0.11	3.99	2.78	1.67	1.06
TRDT	(PE)	412	175	48	28	209	141	17	17	412	301	101	41
	(m.r. %)	2.51	1.06	0.29	0.17	1.28	0.86	0.10	0.10	2.51	1.83	0.61	0.25
GCDT	(PE)	272	**69**	**8**	**5**	225	**124**	**12**	**12**	272	116	20	9
	(m.r. %)	1.66	**0.42**	**0.04**	**0.03**	1.37	**0.76**	**0.07**	**0.07**	1.66	**0.70**	**0.12**	**0.05**

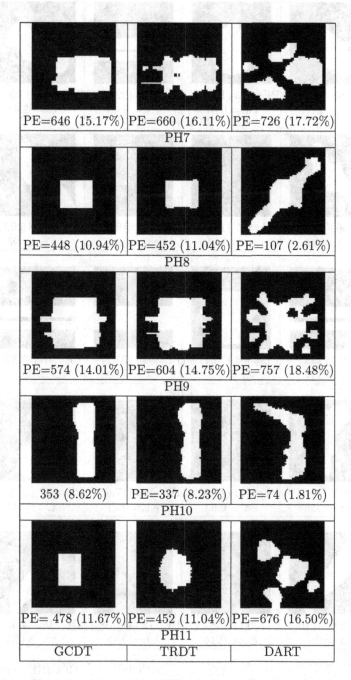

Fig. 6. Reconstructions of the binary test images using data from 2 projection directions.

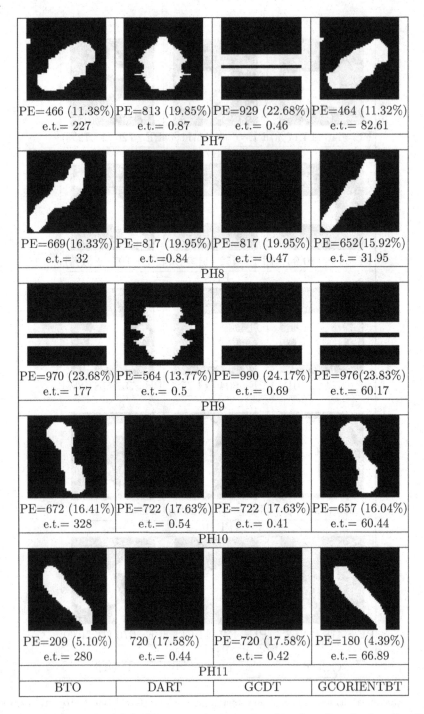

BTO	DART	GCDT	GCORIENTBT

Fig. 7. Reconstructions of the binary test images using data from 1 projection direction, $\alpha = 0°$. The abbreviation e.t. means elapsed time in seconds.

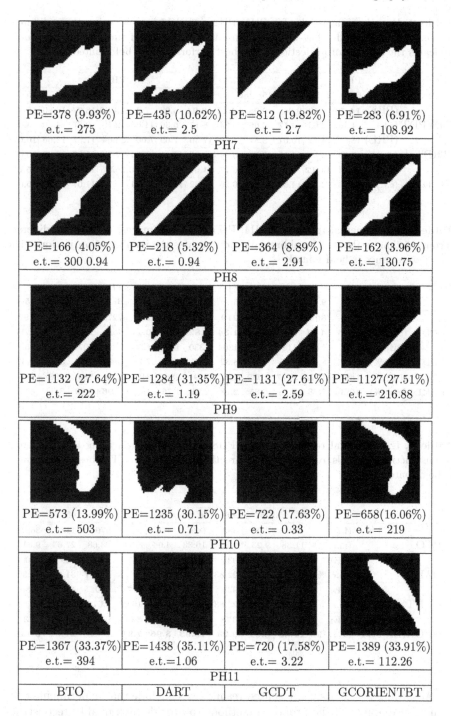

Fig. 8. Reconstructions of the binary test images using data from 1 projection direction, $\alpha = 45°$. The abbreviation e.t. means elapsed time in seconds.

Table 2. Experimental results for PH1, PH2 and PH3 images, using three different reconstruction methods. The abbreviation e.t. means elapsed time in minutes and d indicates the number of projections. The best performance is bold fonted.

d		PH1				PH2				PH3			
		6	9	12	15	6	9	12	15	6	9	12	15
	(PRE)	**14.70**	12.19	9.96	9.08	**14.11**	18.94	6.08	7.71	**19.83**	18.77	18.80	16.43
MWPDT	(e.t.)	1.76	2.63	3.17	4.06	5.34	8.17	6.36	11.62	2.19	2.87	4.30	4.66
	(PRE)	18.66	14.72	10.61	8.87	17.98	17.30	7.09	7.09	23.64	17.87	13.66	10.61
TRDT	(e.t.)	7.73	12.58	14.55	17.77	6.24	10.82	16.01	17.74	7.28	11.07	13.39	16.00
	(PRE)	23.24	**11.12**	**6.52**	**4.39**	26.77	21.04	**6.01**	**6.00**	25.87	**14.96**	**7.59**	**5.60**
GCDT	(e.t.)	7.73	12.58	14.55	17.77	6.25	10.82	16.01	17.74	7.29	11.07	13.40	16.01

Table 3. Experimental results for PH4, PH5 and PH6 images, using three different reconstruction methods. The abbreviation d indicates the number of projections. The best performance is bold fonted.

d		PH4				PH5				PH6			
		6	9	12	15	6	9	12	15	6	9	12	15
	(PE)	1976	**804**	**551**	**399**	219	134	**42**	**28**	727	**473**	**251**	**192**
GCDT	(m.r. %)	12.06	**4.91**	**3.36**	**2.44**	1.34	0.82	**0.26**	**0.17**	4.44	**2.89**	**1.53**	**1.17**
	(PE)	2435	1415	1188	998	1364	1330	1286	1274	889	807	587	552
TRDT	(m.r. %)	14.86	8.64	7.25	6.09	8.32	8.12	7.85	7.78	5.43	4.92	3.58	3.37
	(PE)	**1695**	1242	1177	1089	488	379	288	319	**649**	836	596	707
DART	(m.r. %)	**10.34**	7.58	7.18	6.65	2.98	2.31	1.76	1.95	**3.96**	5.10	3.64	4.32

Table 4. Experimental results for PH7, PH8 and PH9 images, using three different reconstruction methods. α indicates the direction of projections. The best performance is bold fonted.

α		PH7			PH8			PH9		
		0°	45°	90°	0°	45°	90°	0°	45°	90°
	(PE)	466	378	383	669	166	362	970	1132	369
BTO	(m.r. %)	11.38	9.23	9.35	16.33	4.06	8.84	23.68	27.64	9.01
	(PE)	813	435	935	817	218	817	**564**	1284	670
DART	(m.r. %)	19.85	10.62	22.83	19.95	5.32	19.95	**13.77**	31.35	16.36
	(PE)	929	812	935	817	364	817	990	1131	1180
GCDT	(m.r. %)	22.68	19.82	22.83	19.95	8.89	19.95	24.17	27.61	28.81
	(PE)	**464**	**283**	**378**	**652**	**162**	**324**	976	**1127**	**373**
GCORIENTBT	(m.r. %)	**11.32**	**6.91**	**9.23**	**15.93**	**3.96**	**7.91**	23.83	**27.51**	**9.11**

According to the above presented results, we conclude that experiments confirm the capability of the proposed methods to provide high quality reconstructions both for gray-level (GCDT) and binary (GCORIENTBT) images.

Table 5. Experimental results for PH10 and PH11 using three different reconstruction methods. α indicates the direction of projections. The best performance is bold fonted.

α		PH10			PH11		
		$0°$	$45°$	$90°$	$0°$	$45°$	$90°$
BTO	(PE)	672	**573**	597	209	1367	379
	(m.r. %)	16.41	**13.99**	14.58	5.10	33.37	9.25
DART	(PE)	722	1235	414	720	1438	720
	(m.r. %)	17.63	30.15	17.63	17.58	35.11	17.58
GCDT	(PE)	722	722	534	720	**720**	720
	(m.r. %)	17.63	17.63	13.04	17.58	**17.58**	17.58
GCORIENTBT	(PE)	**657**	658	**377**	180	1389	**340**
	(m.r. %)	**16.04**	16.06	**9.20**	4.39	33.91	**8.30**

5   Conclusions

In this paper we presented an approach for solving problem posed by discrete tomography. The approach uses a gradient based method to obtain a smooth reconstruction of an image and then uses graph cuts optimization method for discretization. In cases of lowered projection directions, the method uses orientation as an a priori information. Conducted experiments gave priority in reconstruction quality to the proposed methods compared to the formerly published reconstruction methods. Based on the obtained experimental results and analysis presented in this paper, we have concluded that the combination of a gradient based method with the graph cuts optimization method is suitable for providing high quality reconstructions in discrete tomography.

Acknowledgement. Authors acknowledge the Ministry of Education and Sciences of the R. of Serbia for support via projects OI-174008 and III-44006. T. Lukić acknowledges support received from the Hungarian Academy of Sciences via DOMUS project.

References

1. Batenburg, K.J., Sijbers, J.: DART: a fast heuristic algebraic reconstruction algorithm for discrete tomography. In: Proceedings of International Conference on Image Processing (ICIP), pp. 133–136 (2007)
2. Birchfield, S., Tomasi, C.: Multiway cut for stereo and motion with slanted surfaces. In: Proceedings of the International Conference on Computer Vision, pp. 489–495 (1999)
3. Birgin, E.G., Martínez, J.M., Raydan, M.: Algorithm: 813: SPG - software for convex-constrained optimization. ACM Trans. Math. Softw. **27**, 340–349 (2001)
4. Birgin, E., Martínez, J.: A box-constrained optimization algorithm with negative curvature directions and spectral projected gradients. Computing **15**, 49–60 (2001)
5. Boykov, Y., Jolly, M.P.: Interactive graph cuts for optimal boundary and region segmentation of objects in N-D images. In: Proceedings of the International Conference on Computer Vision, pp. 105–112 (2001)

6. Boykov, Y., Kolmogorov, V.: Computing geodesics and minimal surfaces via graph cuts. In: Proceedings of the International Conference on Computer Vision, pp. 26–33 (2003)

7. Boykov, Y., Kolmogorov, V.: An experimental comparison of min-cut/max-flow algorithms for energy minimization in vision. IEEE Trans. PAMI **26**(9), 1124–1137 (2004)

8. Boykov, Y., Veksler, O., Zabih, R.: Markov random fields with efficient approximations. In: Proceedings of the International Conference on Computer Vision and Pattern Recognition, pp. 648–655 (1998)

9. Boykov, Y., Veksler, O., Zabih, R.: Fast approximate energy minimization via graph cuts. IEEE Trans. PAMI **23**(11), 1222–1239 (2001)

10. Delong, A., Osokin, A., Isack, H.N., Boykov, Y.: Fast approximate energy minimization with label costs. In: Proceedings of the International Conference on Computer Vision and Pattern Recognition, vol. 96(1), pp. 1–27 (2010)

11. Herman, G.T., Kuba, A.: Discrete Tomography: Foundations, Algorithms and Applications. Birkhäuser, Boston (1999)

12. Herman, G.T., Kuba, A.: Advances in Discrete Tomography and Its Applications. Birkhäuser, Boston (2006)

13. Kim, J., Zabih, R.: Automatic Segmentation of Contrast- Enhanced Image Sequences. In: Proceedings of the International Conference on Computer Vision, pp. 502–509 (2003)

14. Kolmogorov, V., Zabih, R.: Visual correspondence with occlusions using graph cuts. In: Proceedings of the International Conference on Computer Vision, pp. 508–515 (2001)

15. Kolmogorov, V., Zabih, R.: What energy functions can be minimized via graph cuts? IEEE Trans. PAMI **26**(2), 147–159 (2004)

16. Kwatra, V., Schoedl, A., Essa, I., Turk, G., Bobick, A.: Graphcut Textures: Image and Video Synthesis Using Graph Cuts. In: Proc. SIGGRAPH 2003. pp. 277–286. ACM Trans. Graphics (2003)

17. Lukić, T.: Discrete tomography reconstruction based on the multi-well potential. In: Aggarwal, J.K., Barneva, R.P., Brimkov, V.E., Koroutchev, K.N., Korutcheva, E.R. (eds.) IWCIA 2011. LNCS, vol. 6636, pp. 335–345. Springer, Heidelberg (2011). https://doi.org/10.1007/978-3-642-21073-0_30

18. Lukić, T., Balázs, P.: Binary tomography reconstruction based on shape orientation. Pattern Recogn. Lett. **79**, 18–24 (2016)

19. Lukić, T., Lukity, A.: A spectral projected gradient optimization for binary tomography. In: Rudas, I.J., Fodor, J., Kacprzyk, J. (eds.) Computational Intelligence in Engineering. SCI, vol. 313, pp. 263–272. Springer, Heidelberg (2010). https://doi.org/10.1007/978-3-642-15220-7_21

20. Lukić, T., Marčeta, M.: Gradient and graph cuts based method for multi-level discrete tomography. In: Brimkov, V.E., Barneva, R.P. (eds.) IWCIA 2017. LNCS, vol. 10256, pp. 322–333. Springer, Cham (2017). https://doi.org/10.1007/978-3-319-59108-7_25

21. Lukić, T., Nagy, B.: Deterministic discrete tomography reconstruction method for images on triangular grid. Pattern Recogn. Lett. **49**, 11–16 (2014)

22. Nagy, B., Lukić, T.: Dense projection tomography on the triangular tiling. Fundam. Inform. **145**, 125–141 (2016)

23. Sonka, M., Hlavac, V., Boyle, R.: Image Processing, Analysis, and Machine Vision. Thomson-Engineering, Toranto (2007)

24. Varga, L., Balázs, P., Nagy, A.: An energy minimization reconstruction algorithm for multivalued discrete tomography. In: Proceedings of 3rd International Symposium on Computational Modeling of Objects Represented in Images, pp. 179–185. Taylor & Francis, Rome (2012)

25. Zisler, M., Petra, S., Schnörr, C., Schnörr, C.: Discrete tomography by continuous multilabeling subject to projection constraints. In: Rosenhahn, B., Andres, B. (eds.) GCPR 2016. LNCS, vol. 9796, pp. 261–272. Springer, Cham (2016). https://doi.org/10.1007/978-3-319-45886-1_21

26. Žunić, J., Rosin, P.L., Kopanja, L.: On the orientability of shapes. IEEE Trans. Image Process. **15**, 3478–3487 (2006)

Dealing with Noise in Cluster Pattern Interface

Valentin E. Brimkov[1,2], Reneta P. Barneva[3(✉)], and Kamen Kanev[4]

[1] SUNY Buffalo State, Buffalo, NY 14222, USA
brimkove@buffalostate.edu
[2] Institute of Mathematics and Informatics, Bulgarian Academy of Sciences,
1113 Sofia, Bulgaria
[3] SUNY Fredonia, Fredonia, NY 14063, USA
barneva@fredonia.edu
[4] Shizuoka University, Hamamatsu 432-8011, Japan
kanev@inf.shizuoka.ac.jp

Abstract. Cluster Pattern Interface (CLUSPI) [6] is an original 2D encoding and a related technology for direct point-and-click interfaces. It allows determining the objects printed on a surface pointed out by the user. The surface is covered by a layer of code composed of tiny dots, which blends with the original surface texture. In practice, however, some of the dots of the code may be damaged or obscured, which creates decoding challenges. In this work we develop theoretical solutions to some of the challenges related to decoding noisy patterns.

Keywords: Cluster pattern interface CLUSPI · Primitive polynomial · Shift-register Σ-sequence · Noisy pattern

1 Introduction

When interacting with humans, we often refer to nearby objects by pointing gestures. The pointing functionality is also used for interacting in VR (Virtual Reality), AR (Augmented Reality), and MR (Mixed Reality) models where both physical and virtual components are integrated and simultaneously presented on the screen. One of the main issues with the AR and MR implementations is the synchronization of the real and the virtual components. While a lot of research aiming at efficient methods for recognition and tracking of physical objects appearing in AR/MR is being done, a highly reliable practical solution is still to come.

In this work we focus on a different approach facilitating the recognition and tracking of the target physical objects by applying specialized digital codes to their surfaces. In this way, instead of complex scene analysis and object recognition, just extracting and decoding of the embedded digital information will suffice for the object identification and tracking.

A well-known technology, widely applied for identifying objects, is based on optical scanners that recognize codes printed on the surface of the objects, such

© Springer Nature Switzerland AG 2020
T. Lukić et al. (Eds.): IWCIA 2020, LNCS 12148, pp. 236–244, 2020.
https://doi.org/10.1007/978-3-030-51002-2_17

as barcodes and QR-codes [13]. Note that the code must be oriented properly with respect to the scanner and be in close proximity to it in order to be recognized. It should also be noted that these codes must have certain minimal size and thus they cover part of the surface on which they are printed. The scanning of the optical code results in extracting an identifier, which correspond to a record stored in a database. This could be the name of the object, some attributes such as its price, or a related URL link.

Obviously, in the frameworks of the VR, AR, and MR models it would be very useful to be able to directly point to a physical object and get its identifier irrespectively of its position and orientation. For this, instead of a single barcode, some encoding that covers the entire object surface must be applied. In addition to the identifier that corresponds to the barcode, this approach would bring information about the orientation of the object with respect to the scanner that can also be extracted and employed, for example, for spatial queries. It is desirable that the encoding consists of small elements and blends with the object surface so that it does not obstruct the other information printed on it and does not disrupt the aesthetic of its appearance.

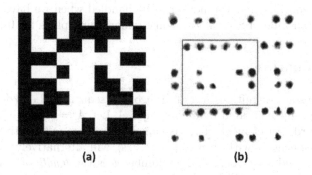

Fig. 1. A sample Data Matrix barcode (a) and a proportionally enlarged D-touch code pattern (b) based on dots. The rectangle denotes the code block size.

A code with such desired properties should obviously be based on dots, instead of squares. (See Fig. 1.) However, most of the existing dot codes [3,12,14,15] still produce dense overlays that make the encoded surface appear grayish and often require specialized reading equipment. With respect to this, we focus on the CLUSPI technology, which has been co-invented by one of the authors of this paper [5,6] and addresses the issues discussed above. It is based on pixels with a specific color, which form 2-dot groups corresponding to the binary pairs (0,0), (0,1), (1,0), and (1,1). A multiplicity of such pairs is then employed for encoding of the absolute coordinates over the surface.

The idea to enumerate the positions on the surface with dots has been used in other coding schemes that cover the surface of the physical object with a relatively thick encoding layers. In order to avoid this, CLUSPI applies two innovative ideas. Firstly, the code of the coordinates of a point on the surface depends on the coordinates of the adjacent points and second, the code representing the coordinates of a point is not written linearly, but rather as a two-dimensional sequence. In this way, the code becomes a sparse layer of dots that cover the surface of the physical object in an unobtrusive way.

For details about the encoding and decoding scheme we refer the readers to some previous publications [2,6,7]. We only mention here that it is based on Σ sequences, which will be explained in the next section.

An important issue of the printed code recognition is the case when part of the code is damaged or obscured. For example, a barcode cannot be recognized properly if entire bars are missing. However, decoding based on partially visible bars is still possible, since the code is printed as a 2D picture, although it is essentially a 1D code. When we minimize the number of the dots in the code to make the layer blend with the surface texture, we face the problem that a small damage to the surface could make it unreadable.

In the sequel we investigate this problem from mathematical perspective and consider solutions.

The paper is structured as follows. In the next section we provide the necessary preliminaries about the Σ-sequences on which the code of CLUSPI is based. In Sect. 3 we consider how the noise can be handled when we have gaps in the pattern. We conclude in Sect. 4 with some remarks and plans for further work.

2   Preliminaries

We start with some definitions from the theory of sequences. Let's denote by A^+ the free semigroup generated by an alphabet A. Let $A^* = A^+ \cup \{\lambda\}$, where λ is the empty word. An element of A^* is called a *sequence* (alternatively, a *string* or *word*). A sequence s with terms $s_1, s_2, \ldots, s_n$ will alternatively be written as $s = s_1 s_2 \ldots s_n$ and $s = s[1 \ldots n]$. The number n is the *length* of s, denoted $|s|$. A sequence of length m will be called an *m-sequence*.

Let $s = uvw$ where u, v, and w are sequences. Then v is a *subsequence* of s with a starting position in s the integer $|u| + 1$, and u and w are the respectively *prefix* and *suffix* of s. $s[j]$ is the j'th term of s, $1 \leq j \leq n$, and $s[p \ldots q]$ is the subsequence of s starting at the p'th position and ending at the q'th position of s, $1 \leq p < q \leq n$. A sequence s that consists of k consecutive copies of a subsequence w is denoted $s = w^k$; w is a *period* of s if s is a prefix of w^k for some $k \geq 1$. Whenever no confusion arises, "period" of a sequence s may also refer to the length $|w|$ of a period w of s. A sequence s may have different periods, as the shortest one is called *the period* of the sequence. A sequence is always a period of itself, called the *trivial* period. A sequence that has only the trivial period is called *aperiodic*. For more details on theory of words see [1].

CLUSPI technology has as a theoretical background a special class of 0/1 sequences, commonly known as Σ-*sequences*. Such a sequence is generated from a polynomial of a certain degree, called a *generator polynomial* and has the property that all subsequences, called patterns, of certain lengths are unique within the sequence. These patterns are the basis of encoding the coordinates on the surface. In what follows we briefly explain the construction of Σ-sequences (see [4] for further details).

Let $P(x) = a_m x^m + a_{m-1} x^{m-1} + \cdots + a_1 x + a_0$, $m \geq 2$, be a polynomial with coefficients 0's or 1's. All operations on such polynomials are performed modulo 2. $P(x)$ is called *irreducible* (or *prime*), if it cannot be factored into a product of a number of polynomials with coefficients equal to 0 or 1 and of degree greater than or equal to one. $P(x)$ is *primitive* if it is irreducible and is a factor of $x^N + 1$, where $N = 2^m - 1$.

For a primitive polynomial $P(x)$ of degree $m \geq 2$, let $x^{\alpha_1}, x^{\alpha_2}, \ldots, x^{\alpha_k}, x^{\alpha_{k+1}}$ be its nonzero monomials, i.e., those with coefficients $a_{\alpha_1} (= a_m) = a_{\alpha_2} = \cdots = a_{\alpha_k} = \alpha_{k+1} = 1$. The indexes $\alpha_1, \alpha_2, \ldots, \alpha_{k+1}$ are called *degree positions*. Since $P(x)$ is irreducible and of degree m, it follows that $a_0 = 1$, i.e., $\alpha_{k+1} = 0$, and $a_m = 1$, i.e., $\alpha_1 = m$. For simplicity, this last coefficient α_{k+1} is always excluded from the description of Σ-sequence construction. Thus the remaining degree positions are $\alpha_1 = m, \alpha_2, \ldots, \alpha_k \geq 1$. We will be concerned with these degree positions, sometimes called *taps*.

A polynomial $P(x)$ is used to build a binary sequence S of length $|S| = 2^m + m - 2$ with the property that every subsequence of S of length m is unique within S. Given a binary sequence $y_1 y_2 \ldots y_m$, define an *Exclusive OR (XOR)* operator *relative to the indexes* the taps $\alpha_1, \alpha_2, \ldots, \alpha_k$ by $XOR_{\alpha_1, \alpha_2, \ldots, \alpha_k}(y_1 y_2 \ldots y_m) = XOR(y_{\alpha_1}, y_{\alpha_2}, \ldots, y_{\alpha_k})$. (Note that the XOR-operator applied to a number of binary values returns 1 if the number of 1's among these values is odd, and 0 otherwise.)

The recursive construction of a shift-register Σ-sequence S starts from an arbitrary binary sequence $a_1 a_2 \ldots a_m$, $a_i = 0$ or 1 for $1 \leq i \leq m$, such that not all a_i's are 0's. This m-tuple is a prefix of S. Then the next element of S is found as $a_{m+1} = XOR_{\alpha_1, \alpha_2, \ldots, \alpha_k}(a_1, a_2, \ldots, a_m)$. More in general, let $a_1, a_2, \ldots, a_m, \ldots, a_j$ be the first j elements of S, where $m \leq j < 2^m + m - 2$. Then

$$a_{j+1} = XOR_{\alpha_1, \alpha_2, \ldots, \alpha_k}(a_{j-m+1}, a_{j-m+2}, \ldots, a_j). \tag{1}$$

The generation of S ends after performing $2^m - 2$ iterations.

It follows that, under the assumption for primitiveness of $P(x)$, the period of the so-constructed sequence S is $2^m - 1$ and every m-subsequence of S is met in S only once; moreover, if another $2^m + m - 1$'th element a_{2^m+m-1} is concatenated to S, then the m-prefix of the obtained sequence $S|a_{2^m+m-1}$ will be the same subsequence of symbols as the m-suffix of S.

Example 1. Consider the primitive polynomial $P(x) = x^3 + x + 1$. In terms of the denotations used above, we have $\alpha_1 = m = 3$ and $\alpha_2 = \alpha_3 = 1$. Starting, e.g., from the binary sequence 001 of length $m = 3$, after six iterations of the described generation process, we obtain the shift-register Σ-sequence 001110100.

3   Handling Noise

As we discussed in Sect. 1, a problem, which often appears in practice is that some of the dots, of which the code is composed, are either damaged or obscured and thus the code cannot be deciphered. In other words, there is noise in the input sequence. In this section we obtain some theoretical results about searching noisy patterns in Σ-sequences.

3.1 Decoding Patterns with Gaps

As we mentioned in Sect. 1, CLUSPI code is based on folded patterns. At one reading instance, the camera captures an input in the form of a rectangle populated by binary elements. We can think of it as a row-wise matrix representation of a pattern of length $m = p \times q$, where p and q are integer numbers.

In order to decode the coordinates encoded by the pattern, it is searched for occurrence in the Σ-sequence and the position of its (unique) occurrence is the key of the coordinates.

Obviously, the length of the pattern should be at least m so that it has a unique occurrence in the Σ-sequence. However, in practice, the length of the pattern is longer as the camera captures more points. This fact may be used to solve the important problem of having some of the dots in the pattern obscured or noisy.

As an illustration, let us consider the sequence in Example 1. The pattern 1 * 0, where * denotes the unreadable element, has two possible occurrences: as 110 and as a 100.

In other words, if the pattern has length m, but one of its elements is unreadable – that is, the pattern has a *gap* its occurrence in the Σ-sequence won't be unique anymore. However, in practice the captured pattern has length longer than m and these additional sequence elements could be used to ensure the uniqueness of the pattern in the Σ-sequence.

A major question is: if we assume that the pattern has no more than k gaps, what the length of the captured sequence should be, so that we ensure uniqueness.

The following theorem gives an answer to this question.

Theorem 1. *Let S be a Σ-sequence generated by a primitive polynomial of degree $m \geq 2$. Let P be an m-pattern that has gaps (*'s) at positions $j_1^*, j_2^*, \ldots, j_k^*$. Let $j_1' = 1, j_2', \ldots, j_k'$ be the closest degree positions to the left of the j_i^*'s $(1 \leq i \leq k)$. Denote $J = \min_{i=1,\ldots,k}\{j_i^* - j_i' \ : \ S[k_1+j_i^*-1] \neq S[k_2+j_i^*-1]\}$, where k_1 and k_2 are the positions within S where two different occurrences of P start. Then all patterns that have P as their prefix and are of length greater than $m + J - 1$, are unique.*

Proof. Consider two occurrences $P_1 = S[k_1, k_1+m-1]$ and $P_2 = S[k_2, k_2+m-1]$ of pattern P in S, with k_1 and k_2 being the positions within S where P_1 and P_2 start.

If $k = 1$, as already discussed above, the gaps in P_1 and P_2 must have different unknown values.

If $k > 1$, it is possible that gaps at identical positions in P_1 and P_2 had featured identical (unknown, or "deleted") values. However, since by construction of S each m-pattern in S is unique, for at least one i, $1 \le i \le k$, the values at corresponding gap positions in P_1 and P_2 must be different.

For an arbitrary such i, let us perform a series of operations $O_1, O_2, \ldots$ of type (1) to both patterns, thus extending them to the right. One can observe that after performing an operation O_i, the corresponding elements which are added to P_1 and to P_2, respectively, will be the same as long as no degree position interferes with the values at positions $k_1 + j_i^* - 1$ and $k_2 + j_i^* - 1$. This is because the newly computed values are obtained by performing the same operation O_i of type (1) on identical inputs.

After performing a number of operations causing a number of shifts, $k_1 + j_i^* - 1$ and $k_2 + j_i^* - 1$ will simultaneously become degree positions. Since $S[k_1 + j_i^* - 1] \ne S[k_2 + j_i^* - 1]$ (by the choice of index i), the next values in the sequence S will certainly be different. It is clear that such a difference will first occur for the value of i for which J is obtained. □

The following corollary provides a simpler (although, in general, less precise) bound on the length of the unique patterns with gaps.

Corollary 1. *Under the terms of Theorem 1, denote $J_0 = \min_{i=1,\ldots,k}\{j_i^* - j_i'\}$. Then all patterns that have P as their prefix and are of length greater than $m + J_0 - 1$, are unique.*

Follows from the obvious inequality $J_0 \ge J$.

Theorem 1 implies a simple way of identifying the shortest unique pattern that contains P as a prefix. For this, after detecting the positions of the gaps in P (which requires $O(m)$ arithmetic operations), one has to locate the closest degree positions to the left of all gaps. This can be accomplished with other $O(m)$ arithmetic operations, i.e., one should perform $O(m)$ operations overall. Then, the required dimension is implied by Theorem 1.

3.2   Reconstructing Patterns with Gaps

In the previous section we showed that if a pattern with gaps is sufficiently long, then it may be unique within the Σ-sequence. In this section we show that if a pattern with a certain number of gaps is sufficiently long, then the unknown values can be recovered.

The shift-register Σ-sequence S constructed in Sect. 2 has a number of properties that make it a useful tool for various applications. Here we list one of them for future references.

Lemma 1. *[4] The set of taps for any Σ-sequence has an even number of elements.*

Here are two more facts to be used in the sequel.

Lemma 2. *Let P be an $(m+1)$-subsequence of S with an unknown value $P[i]$ at a tap-position i, $1 \le i < m + 1$. Then $P[i]$ can be determined with $O(m)$ operations.*

Proof. By construction of S (see Eq. (1)), if the number of 1's at tap-positions different than i is even, then $P[i] = 1$ if $P[m+1] = 1$ and $P[i] = 0$ if $P[m+1] = 0$. If the number of 1's at those tap-positions is odd, then $P[i] = 1$ if $P[m+1] = 0$ and $P[i] = 0$ if $P[m+1] = 1$.   □

Lemma 3. *Let P with $|P| > m$ be a subsequence of S which contains gaps at certain positions. If P contains an m-subsequence w all elements of which are known, then the whole subsequence P can be restored with $O(|P|)$ operations.*

Proof. By Eq. (1), any m consecutive values determine the value at the position next to them, thus one can compute all unknown values to the right of w. The unknown values to the left of w can be found using Lemma 2.   □

Now we can prove the following theorem.

Theorem 2. *Let P be a subsequence of a Σ-sequence generated by a primitive polynomial of degree $m \geq 2$, and let P have g gaps, $1 \leq g \leq |P|$. If $|P| \geq \omega(g, m) = (g+1)m - 1$, then the whole P can be restored with $O(|P|)$ operations.*

Proof. Consider the extreme (worst) case where $|P| = \omega(g, m)$. Partition P into $g + 1$ subsequences:

$$P = P_0 P_1 P_2 \ldots P_g, \text{ where } |P_0| = m - 1, |P_1| = m, |P_2| = m, \ldots, |P_g| = m.$$

If P_0 contains at least one gap, by the pigeonhole principle there will be a P_i, $i \neq 0$, without a gap. Then, by Lemma 3, the entire pattern P can be restored with $O(|P|)$ operations.

Let P_0 have no gaps. If $P_1[1] \neq *$, then the subsequence $P_0 P_1[1]$ is of length m and has no gaps. Once again, by Lemma 3 the entire pattern P can be recovered.

Now let $P_1[1] = *$. If P_1 has another gap, then some P_i, $i \neq 0, 1$, will contain no gap and we fall within a case already considered. Therefore, suppose that $P[1] = *$ is the only gap in P_1.

By Lemma 1, a shift-register sequence has an even number of taps, i.e., at least two, for any $m \geq 2$. Hence, if the leftmost tap is placed at the first position of P_0 (and thus also of P), the position of the rightmost tap will be greater than or equal to two. Therefore, after no more than $m - 2$ shifts, $P_1[1]$ will feature the rightmost tap position since there is at least one more to the left of it which is with a known value. Then the result stated follows from Lemma 2.   □

The above theorem provides a necessary condition for a pattern length so that a pattern with a number of gaps to be restorable. This is a "worst-case" bound. The following example shows that it is also achievable.

Example 2. Consider the primitive polynomial $P(x) = x^2 + x + 1$. We have $m = 2$. Starting from sequence 01, this polynomial generates the Σ-sequence 0110. For $g = 1$, the bound of Theorem 2 is $\omega(1, 2) = 3$. Obviously, a gap in any 3-pattern can be filled in. This, however, is not possible for a pattern of length two.

It is not hard to realize that the possibility to successfully determine a gap value depends on the specific pattern and polynomial used. This is illustrated by the following example.

Example 3. Consider the primitive polynomial $P(x) = x^5 + x^2 + 1$. Starting from the 5-pattern 00001, it generates the Σ-sequence

$$S = 00001010111011000111110\underline{0110100}10000.$$

Consider the underlined pattern $P = 0110100$ and first assume that it has a gap in its third position, i.e., let $P' = 01*0100$. Clearly, the value of $*$ cannot be found. Consider now the same pattern with two gaps at its first and last positions, i.e., let $P'' = 0*101*0$. It's easy to see that the values for both gaps can be determined, although the pattern has a length 7 which is twice less than $\omega(2,5) = 14$.

4 Concluding Remarks

The Cluster Pattern Interface technology has numerous applications [8–11]. In this paper we presented some theoretical results related to coping with the problem of a noisy input. Some open questions remain as to how to optimally calibrate the camera for a given Σ-sequence, so that on the one hand it captures enough dots to be able to recognize even noisy patterns, but on the other hand it can be positioned at a reasonable distance. Another direction of the theoretical results is aiming at utilizing the information about the angle of the camera towards the encoded surface to implement additional functionality. We are also planning to extend the results into higher dimensions.

Acknowledgements. The work was partly supported by a Cooperative Research Project of RIE, Shizuoka University, Japan.

References

1. Apostolico, A., Chrochemore, M.: Pattern Matching Algorithms. Oxford University Press, New York (1997)
2. Barneva, R.P., Brimkov, V.E., Kanev, K.K.: Theoretical issues of cluster pattern interfaces. In: Wiederhold, P., Barneva, R.P. (eds.) IWCIA 2009. LNCS, vol. 5852, pp. 302–315. Springer, Heidelberg (2009). https://doi.org/10.1007/978-3-642-10210-3_24
3. Shim, J.-Y., Kim, S.-W.: Design of circular dot pattern code (CDPC) for maximum information capacity and robustness on geometric distortion/noise. Multimedia Tools Appl. **70**(3), 1941–1955 (2012). https://doi.org/10.1007/s11042-012-1222-x
4. Golomb, S.W.: Shift Register Sequences. Holden-Day Inc., San Francisco, Cambridge, London, Amsterdam (1967)
5. Kanev, K., Kimura, S.: Digital Information Carrier. Patent Registration No 3635374, Japan Patent Office (2005)

6. Kanev, K., Kimura, S.: Digital Information Carrier. Patent Registration No 4368373, Japan Patent Office (2009)

7. Kanev, K., Kimura, S.: Direct point-and-click functionality for printed materials. J. Three Dimension. Images **20**(2), 51–59 (2006)

8. Kanev, K., Kimura, S., Orr, T.: A framework for collaborative learning in dynamic group environments. Int. J. Distance Educ. Technol. **7**(1), 58–77 (2009)

9. Kanev, K., Mirenkov, N., Brimkov, B., Dimitrov, K.: Semantic surfaces for business applications. In: Dicheva, D., et al. (eds.) Software, Services, and Semantic Technologies, pp. 36–43. Demetra Publishing, Sofia (2009)

10. Kanev, K., Mirenkov, N., Hasegawa, A.: Newsputers: digitally enhanced printouts supporting pervasive multimodal interactions. In: Proceedings of the First IEEE International Conference on Ubi-Media Computing (U-Media 2008), Lanzhou, China, pp. 1–7 (2008)

11. Kanev, K., Mirenkov, N., Urata, A.: Position sensing for parking simulation and guidance. J. Three Dimension. Images **21**(1), 66–69 (2007)

12. Mohan, A., Woo, G., Hiura, S., Smithwick, Q., Raskar, R.: Bokode, Imperceptible visual tags for camera-based interaction from a distance. ACM Trans. Graph. **28**(3), 98:1–98:8 (2009)

13. QR-code Standardization. https://www.qrcode.com/en/. Accessed 20 April 2020

14. Signer, B., Fundamental Concepts for Interactive Paper and Cross-Media Information Spaces. Doctoral Thesis, Swiss Federal Institute of Technology, Zurich, May 2008

15. Silberman, S., The Hot New Medium: Paper. Wired Magazine, April 1, 2001

Local Q-Convexity Histograms
for Shape Analysis

Judit Szűcs[✉] and Péter Balázs

Department of Image Processing and Computer Graphics,
University of Szeged, Szeged, Hungary
{jszucs,pbalazs}@inf.u-szeged.hu

Abstract. In this paper we propose a novel local shape descriptor based on Q-convexity histograms. We investigate three different variants: (1) focusing only on the background points, (2) examining all the points and (3) omitting the zero bin. We study the properties of the variants on a shape and on a texture dataset. In an illustrative example, we compare the classification accuracy of the introduced local descriptor to its global counterpart, and also to a variant of Local Binary Patterns which is similar to our descriptor in the sense that its histogram collects frequencies of local configurations. We show that our descriptor can reach in many cases higher classification accuracy than the others.

Keywords: Quadrant-convexity · Local shape descriptor · Classification · Shape analysis

1 Introduction

The measure of convexity is a frequently used shape descriptor in digital image analysis. Numerous convexity measures have been proposed in the past years based on different approaches. There are, among others, boundary-based [15], area-based [12], direction-based [9], as well as probability-based methods [11] to express the degree of convexity of a shape. Recently, in [2,3] the authors introduced a convexity measure that relies on the concept of so-called Quadrant-convexity (shortly, Q-convexity) [4,5]. The measure is global in the sense that it describes the degree of Q-convexity of the entire image by a single scalar value.

As many studies revealed, histograms built on local features can provide much richer information on the geometry and structural properties of the shape than single scalar descriptors do (see, e.g., the Histogram of Oriented Gradients [7], the Speeded Up Robust Features [1] or variants of the Local Binary Patterns [10]). In this paper we extend the abovementioned global Q-convexity measure to histograms collecting Q-convexity values calculated under all possible positions of an image window of predefined size. In Sect. 2 we give the necessary notions and definitions. Then, in Sect. 3 we present an experimental analysis of the introduced descriptor and show its effectiveness in a classification problem. Finally, in Sect. 4 we conclude our results.

© Springer Nature Switzerland AG 2020
T. Lukić et al. (Eds.): IWCIA 2020, LNCS 12148, pp. 245–257, 2020.
https://doi.org/10.1007/978-3-030-51002-2_18

2   Definitions and Notation

We first recall the global Q-convexity measure from [2,3]. Consider a two-dimensional non-empty *lattice set* F, i.e., a finite subset $F \subset \mathbb{Z}^2$ defined up to translation. Let $\mathcal{R}$ be the smallest discrete rectangle covering F and suppose it is of size $m \times n$. Without loss of generality we can assume that the bottom-left corner of $\mathcal{R}$ is in the origin $(0,0)$, i.e., $\mathcal{R} = \{0,\dots,m-1\} \times \{0,\dots,n-1\}$. Alternatively, F can be viewed as a binary image, i.e., as a union of white unit squares (foreground pixels) corresponding to points of F, and $\mathcal{R} \setminus F$ being the union of black unit squares (background pixels).

Each position (i,j) in the rectangle $\mathcal{R}$ together with the horizontal and vertical directions determines the following four quadrants:

$$Z_0(i,j) = \{(l,k) \in \mathcal{R} : 0 \le l \le i,\ 0 \le k \le j\},$$
$$Z_1(i,j) = \{(l,k) \in \mathcal{R} : i \le l \le m-1,\ 0 \le k \le j\},$$
$$Z_2(i,j) = \{(l,k) \in \mathcal{R} : i \le l \le m-1,\ j \le k \le n-1\},$$
$$Z_3(i,j) = \{(l,k) \in \mathcal{R} : 0 \le l \le i,\ j \le k \le n-1\}.$$

The number of object points (foreground pixels) of F in $Z_p(i,j)$ is denoted by $n_p(i,j)$, for $p = 0,\dots,3$, i.e.,

$$n_p(i,j) = card(Z_p(i,j) \cap F) \quad (p = 0,\dots,3). \tag{1}$$

We say that a lattice set F is Q-convex if for each (i,j), $(n_0(i,j) > 0 \wedge n_1(i,j) > 0 \wedge n_2(i,j) > 0 \wedge n_3(i,j) > 0)$ implies $(i,j) \in F$. If F is not Q-convex, then there exists a position (i,j) violating the Q-convexity property, i.e. $n_p(i,j) > 0$ for all $p = 0,\dots,3$ and still $(i,j) \notin F$. Figure 1 illustrates the above concepts.

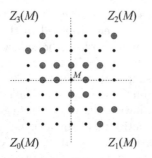

Fig. 1. A lattice set F (left) together with the four quadrants around the point M, and the corresponding binary image representation (right). F is not Q-convex since $n_p(M) > 0$ for all $p = 0,\dots,3$ and still $M \notin F$.

Now, let (i,j) be an arbitrary point of $\mathcal{R}$. The Q-*concavity contribution* of (i,j) w.r.t. F is defined as

$$\varphi_F(i,j) = n_0(i,j)n_1(i,j)n_2(i,j)n_3(i,j)(1 - f(i,j)), \tag{2}$$

where $f(i, j) = 1$ if $(i, j) \in F$, otherwise $f(i, j) = 0$. The sum of the contributions of Q-concavities for each point in $\mathcal{R}$ is expressed as

$$\varphi_F = \sum_{(i,j) \in \mathcal{R}} \varphi_F(i, j). \tag{3}$$

Our idea is to refine (2) and (3) to achieve a local and thus more informative measure of Q-concavity. For this purpose, consider a $(2w+1) \times (2w+1)$ window. The quadrants around (i, j) restricted to this window size can then be defined as

$$Z_0^w(i, j) = \{(l, k) \in \mathcal{R} : (i - w) \leq l \leq i, \ (j - w) \leq k \leq j\},$$
$$Z_1^w(i, j) = \{(l, k) \in \mathcal{R} : i \leq l \leq (i + w), \ (j - w) \leq k \leq j\},$$
$$Z_2^w(i, j) = \{(l, k) \in \mathcal{R} : i \leq l \leq (i + w), \ j \leq k \leq (j + w)\},$$
$$Z_3^w(i, j) = \{(l, k) \in \mathcal{R} : (i - w) \leq l \leq i, \ j \leq k \leq (j + w)\}.$$

The number of object points in Z_p^w is

$$n_p^w(i, j) = card(Z_p^w(i, j) \cap F) \ (p = 0, \dots, 3), \tag{4}$$

and the local Q-concavity contribution at the point (i, j) is

$$\varphi_F^w(i, j) = n_0^w(i, j) n_1^w(i, j) n_2^w(i, j) n_3^w(i, j)(1 - f(i, j)). \tag{5}$$

Finally, the *local Q-convexity histogram* of F is a mapping $hist_{F,w} : \mathbb{Z} \to \mathbb{Z}$ which we can define in different ways. The first approach focuses on the background points in $\mathcal{R}$. In this case

$$hist_{F,w,background}(r) = |(i, j) \in \mathcal{R} \setminus F : \varphi^w(i, j) = r|, \tag{6}$$

i.e., we take each background point, calculate its local Q-concavity value, and increase (by 1) the value of the corresponding bin. Alternatively, we can take into account all the points in $\mathcal{R}$. Then,

$$hist_{F,w,all}(r) = |(i, j) \in \mathcal{R} : \varphi^w(i, j) = r|. \tag{7}$$

A third approach focuses exclusively on the points that truly violate local Q-convexity. Then we get

$$hist_{F,\hat{w},nonzero}(r) = |(i, j) \in \mathcal{R} : \varphi^w(i, j) = r > 0|, \tag{8}$$

i.e., in this case the 0 bin is omitted.

Let F be an arbitrary binary image. With elementary combinatorics we get that the theoretical maximum of $\varphi_F^w(i, j)$ is $((w + 1)^2 - 1)^4$. Moreover, $\varphi_F^w(i, j)$ is either equal to 0 or it is a product of four positive integers from the interval $[1, (w + 1)^2 - 1]$. Owing to the associative property of multiplication, using the formula of k-combinations with repetition we get an upper bound on the possible number of the necessary bins equals to

$$\binom{((w + 1)^2 - 1) + 4 - 1}{4} + 1 = \binom{(w + 1)^2 + 2}{4} + 1. \tag{9}$$

Nevertheless, many of the bins are inherently empty, e.g. those having indices obtained by multiplying five or more primes.

Figure 2 shows two binarized retina images of size 1000×1000 together with their local Q-convexity histograms $hist_{F,2,nonzero}$ processed for a 5×5 window. Since the maximum of $\varphi^w(i,j)$ can be 4096, we allocated 4096 bins to store the values (remember that the 0 bin is omitted). The histograms $hist_{F,2,background}$ and $hist_{F,2,all}$ would look the same except that in these cases the 0 bin is also presented. For the image in Fig. 2a the value of this bin is 937 885 and 991 253, in case of $hist_{F,2,background}$ and $hist_{F,2,all}$, respectively. Concerning the image of Fig. 2c the value of the 0 bin is 918 465 and 991 167, in case of $hist_{F,2,background}$ and $hist_{F,2,all}$, respectively. Thus, in these cases the 0 bin gives a significant peak in the histograms, and all the other bins become negligible.

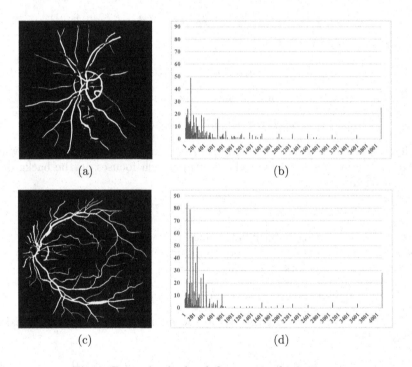

(a)   (b)

(c)   (d)

Fig. 2. Examples for local Q-convexity histograms.

For an other example, consider the bullseye image of Fig. 3 (of size 256×256), and let the window be, again, of size 5×5. In this case all the bins excluding the 0 bin will have 0 value, independently from the type of the histogram calculated. This reveals an interesting behavior of the descriptor: the shape seems to be locally Q-convex in each window position, even though, clearly, it is globally not Q-convex. For completeness, the value of the 0 bin for this image is 43 492 (the number of background points) and 63 504 (the number of positions where the window entirely fit into the image), in case of $hist_{F,2,background}$ and

$hist_{F,2,all}$, respectively. Naturally, for sufficiently large window sizes the shape will be neither locally nor globally Q-convex.

Fig. 3. A globally non-Q-convex shape that is locally Q-convex in each window position, e.g., with window size 5×5.

3   Experimental Results

3.1   Histogram Variants and the Effect of Window Size

To compare the three histogram variants introduced in Sect. 2 we took 22 binary images of size 128×128 from [14] (see Fig. 4) and computed their local Q-convexity histograms. Then, we calculated the Euclidean distances of the histogram vectors between all possible (distinct) image pairs. Table 1 collects the maximal, minimal, and mean distance between the image pairs, as well as the standard deviation of the distances for different window sizes and for all three histogram approaches. From the entries of the table we deduce the following. The minimum and maximum values of the approach that takes only the background points into account (Eq. 6, columns *background* in the table) stretches a wider interval than those of the other two approaches and, in the same time, with greater deviation. Furthermore, the histograms of Eqs. 7 and 8 (columns *all* and *nonzero*, respectively) cannot distinguish two locally Q-convex images, whereas that of Eq. 6 still can measure some difference, based on the number of background points. Concerning this latter approach, we notice that, in general, greater window sizes provide greater difference between the minimum and maximum values, and thus also greater mean and deviance values.

In a second experiment, we investigated how different textures can be separated by the local Q-convexity histograms. We took 6 texture images (see Fig. 5) and cut 50-50 random patches of size 128×128 of them. Table 2 shows the mean Euclidean distance between the representatives of different texture classes of Fig. 5 (in other words, the mean interclass distance), whereas Table 3 gives the intraclass standard deviation of the 6 classes. It is clearly seen that for the *background* approach the interclass Euclidean distances are typically greater than in the case of the other two histogram variants. We also observe that taking an

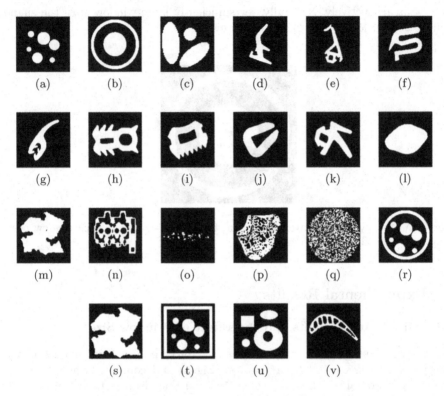

Fig. 4. Images taken from [14].

Table 1. Statistics of Euclidean distances for images of Fig. 4.

	3×3 background	3×3 all	3×3 nonzero	5×5 background	5×5 all	5×5 nonzero
Min	16	0	0	15	0	0
Max	8 806	919	280	8 798	3 499	421
Mean	2 401	112	37	2 647	421	53
Deviation	1 749	265	81	1 924	999	118

	7×7 background	7×7 all	7×7 nonzero	9×9 background	9×9 all	9×9 nonzero
Min	12	0	0	34	0	0
Max	10 635	5 533	365	11 251	6 209	244
Mean	2 859	700	50	2 939	840	40
Deviation	2 210	1 568	101	2 318	1 742	66

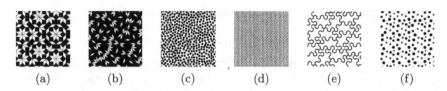

$$\text{(a)} \qquad \text{(b)} \qquad \text{(c)} \qquad \text{(d)} \qquad \text{(e)} \qquad \text{(f)}$$

Fig. 5. Texture image classes.

arbitrary image pair of Table 2 and checking the corresponding intraclass deviation of their classes in Table 3, in general, this latter value is relatively smaller (compared to the distance value, e.g., by taking the ratio of the two) for the *background* variant than for the other two methods. Based on these two experiments, in the sequel we prefer to use the histogram based on the background points.

3.2 Classification of Retina Images

Experimental Setting. As a case study, to investigate the classification power of the introduced shape descriptor, we repeated an experiment of [3,6] with exactly the same dataset and classifier. In [3] the authors concluded that in this experiment the global Q-convexity measure (using just two directions) can ensure comparable or in some cases even better results than that obtained by [6] employing many directions. Thus, the question is, if the local Q-convexity histogram can even outperform the global Q-convexity measure.

We used public datasets of fundus photographs of the retina. The CHASEDB1 [8] dataset is composed of 20 binary images with centered optic disks (see Fig. 2a for one of them), while the DRIVE [13] dataset contains 20 images where the optic disk is shifted from the center (see Fig. 2c). All the images are of the same size 1000×1000.

Following the strategy of [6], different types of random noise were added to the images: Gaussian and Speckle noise were added with 10 increasing variances $\sigma^2 \in [0,2]$, while salt & pepper noise was added with 10 increasing amounts in $[0,0.1]$. We tried to classify the images into two classes (CHASEDB1 and DRIVE) based on their local Q-convexity histograms using the background points (Eq. 6), by the 5-nearest-neighbor classifier with inverse Euclidean distance. To avoid overfitting, we used leave-one-out cross validation.

Normalization. Since our aim was to compare the performance of our descriptor to the global Q-convexity measure published in [3], we modified the formula of local Q-concavity contribution at the point (i,j) (given in Eq. 5) to

$$\varphi_{F,norm}^w(i,j) = \frac{n_0^w(i,j)n_1^w(i,j)n_2^w(i,j)n_3^w(i,j)(1-f(i,j))}{(\alpha+h_i+v_j)^4}, \tag{10}$$

Table 2. Mean interclass Euclidean distances of the texture classes in Fig. 5.

Figure	Figure	3×3 background	3×3 all	3×3 nonzero	5×5 background	5×5 all	5×5 nonzero
5f	5a	5 861	374	191	5 489	439	177
5f	5b	2 524	26	18	2 332	169	83
5f	5c	77	1	1	996	992	267
5f	5d	9 105	240	97	8 809	209	60
5f	5e	511	257	181	2 078	2 689	722
5e	5a	5 386	302	269	7 469	2 306	724
5e	5b	2 052	231	162	4 340	2 550	728
5e	5c	584	259	182	1 250	1 806	766
5e	5d	8 629	216	213	10 781	2 497	722
5d	5a	3 251	193	164	3 327	263	168
5d	5b	6 582	224	100	6 480	134	124
5d	5c	9 182	241	96	9 773	802	271
5c	5a	5 937	375	191	6 454	637	315
5c	5b	2 600	27	19	3 302	856	282
5b	5a	3 340	359	193	3 162	328	207

Figure	Figure	7×7 background	7×7 all	7×7 nonzero	9×9 background	9×9 all	9×9 nonzero
5f	5a	5 525	340	129	5 595	130	105
5f	5b	1 852	567	176	278	2 225	263
5f	5c	2 602	2 506	311	2 238	2 218	173
5f	5d	8 710	112	54	8 636	186	58
5f	5e	2 750	3 373	597	2 261	2 846	367
5e	5a	8 229	3 065	604	7 834	2 770	372
5e	5b	4 571	2 851	631	2 368	768	462
5e	5c	680	1 070	673	399	730	399
5e	5d	11 410	3 276	594	10 873	3 021	362
5d	5a	3 189	246	115	3 044	269	87
5d	5b	6 870	489	211	8 549	2 403	285
5d	5c	11 297	2 409	311	10 869	2 393	165
5c	5a	8 113	2 197	331	7 827	2 141	186
5c	5b	4 441	1 981	359	2 346	328	328
5b	5a	3 688	330	242	5 510	2 152	298

Table 3. Intraclass deviation of the texture classes of Fig. 5.

Figure	3×3 background	3×3 all	3×3 nonzero	5×5 background	5×5 all	5×5 nonzero
5a	363	81	52	267	98	82
5b	74	6	4	58	20	12
5c	124	5	3	121	86	19
5d	301	46	26	253	41	19
5e	52	24	16	75	84	69
5f	125	3	2	123	6	4

Figure	7×7 background	7×7 all	7×7 nonzero	9×9 background	9×9 all	9×9 nonzero
5a	363	63	46	394	59	37
5b	61	19	17	48	36	30
5c	74	109	24	17	99	17
5d	342	62	12	294	58	4
5e	53	63	53	30	50	30
5f	130	12	4	140	36	7

where α denotes the number of points belonging to F in the current window, h_i is the number of points of F in the ith row, and v_j is number of points of F in the jth column, restricted for the window.

Remark 1. In [3], the normalizing denominator for the local Q-concavity contribution at the point (i, j) is $((\alpha + h_i + v_j)/4)^4$. However, the constant factor $\frac{1}{4}$ can be omitted in case of local Q-convexity histograms as only the relative values of the bins are important.

Quantization. The theoretical maximum value of $\varphi_{F,norm}^{w}(i, j)$ is

$$\max \varphi_{F,norm}^{w} = \frac{1}{4^4} = 0.00390625, \tag{11}$$

independently on the size of the window (see also Remark 1). Since $\varphi_{F,norm}^{w}(i, j)$ is not an integer, the bins of the histogram are not straightforward to index. To overcome this problem we do quantization. Let q be the number of quantization levels, i.e., the number of bins to occur in the histogram. Then we get the bin indices by the following formula:

$$bin\_index = \left\lfloor \varphi_{F,norm}^{w}(i, j) \middle/ \frac{\max \varphi_{F,norm}^{w}}{q - 1} \right\rfloor. \tag{12}$$

Results. Tables 4 and 5 show the classification accuracy for different window sizes with quantization levels $q = 2$ and $q = 10$, respectively, together with the classification results of the global Q-convexity descriptor of [3]. For further reference the accuracy achieved by Shift Local Binary Patterns (SLBP) is also presented. SLBP [10] is a visual descriptor mostly used for classification in computer vision. It illustrates the relationship between the pixels and their

Table 4. Classification accuracy of retina images for $q = 2$.

Noise type	Noise level	SLBP	3×3	5×5	7×7	9×9	Global
Gaussian	0	1	0.9	0.9	0.725	0.725	0.925
Gaussian	1	0.825	0.825	0.85	0.775	0.775	0.775
Gaussian	2	0.825	0.9	0.8	0.875	0.85	0.575
Gaussian	3	0.9	0.875	1	0.95	0.95	0.425
Gaussian	4	0.775	0.7	0.75	0.725	0.875	0.65
Gaussian	5	0.675	0.7	0.8	0.75	0.85	0.475
Gaussian	6	0.725	0.75	0.825	0.8	0.725	0.65
Gaussian	7	0.65	0.725	0.675	0.775	0.75	0.725
Gaussian	8	0.65	0.7	0.7	0.575	0.625	0.675
Gaussian	9	0.65	0.525	0.6	0.675	0.275	0.35
S&P	0	1	0.975	0.95	0.875	0.825	0.925
S&P	1	1	0.925	0.875	0.85	0.825	0.9
S&P	2	1	0.95	0.875	0.95	0.95	0.925
S&P	3	1	0.85	0.775	0.85	0.925	0.925
S&P	4	1	0.85	0.825	0.875	0.875	0.9
S&P	5	1	0.9	0.8	0.85	0.875	0.9
S&P	6	1	0.75	0.9	0.925	0.95	0.925
S&P	7	1	0.9	0.8	0.875	0.85	0.9
S&P	8	1	0.85	0.925	0.925	0.975	0.875
S&P	9	1	0.8	0.775	0.875	0.875	0.9
Speckle	0	1	0.825	0.925	0.725	0.725	0.925
Speckle	1	0.775	0.875	0.775	0.8	0.775	0.9
Speckle	2	0.85	0.875	0.875	0.875	0.825	0.925
Speckle	3	0.925	0.95	0.9	0.875	0.875	0.925
Speckle	4	0.875	0.875	0.825	0.85	0.75	0.95
Speckle	5	0.9	0.925	0.875	0.825	0.825	0.925
Speckle	6	0.95	0.95	0.95	0.85	0.825	0.95
Speckle	7	0.9	0.975	0.95	0.85	0.825	0.925
Speckle	8	0.925	0.975	0.95	0.9	0.925	0.925
Speckle	9	0.9	0.925	0.9	0.9	0.875	0.95

neighbors (most often 8-neighbors) with a histogram (having 256 bins in case of 8-neighbors). We chose this method as a reference being its approach similar to that of our local Q-convexity histograms. In the tables, the best values achieved by the Q-convexity based approaches (i.e., not considering SLBP) are highlighted.

Table 5. Classification accuracy of retina images for $q = 10$.

Noise type	Noise level	SLBP	3×3	5×5	7×7	9×9	Global
Gaussian	0	1	1	1	0.975	0.95	0.925
Gaussian	1	0.825	0.8	0.7	0.775	0.975	0.775
Gaussian	2	0.825	0.85	0.85	0.85	0.875	0.575
Gaussian	3	0.9	0.85	0.95	0.875	0.875	0.425
Gaussian	4	0.775	0.775	0.75	0.7	0.85	0.65
Gaussian	5	0.675	0.7	0.75	0.8	0.8	0.475
Gaussian	6	0.725	0.75	0.775	0.85	0.725	0.65
Gaussian	7	0.65	0.7	0.55	0.7	0.775	0.725
Gaussian	8	0.65	0.425	0.575	0.7	0.65	0.675
Gaussian	9	0.65	0.625	0.6	0.725	0.45	0.35
S&P	0	1	1	1	0.875	0.9	0.925
S&P	1	1	0.95	0.975	0.875	0.95	0.9
S&P	2	1	0.9	0.975	0.95	0.95	0.925
S&P	3	1	0.875	0.95	0.975	0.95	0.925
S&P	4	1	0.925	0.925	0.95	0.95	0.9
S&P	5	1	0.9	0.975	0.975	0.975	0.9
S&P	6	1	0.95	0.975	1	0.975	0.925
S&P	7	1	0.975	0.975	0.95	0.975	0.9
S&P	8	1	0.925	0.95	0.95	0.95	0.875
S&P	9	1	0.95	0.975	0.975	0.975	0.9
Speckle	0	1	1	1	0.975	0.925	0.925
Speckle	1	0.775	0.9	0.625	0.7	0.925	0.9
Speckle	2	0.85	0.75	0.85	0.725	0.875	0.925
Speckle	3	0.925	0.95	0.925	0.875	0.825	0.925
Speckle	4	0.875	0.85	0.725	0.7	0.775	0.95
Speckle	5	0.9	0.925	0.85	0.825	0.675	0.925
Speckle	6	0.95	0.925	0.9	0.825	0.7	0.95
Speckle	7	0.9	0.9	0.875	0.925	0.85	0.925
Speckle	8	0.925	0.95	0.95	0.825	0.825	0.925
Speckle	9	0.9	0.9	0.875	0.775	0.825	0.95

We observe that in presence of Gaussian noise, in almost all cases, the local, histogram-based method performs significantly better than the one based on the global Q-convexity measure, and also better than SLBP, both for $q = 2$ and $q = 10$. For salt & pepper noise, SLBP is the best, however, especially for $q = 10$ the local method is almost as good as SLBP. We stress that SLBP is a 256-dimensional descriptor, whereas the histogram-based one uses only 10-dimensional vectors (when $q = 10$). In case of Speckle noise, the global Q-convexity measure seems to be the best choice, although in some cases, and especially for smaller window sizes the local approach as well as SLBP ensures comparable accuracy.

Finding the proper window size and the appropriate number of quantization levels for a classification problem is, of course, challenging for which one can utilize feature selection methods from the field of machine learning. However, this topic is out of scope of the paper.

4   Conclusions

In this research we introduced a Quadrant-convexity based local shape descriptor which uses predefined windows to create histograms. We studied three different variants of the approach (*background, all, nonzero*) on two datasets and found that the histogram based on the background points is the most suitable for classification. We conducted an illustrative experiment to compare the classification accuracy of our proposal to SLBP and a global Quadrant-convexity descriptor. We deduced that our descriptor can ensure comparable accuracy and, in many cases, it outperforms the others.

Currently, we have no general strategy to choose the proper window size for a problem, which, of course may be different from task to task. One simple strategy in classification tasks would be to try out different window sizes on the train set and choose the one ensuring the highest classification accuracy on a validation set. In addition, one could also combine histograms belonging to different window sizes to gain more informative shape descriptors.

Acknowledgments. Judit Szűcs was supported by the UNKP-19-3-SZTE-291 New National Excellence Program of the Ministry for Innovation and Technology, Hungary. This research was supported by the project "Integrated program for training new generation of scientists in the fields of computer science", no. EFOP-3.6.3-VEKOP16-2017-00002. This research was supported by grant TUDFO/47138-1/2019-ITM of the Ministry for Innovation and Technology, Hungary.

References

1. Bay, H.B., Ess, A., Tuytelaars, T., Luc, V.G.: SURF: speeded up robust features. Comput. Vis. Image Underst. **110**(3), 346–359 (2008)
2. Brunetti, S., Balázs, P., Bodnár, P.: Extension of a one-dimensional convexity measure to two dimensions. In: Brimkov, V.E., Barneva, R.P. (eds.) IWCIA 2017. LNCS, vol. 10256, pp. 105–116. Springer, Cham (2017). https://doi.org/10.1007/978-3-319-59108-7_9

3. Brunetti, S., Balázs, P., Bodnár, P., Szűcs, J.: A spatial convexity descriptor for object enlacement. In: Couprie, M., Cousty, J., Kenmochi, Y., Mustafa, N. (eds.) DGCI 2019. LNCS, vol. 11414, pp. 330–342. Springer, Cham (2019). https://doi.org/10.1007/978-3-030-14085-4_26

4. Brunetti, S., Daurat, A.: An algorithm reconstructing convex lattice sets. Theoret. Comput. Sci. **304**(1–3), 35–57 (2003)

5. Brunetti, S., Daurat, A.: Reconstruction of convex lattice sets from tomographic projections in quartic time. Theoret. Comput. Sci. **406**(1–2), 55–62 (2008)

6. Clement, M., Poulenard, A., Kurtz, C., Wendling, L.: Directional enlacement histograms for the description of complex spatial configurations between objects. IEEE Trans. Pattern Anal. Mach. Intell. **39**(12), 2366–2380 (2017)

7. Dalal, N., Triggs, B.: Histograms of oriented gradients for human detection. In: 2005 IEEE Computer Society Conference on Computer Vision and Pattern Recognition (CVPR 2005), vol. 1, pp. 886–893, June 2005

8. Fraz, M.M., Remagnino, P., Hoppe, A., Uyyanonvara, B., Rudnicka, A.R., Owen, C.G., Barman, S.A.: An ensemble classification-based approach applied to retinal blood vessel segmentation. IEEE Trans. Biomed. Eng. **59**(9), 2538–2548 (2012)

9. Gorelick, L., Veksler, O., Boykov, Y., Nieuwenhuis, C.: Convexity shape prior for binary segmentation. IEEE Trans. Pattern Anal. Mach. Intell. **39**(2), 258–271 (2016)

10. Kylberg, G., Sintorn, I.-M.: Evaluation of noise robustness for local binary pattern descriptors in texture classification. EURASIP J. Image Video Process. **2013**(1), 1–20 (2013). https://doi.org/10.1186/1687-5281-2013-17

11. Rahtu, E., Salo, M., Heikkila, J.: A new convexity measure based on a probabilistic interpretation of images. IEEE Trans. Pattern Anal. Mach. Intell. **28**(9), 1501–1512 (2006)

12. Sonka, M., Hlavac, V., Boyle, R.: Image Processing, Analysis, and Machine Vision. Cengage Learning, Stamford (2014)

13. Staal, J., Abràmoff, M.D., Niemeijer, M., Viergever, M.A., Van Ginneken, B.: Ridge-based vessel segmentation in color images of the retina. IEEE Trans. Med. Imaging **23**(4), 501–509 (2004)

14. Varga, L.G., Nyl, L.G., Nagy, A., Balzs, P.: Local and global uncertainty in binary tomographic reconstruction. Comput. Vis. Image Understand. **129**, 52–62 (2014), http://www.sciencedirect.com/science/article/pii/S1077314214001179; Special section: Advances in Discrete Geometry for Computer Imagery

15. Zunic, J., Rosin, P.L.: A new convexity measure for polygons. IEEE Trans. Pattern Anal. Mach. Intell. **26**(7), 923–934 (2004)

k-Attempt Thinning

Kálmán Palágyi$^{(\boxtimes)}$ and Gábor Németh

Department of Image Processing and Computer Graphics,
University of Szeged, Szeged, Hungary
{palagyi,gnemeth}@inf.u-szeged.hu

Abstract. Thinning is a frequently used approach to produce all kinds of skeleton-like shape features in a topology-preserving way. It is an iterative object reduction: some border points of binary objects are deleted, and the entire process is repeated until stability is reached. In the conventional implementation of thinning algorithms, we have to investigate the deletability of all border points in each iteration step. In this paper, we introduce the concept of k-attempt thinning ($k \geq 1$). In the case of a k-attempt algorithm, if a border point 'survives' at least k successive iterations, it is 'immortal' (i.e., it belongs to the produced feature). We propose a computationally efficient implementation scheme for k-attempt thinning. It is shown that an existing parallel thinning algorithm is 5-attempt, and the advantage of the new implementation scheme over the conventional one is also illustrated.

Keywords: Digital topology · Skeletonization · Thinning

1 Introduction

A *binary picture* [8] on a grid is a mapping that assigns a color of *black* or *white* to each grid element that is called a *point*. A *reduction* [3] transforms a binary picture only by changing some black points to white ones, which is referred to as *deletion*. *Parallel reductions* can delete a set of black points simultaneously, while *sequential reductions* traverse the black points of a picture, and focus on the actually visited point for possible deletion at a time [3].

For digital pictures, *skeletonization* means extraction of *skeleton-like shape features* from digital binary objects [19]. In 2D, two kinds of features are taken into consideration: the *centerline* that approximates the *continuous skeleton* [1], and the *topological kernel* that is a minimal set of points being topologically equivalent [8] to the original object (i.e., if we remove any further point from it, then the topology is not preserved). In the 3D case, there are three types of skeleton-like shape features: the *centerline* (that is a concise representation of tubular and tree-like objects), the *medial surface* (that provides an approximation to the continuous 3D skeleton, since it can contain 2D surface patches), and the *topological kernel* (that is useful in representing or checking the topological structure of the object to be described).

© Springer Nature Switzerland AG 2020
T. Lukić et al. (Eds.): IWCIA 2020, LNCS 12148, pp. 258–272, 2020.
https://doi.org/10.1007/978-3-030-51002-2_19

Thinning [3,10,20] is a skeletonization method: border points of binary objects that satisfy certain topological and geometric constraints are deleted in iteration steps. The entire process is then repeated until stability is reached. Thinning is the fastest skeletonization technique, preserve the topology [6,8], and can produce all kinds of skeleton-like shape features in 2D [13] and 3D [14]. One may think that thinning is an obsolete approach, but it remains a frequently used skeletonization method in numerous applications, see for example [5,12].

In the conventional implementation of thinning, deletability of all border points are to be investigated in each iteration step. It is a time-consuming process for objects that contain also 'thinner' and 'thicker' parts (or images with 'thinner' and 'thicker' objects), since some elements in the produced features are formed within 'a few' iterations, but the iterative object reduction is to be continued until stability is reached. That is why we have investigated the *fixpoints* (i.e., survival points whose rechecking is not needed in the remaining iterations) of some special iterated reductions [15,16].

In this paper, we propose another way to speed up the process by introducing the notion of *k-attempt thinning* ($k \geq 1$). In a *k*-attempt algorithm, if a border point is not deleted in at least k successive iterations, it cannot be deleted later (i.e., it belongs to the produced feature). We give a computationally efficient implementation scheme for *k*-attempt thinning, and it is shown that the parallel thinning algorithm proposed by Eckhardt and Maderlechner [2] is 5-attempt. Lastly the advantage of the new implementation scheme over the conventional one is instantiated.

2   Basic Notions and *k*-Attempt Thinning

Next, we apply the fundamental concepts of digital topology as reviewed by Kong and Rosenfeld [8]. Despite the fact that there are other approaches based on cellular/cubical complexes [9] or polytopal complexes [7], here we shall consider the 'conventional paradigm' of digital topology.

An (m,n) *(binary digital) picture* on a grid $\mathcal{V}$ is a quadruple $(\mathcal{V}, m, n, B)$ [8], where $B \subseteq \mathcal{V}$ denotes the set of *black points*; each point in $\mathcal{V} \setminus B$ is said to be a *white point*; adjacency relations m and n are assigned to B and $\mathcal{V} \setminus B$, respectively. In order to avoid connectivity paradoxes, it is generally assumed that $m \neq n$ [8,11]. Since all studied relations are reflexive and symmetric, their transitive closure form equivalence relations, and their equivalence classes are called *components*. A *black component* or an *object* is an m-component of B, while a *white component* is a n-component of $\mathcal{V} \setminus B$.

For practical purposes, we assume that all pictures are *finite* (i.e. they contain finitely many black points). In a finite picture there is a unique infinite white component, which is called the *background*. A finite white component is said to be a *cavity*. A cavity is *isolated* if it is formed by just one (white) point.

A point $p \in B$ is an *interior point* if all further points being n-adjacent to p are in B. A black point is a *border point* if it is not an interior point. Let δB denote the set of all border points in B.

Since the thinning algorithm taken into consideration in Sect. 4 acts on $(8, 4)$
pictures on the (2D) regular square grid $\mathcal{S}$, 8- and 4-adjacency relations are given
by Fig. 1.

$\bullet$	N	$\bullet$
W	p	E
$\bullet$	S	$\bullet$

Fig. 1. The adjacency relations studied on the square grid $\mathcal{S}$. The four points N, E,
S, and W are 4-*adjacent* to the central point p, and they are its 4-*neighbors*. The
4-neighbors and the four points marked '$\bullet$' are 8-*adjacent* to p, and they are its 8-
neighbors. Note that two points (i.e., square) are 4-adjacent if they share an edge,
and they are 8-adjacent if they share an edge or a vertex. The two *opposite pairs of
4-neighbors* of p are (N, S) and (E, W).

A *reduction* [4] transforms a binary picture only by changing some black
points to white ones, which is referred to as *deletion*. Hence, if a reduction
with deletion rule $R : 2^{\mathcal{V}} \to 2^{\mathcal{V}}$ transforms picture $(\mathcal{V}, m, n, B)$ to picture
$(\mathcal{V}, m, n, R(B))$, then $R(B) \subseteq B$.

Let us define deletion rules of *iterated reductions* as follows:

$$R_j(B) = \begin{cases} B & \text{if } j = 0 \\ R(R_{j-1}(B)) & \text{if } j \geq 1 \end{cases}.$$

We are now ready to define the main concept of this work:

Definition 1. *Let R be the deletion rule of a thinning algorithm (i.e., iterated
reduction). This algorithm is k-attempt ($k \geq 1$) if k is the smallest number such
that for any set of black points B and $p \in \delta B$, $p \in R_k(B)$, implies $p \in R_{k+1}(B)$.*

In other words, if a border point 'survives' at least k successive iterations of
a k-attempt thinning algorithm, it is 'immortal' (i.e., it is an element of the final
result).

3   Implementations for Thinning Algorithms

Palágyi proposed a general and computationally efficient implementation scheme
for arbitrary sequential and parallel thinning algorithms [17,18]. This method
utilizes that only border points in the current picture are to be examined in
each iteration (i.e., we do not have to evaluate the deletion rules for interior
points). It uses a list for storing the border points in the current picture, thus
the repeated scans/traverses of the entire array (that stores the actual picture)
are avoided. The pseudocode of collecting border points in the input picture (i.e.,
the initialization step of the thinning process) is described by Algorithm 1.

Algorithm 1. Collecting border points

---

1 **Input**: array A storing the picture to be thinned
2 **Output**: list *border_list* storing the border points in that picture
3 *border_list* ← < empty list >
4 **foreach** element p in array A **do**
5   **if** $A[p] = 1$ and p is a border point **then**
6   *border_list* ← *border_list* + < p >
7          $A[p]$ ← 2

In input array A, the value '1' corresponds to black points in the picture to be thinned, and the value '0' is assigned to white ones. In order to avoid storing more than one copy of a border point in *border_list*, a three-color picture is assumed in which the value '2' corresponds to border points to be checked in the forthcoming (1st) iteration.

Algorithm 2 describes one iteration of arbitrary parallel thinning algorithm.

Algorithm 2. 'Conventional' parallel thinning iteration

---

1 **Input**: array A storing the (input or interim) (m, n) picture and
2   list *border_list* storing the border points in that picture
3 **Output**: array A containing the result of the parallel reduction and
4   the updated *border_list*
5 // collecting deletable points
6 *deletable_list* ← < empty list >
7 **foreach** point p in *border_list* **do**
8   **if** p is 'deletable' **then**
9   *border_list* ← *border_list* − < p >
10   *deletable_list* ← *deletable_list* + < p >

11 **foreach** point p in *deletable_list* **do**
12   // deletion
13     $A[p]$ ← 0
14   // updating the list of border points
15   **foreach** point q being n-adjacent to p **do**
16   **if** $A[q] = 1$ **then**
17             $A[q]$ ← 2
18   *border_list* ← *border_list* + < q >

It is used a second list for storing the 'deletable' points of the current iteration by Algorithm 2. Note that both the input and the output pictures of an iteration can be stored in a single array, and the evaluation and the deletion phases are separated: first all 'deletable' points are added to the *deletable_list*, then they are deleted and *border_list* is updated accordingly.

If a border point is deleted, all interior points that are n-adjacent to it become border points. These brand new border points of the resulted picture are added to the *border_list*.

The thinning process terminates when no more points can be deleted (i.e., stability is reached). When it is completed, all points having a nonzero value in array A belong to the produced skeleton-like feature.

Note that one iteration of sequential thinning algorithms can be implemented accordingly. In the sequential case, we do not need to use the second list (i.e., *deletable_list*).

Algorithm 3 describes one iteration of k-attempt parallel thinning. Note that one iteration of sequential k-attempt thinning can be implemented accordingly.

Algorithm 3. k-attempt parallel thinning iteration

```
 1 Input: array A storing the (input or interim) (m, n) picture,
 2          number of attempts k, and
 3          list border_list storing the border points in that picture
 4 Output: array A containing the result of the parallel reduction and
 5          the updated border_list
 6 deletable_list ← < empty list >
 7 foreach point p in border_list do
 8     if p is 'deletable' then
 9         // deletable point found
10         border_list ← border_list − < p >
11         deletable_list ← deletable_list + < p >
12     else
13         if A[p] = k + 2 then
14             // 'safe' point found
15             border_list ← border_list − < p >
16         else
17             // to be evaluated next time
18             A[p] ← A[p] + 1

19 foreach point p in deletable_list do
20     // deletion
21     A[p] ← 0
22     // updating the list of border points
23     foreach point q being n-adjacent to p do
24         if A[q] = 1 then
25             A[q] ← 2
26             border_list ← border_list + < q >
```

Similarly to Algorithm 2, Algorithm 3 uses two lists, *border_list* for storing the potentially deletable border points and *deletable_list* for storing the 'deletable' points of the current iteration. It concerns $(k + 3)$-*color* pictures in which value

'0' corresponds to white points, value '1' is assigned to interior (black) points, values '2', ..., '$k+1$' correspond to potentially deletable border (black) points, and value '$k+2$' is assigned to border points that cannot be deleted in the remaining iteration steps.

The significant difference between Algorithm 2 and Algorithm 3 is illustrated in Fig. 2.

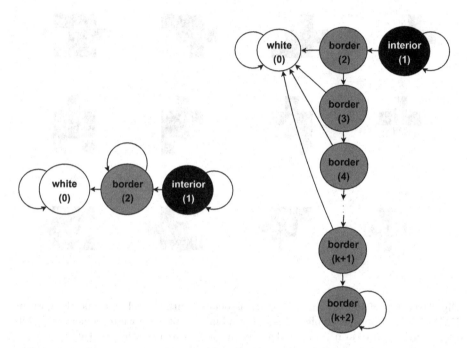

Fig. 2. Transition graphs associated with the 'conventional' (left) and the *k*-attempt (right) thinning schemes. Numbers in parentheses correspond to the values present in array that stores the (input or interim) picture, see Algorithm 2 and Algorithm 3.

4   An Existing 5-Attempt Thinning Algorithm

In this section, we show that the parallel thinning algorithm proposed by Eckhardt and Maderlechner [2] called **EM1993** is 5-attempt. This algorithm falls into the category of *fully parallel* [3] since it uses the same parallel reduction in each iteration, and it acts on $(8, 4)$ pictures (on the square grid $\mathcal{S}$).

The deletion rule of algorithm **EM1993** is given by the set of 16 matching templates depicted in Fig. 3. A (black) point is said to be *deletable* if at least one template (in Fig. 3) matches it. Otherwise, it is called *non-deletable*.

Let us notice some useful properties of the deletion rule of algorithm **EM1993** and the matching templates associated with it (see Fig. 3).

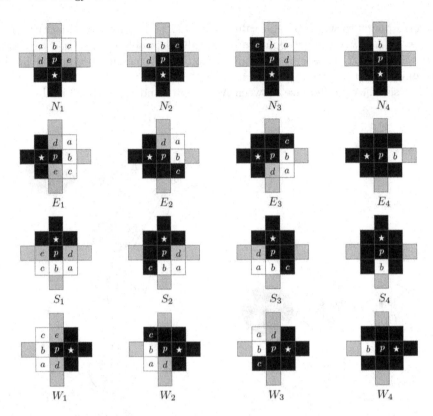

Fig. 3. The set of 16 matching templates associated with the deletion rule of algorithm **EM1993**. Notations: each black position is a black point; each white element is a white point; each (don't care) position depicted in grey matches either a black or a white point; p indicates the central point of a pattern; each position marked '★' is an interior point. (Note that positions marked a, b, c, d, and e help us to prove the properties of the algorithm.)

Proposition 1. *Only border points are matched by templates in Fig. 3.*

Proposition 2. *If a point is matched by a template in Fig. 3, it has an opposite pair of 4-neighbors (see Fig. 1) that contains a white point and an interior point.*

Proposition 3. *If point p is deletable, it is not matched by the four configurations depicted in Fig. 4.*

Proposition 4. *Matching templates E_i, S_i, and W_i are rotated versions of N_i ($i = 1, 2, 3, 4$), where the rotation angles are 90°, 180°, and 270°.*

Proposition 5. *Matching templates N_3, E_3, S_3, and W_3 are reflected versions of N_2, E_2, S_2, and W_2, respectively.*

Fig. 4. Restoring configurations assigned to Proposition 3.

Let us now introduce the concept of a set of blocker points associated with a non-deletable point.

Definition 2. *Let us assume that a black point p is non-deletable in picture $(\mathcal{S}, 8, 4, B \cup D)$, but it is deletable in picture $(\mathcal{S}, 8, 4, B)$, where D is a nonempty (sub)set of deletable points in picture $(\mathcal{S}, 8, 4, B \cup D)$. Then D is called a* blocker set *associated with p.*

By Propositions 1–5, the following three lemmas characterize blocker sets associated with border points.

Lemma 1. *If a deletable point is matched by template N_1 (see Fig. 3) in the current iteration, then it was not a non-deletable border point in the previous iteration.*

Proof. Let us assume that p is matched by template N_1 in the current iteration, and p was a non-deletable border point in the previous iteration. Then the deletion of p was blocked by a set of deletable (black) points $D \subseteq \{a, b, c\}$ in the previous iteration.

The following cases are to be investigated:

- Assume that $b \in D$.
  Since p was a border point in the previous iteration, b was not matched by templates N_i and S_i ($i = 1, 2, 3, 4$), by Proposition 2. Since a and c were white points or border points in the previous iteration, b was not matched by templates E_i and W_i ($i = 1, 2, 3, 4$), by Proposition 2. Thus b was non-deletable in the previous iteration, and we arrive at a contradiction.
- Assume that $D = \{a\}$.
  Since a was deletable and b was white in the previous iteration, d was black by Proposition 3. Consequently, p was matched by template N_3 in the previous iteration. Since p was non-deletable, we arrive at a contradiction.
- Assume that $D = \{c\}$.
  Since c was deletable and b was white in the previous iteration, e was black by Proposition 3. Consequently, p was matched by template N_2 in the previous iteration. Since p was non-deletable, we arrive at a contradiction.
- Assume that $D = \{a, c\}$.
  Since a and c were deletable and b was white in the previous iteration, d and e were black by Proposition 3. Consequently, p was matched by template N_4 in the previous iteration. Since p was non-deletable, we arrive at a contradiction.

□

Lemma 2. *If a deletable point p is matched by template N_2 (or N_3) in the current iteration, and p was a non-deletable border point in the previous iteration, then the only blocker set associated with p is $\{b\}$ (see Fig. 3).*

Proof. Let us assume that p is matched by template N_2 in the current iteration, and p was a non-deletable border point in the previous iteration. Then the deletion of p was blocked by a set of deletable (black) points $D \subseteq \{a, b\}$ in the previous iteration.

The following cases are to be investigated:

- Assume that $D = \{a\}$.
  Since a was deletable and b was white in the previous iteration, d was black by Proposition 3. Consequently, p was matched by template N_4 in the previous iteration. Since p was non-deletable, we arrive at a contradiction.
- Assume that $D = \{a, b\}$.
  Since p was a border point in the previous iteration, b was not matched by templates N_i and S_i ($i = 1, 2, 3, 4$), by Proposition 2. Since deletable point a was a border point in the previous iteration, b was not matched by templates E_i and W_i ($i = 1, 2, 3, 4$), by Proposition 2. Thus b was non-deletable, and we arrive at a contradiction.
- Assume that $D = \{b\}$.
  Since a was white in the previous iteration, b could be matched by templates W_i ($i = 1, 2, 3, 4$) (if c was an interior point).

Consequently, $D = \{b\}$ may be the only blocker set.

By Proposition 5, if a deletable point p is matched by template N_3, $D = \{b\}$ may be the only blocker set. $\qquad\qquad\square$

Lemma 3. *If a deletable point is matched by template N_4 (see Fig. 3) in the current iteration, then it was not a non-deletable border point in the previous iteration.*

Proof. Let us assume that p is matched by template N_1 in the current iteration, but the deletion of the border point p was blocked by the singleton subset of deletable (black) points $D = \{b\}$ in the previous iteration.

In this case p was an interior point. Since p was a border point, we arrive at a contradiction. $\qquad\qquad\square$

We can summarize Lemmas 1–3 as follows:

Corollary 1. *Only points matched by templates N_i, E_i, S_i, and W_i, ($i = 2, 3$) could be blocked in the previous iteration, and the blocker set associated with a border point is always singleton (i.e., $D = \{b\}$, see Fig. 3).*

We are now ready to state our main theorem.

Theorem 1. *Thinning algorithm **EM1993** is 5-attempt.*

Proof. We need to show that if a black point p is deletable in the current iteration (i.e., the j-th thinning phase), p could be a non-deletable border point only in the $(j-1)$-th, the $(j-2)$-th, the $(j-3)$-th, and the $(j-4)$-th iterations. In other words, a longest chain of blocker sets can be expressed as follows:

$$\{p\} \ \Leftarrow \ \{q\} \ \Leftarrow \ \{r\} \ \Leftarrow \ \{s\} \ \Leftarrow \ \{t\} \ \Leftarrow \ \emptyset$$

in which the deletion of p can be blocked by set $\{q\}$ in the $(j-1)$-th iteration, ..., and the deletion of t cannot be blocked in the $(j-4)$-th iteration. (Recall that blocker sets associated with border points are singleton by Corollary 1.)

Let us construct the above mentioned chain of blocker sets.

- Without loss of generality, we can assume that p is matched by template N_2 in the j-th iteration.
  By Propositions 1–5 and Corollary 1, $\{q\}$ could be the only blocker set associated with p (see Fig. 5a).
- Point q was a deletable border point in the $(j-1)$-th iteration, and it is assumed that p was a (non-deletable) border point in that thinning phase. Thus w was white (otherwise p would be an interior point), and it can be readily seen that q could be matched only by template W_2 (see Fig. 5b).
  In addition, point x was white (otherwise p would be matched by template W_3 or template W_4).
  Let us continue to construct the chain of blocker sets. Since q could be matched only by template W_2 in the $(j-1)$-th iteration, the only blocker set associated with q could be $\{r\}$ by Corollary 1 (see Fig. 5b).
- Let us deal with the blocker set $\{r\}$.
  If r was a deletable (black) point in the $(j-2)$-th iteration, it can be readily seen that r could be matched by template S_3 or template S_4 (see Fig. 5c), or template W_2 (see Fig. 5d).
  If r was matched by template S_3 or template S_4 (see Fig. 5c), $\{w\}$ was not the blocker set associated with r. Thus the chain of blocker sets ends.
  If r was matched by template W_2 in the $(j-2)$-th iteration (see Fig. 5d), the only blocker set associated with r could be $\{s\}$ by Corollary 1. In this case, point y was white (otherwise r was matched by template S_3 in the $(j-3)$-th iteration (see Fig. 5d).
- Let us continue the chain of blocker sets from $\{s\}$.
  If the deletion of s was blocked in the $(j-4)$-th iteration, s could be matched only by template S_2, and $\{t\}$ was the only blocker set associated with s by Corollary 1 (see Fig. 5e).
- Since w was white in the $(j-4)$-th iteration, t was not matched by templates W_i ($i = 1, 2, 3, 4$). Since s was black in the $(j-4)$-th iteration, t was not matched by templates N_i ($i = 1, 2, 3, 4$). If t was a deletable point in the $(j-4)$-th iteration, point t was black by Proposition 3 (see Fig. 5f). Hence t could be matched only by template E_4, and the only white 8-neighbor of t was w. Since $\{w\}$ may not be the blocker set associated with t, the chain of blocker sets ends with t.

$\square$

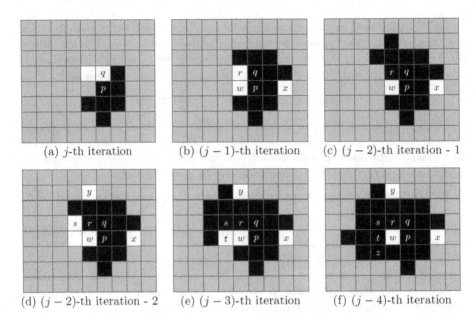

(a) j-th iteration   (b) $(j-1)$-th iteration   (c) $(j-2)$-th iteration - 1

(d) $(j-2)$-th iteration - 2   (e) $(j-3)$-th iteration   (f) $(j-4)$-th iteration

Fig. 5. Input pictures of five iterations assigned to proof of Theorem 1.

Since longest chains of blocker sets end with black points that are 4-adjacent to isolated cavities (see the proof of Theorem 1), we can state the followings:

Corollary 2. *Thinning algorithm* **EM1993** *is 4-attempt for pictures that are free from isolated cavities.*

5   The Advantage of the New Implementation Scheme

The 'conventional' implementation scheme (see Algorithm 2) and the proposed 'advanced' one (for k-attempt thinning, see Algorithm 3) were compared on the 5-attempt thinning algorithm **EM1993** for numerous objects of different shapes. For reasons of scope, we present the results only for six test images, see Fig. 6.

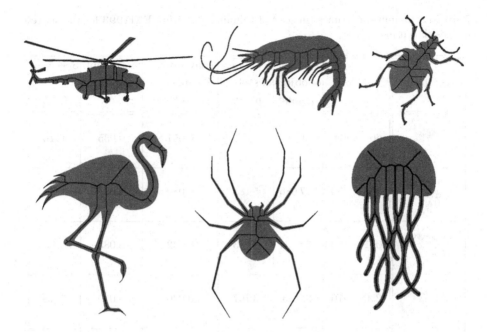

Fig. 6. Centerlines produced by algorithm **EM1993** for six test pictures. The extracted features are superimposed on the original objects.

The advantage of the proposed 'advanced' implementation scheme over the 'conventional' one is illustrated in Table 1 for the selected six test images. Both implementations under comparison were run on a usual PC under Linux (Fedora 30–64 bit), using a 3.30 GHz 4x Intel Core i5-2500 CPU. (Note that just the iterative thinning process itself was considered here; reading the input image and writing the output image were not taken into account but the processing involved is not excessive.)

In spite of that the 'conventional' implementation scheme is computationally efficient, the 'advanced' one speeds up the process considerably. Note that less number of attempts involves the greater speed-up.

Table 1. Computation times (in sec.) of thinning algorithm **EM1993** for the six test images shown in Fig. 6.

Test image	Size	Number of object points	Number of skeletal points	'conventional' comp. time **C**	'5-attempt' comp. time **A**	Speed-up **C/A**
	667 × 490	80 193	2 782	0.051	0.035	**1.46**
	450 × 350	117 086	5 183	0.032	0.021	**1.52**
	1011 × 1000	154 104	4 963	0.132	0.085	**1.55**
	290 × 476	40 196	3 767	0.030	0.019	**1.58**
	242 × 616	33 856	3 139	0.024	0.014	**1.71**
	725 × 481	43 134	3 797	0.039	0.021	**1.86**

6   Conclusions

In this paper, we introduced the concept of k-attempt thinning. In the case of a k-attempt algorithm, if a border point is not deleted in at least k successive iterations, it belongs to the produced skeleton-like feature. We proposed a computationally efficient implementation scheme for k-attempt thinning, show that an existing parallel thinning algorithm is 5-attempt, and the advantage of the proposed 'advanced' implementation (for k-attempt algorithms) over the 'conventional' one is also illustrated.

Future research should be devoted to estimate the performance of the 'advanced' implementation, to prove that further existing thinning algorithms are k-attempt (for some k), and to construct new (geometrically and topologically correct) 2D and 3D k-attempt thinning algorithms (for small attempts).

Acknowledgments. This research was supported by the project "Integrated program for training new generation of scientists in the fields of computer science", no EFOP-3.6.3-VEKOP-16-2017-00002. This research was supported by grant TUDFO/47138-1/2019-ITM of the Ministry for Innovation and Technology, Hungary.

References

1. Blum, H.: A transformation for extracting new descriptors of shape. In: Wathen-Dunn, W. (ed.) Models for the Perception of Speech and Visual form, pp. 362–380. MIT Press, Cambridge (1967)

2. Eckhardt, U., Maderlechner, G.: Invariant thinning. Int. J. Pattern Recogn. Artif. Intell. **7**(5), 1115–1144 (1993)

3. Hall, R.W.: Parallel connectivity-preserving thinning algorithms. In: Kong, T.Y., Rosenfeld, A. (eds.) Topological Algorithms for Digital Image Processing, vol. 19, pp. 145–179. Elsevier Science, Amsterdam (1996)

4. Hall, R.W., Kong, T.Y., Rosenfeld, A.: Shrinking binary images. In: Kong, T.Y., Rosenfeld, A. (eds.) Topological Algorithms for Digital Image Processing, pp. 31–98. Elsevier Science, Amsterdam (1996)

5. Holmgren, M., Wahlin, A., Dunas, T., Malm, J., Eklund, A.: Assessment of cerebral blood flow pulsatility and cerebral arterial compliance with 4D flow MRI. J. Magn. Reson. Imaging (2020). https://doi.org/10.1002/jmri.26978

6. Kong, T.Y.: On topology preservation in 2-D and 3-D thinning. Int. J. Pattern Recogn. Artif. Intell. **9**(05), 813–844 (1995). https://doi.org/10.1142/S0218001495000341

7. Kong, T.Y.: Critical kernels, minimal nonsimple sets, and hereditarily simple sets in binary images on n-dimensional polytopal complexes. In: Saha, P.K., Borgefors, G., Sanniti di Baja, G. (eds.) Skeletonization: Theory, Methods and Applications, pp. 211–256. Academic Press, San Diego (2017)

8. Kong, T.Y., Rosenfeld, A.: Digital topology: introduction and survey. Comput. Vis. Graph. Image Process. **48**, 357–393 (1989). https://doi.org/10.1016/0734-189X(89)90147-3

9. Kovalevsky, V.A.: Geometry of Locally Finite Spaces. Publishing House, Berlin (2008). https://doi.org/10.1142/S0218654308001178

10. Lam, L., Lee, S.-W., Suen, C.Y.: Thinning methodologies - a comprehensive survey. IEEE Trans. Pattern Anal. Mach. Intell. **14**(9), 869–885 (1992). https://doi.org/10.1109/34.161346

11. Marchand-Maillet, S., Sharaiha, Y.M.: Binary Digital Image Processing: A Discrete Approach. Academic Press (2000). https://doi.org/10.1117/1.1326456

12. Matejek, B., Wei, D., Wang, X., Zhao, J., Palágyi, K., Pfister, H.: Synapse-aware skeleton generation for neural circuits. In: Shen, D. (ed.) MICCAI 2019. LNCS, vol. 11764, pp. 227–235. Springer, Cham (2019). https://doi.org/10.1007/978-3-030-32239-7_26

13. Németh, G., Kardos, P., Palágyi, K.: 2D parallel thinning and shrinking based on sufficient conditions for topology preservation. Acta Cybernetica **20**, 125–144 (2011). https://doi.org/10.14232/actacyb.20.1.2011.10

14. Palágyi, K., Németh, G., Kardos, P.: Topology preserving parallel 3D thinning algorithms. In: Brimkov, V.E., Barneva, R.P. (eds.) Digital Geometry Algorithms: Theoretical Foundations and Applications to Computational Imaging, pp. 165–188. Springer, Heidelberg (2012). https://doi.org/10.1007/978-94-007-4174-4_6

15. Palágyi, K., Németh, G.: Fixpoints of iterated reductions with equivalent deletion rules. In: Barneva, R.P., Brimkov, V.E., Tavares, J.M.R.S. (eds.) IWCIA 2018. LNCS, vol. 11255, pp. 17–27. Springer, Cham (2018). https://doi.org/10.1007/978-3-030-05288-1_2

16. Palágyi, K., Németh, G.: Endpoint-based thinning with designating safe skeletal points. In: Barneva, R.P., Brimkov, V.E., Kulczycki, P., Tavares, J.R.S. (eds.) CompIMAGE 2018. LNCS, vol. 10986, pp. 3–15. Springer, Cham (2019). https://doi.org/10.1007/978-3-030-20805-9_1

17. Palágyi, K., Tschirren, J., Hoffman, E.A., Sonka, M.: Quantitative analysis of pulmonary airway tree structures. Comput. Biol. Med. **36**, 974–996 (2006). https://doi.org/10.1016/j.compbiomed.2005.05.004

18. Palágyi, K.: A 3D fully parallel surface-thinning algorithm. Theor. Comput. Sci. **406**, 119–135 (2008). https://doi.org/10.1016/j.tcs.2008.06.041

19. Saha, P.K., Borgefors, G., Sanniti di Baja, G.: A survey on skeletonization algorithms and their applications. Pattern Recogn. Lett. **76**, 3–12 (2016)

20. Suen, C.Y., Wang, P.S.P. (eds.): Thinning Methodologies for Pattern Recognition. Series in Machine Perception and Artificial Intelligence, vol. 8. World Scientific (1994). https://doi.org/10.1142/9789812797858_0009

Fuzzy Metaheuristic Algorithm for Copy - Move Forgery Detection in Image

Nataša S. Milosavljević[1,2]([envelope]) and Nebojša M. Ralević[1,2]

[1] Faculty of Agriculture, University of Belgrade, Belgrade, Serbia
natasam@agrif.bg.ac.rs
[2] Faculty of Technical Sciences, University of Novi Sad, Novi Sad, Serbia
nralevic@uns.ac.rs

Abstract. One of the most common methods for counterfeiting digital images is copy-move forgery detection (CMFD). It implies that part of the image is copied and pasted to another part of the same image. The purpose of such changes is to hide certain image content, or to duplicate image content. The aim of this paper is to propose a new clustering algorithm for edited images. The image is divided into non-overlapping blocks. New fuzzy metric is used to calculate the distance between the blocks. In this research the metaheuristic method of the variable neighbourhood search (VNS) is used for the classification of the block. The aim of the classification is that the division should be on the changed and unchanged blocks. The results of this research are compared with the latest results from the literature dealing with this problem and it is shown that the proposed algorithm gives better results. Publicly available image databases were used. The proposed algorithm was implemented in the Python programming language.

Keywords: Image processing · Clustering · Image forensics · Fuzzy metrics · Copy-move forgery detection · VNS

1 Introduction

The methods for detecting the intentionally induced changes in the image content can be divided into two categories: active and passive, depending on the presence of additional information. The active methods involve the presence of additional, retrofitted information in the content of an image, such as a watermark or a digital signature. However, the active methods mean that the additional information is incorporated into the image when generating the image, or subsequently by an authorized author. The passive methods, on the other hand, allow changes to be detected in the images without the presence of additional retrofitted information in the image. The passive methods reveal the manipulations of images by extracting the natural characteristics of the images, as well as the characteristics of the optical devices used to generate an image (noise characteristics). These methods can be further divided into two categories: dependent and independent. The dependent methods mean that a part of an image is copied and

© Springer Nature Switzerland AG 2020
T. Lukić et al. (Eds.): IWCIA 2020, LNCS 12148, pp. 273–281, 2020.
https://doi.org/10.1007/978-3-030-51002-2_20

pasted to a part within the same image or from another image. The other types of image manipulations, such as compression, sub-selection, blurring, and the like, are independent methods. As for the detection of the device that generated the image, it is based on the regularities of the optical sensors of the device. [1,2]

The aim of this research is to propose a new algorithm that will give better precision than existing methods in this field as well as a lower percentage of "false detection", i.e. more specifically detecting unmodified parts of an image as well as the altered ones.

The paper is divided into several sections. Section 2 provides the methodology used in this research. The results of the research as well as the conclusion are given in the following Sects. 3 and 4.

2   Methodology

The proposed method is block-based method because it works on non-overlapping blocks of an image of interest. The method is based on a fuzzy metric that calculates the distances between blocks, and then applies the VNS to divide those blocks into the ones where the change occurred and those unchanged.

2.1   Fuzzy Metrics

In the classical sense, the distances are most commonly defined by the functions that were metrics, pseudo-metrics, semi-metrics and similarities, and they were defined in the following way:

Definition 1. *If $X \neq \emptyset$, function $d : X^2 \to \mathbb{R}_0^+$ to which the following characteristics apply:*

1. $(\forall x \in X) \, d(x, x) = 0$,
2. $(\forall x, y \in X) \, d(x, y) = d(y, x)$,

we say that the distance, and the ordered pair (X, d) is a space with distance. If only property 1. is valid, it is a quasi-distance. If

3. $(\forall x, y \in X) \, d(x, y) = 0 \Rightarrow x = y$,
4. $(\forall x, y, z \in X) \, d(x, z) \leqslant d(x, y) + d(y, z)$,

d is a metric and the ordered pair (X, d) is a metric space. If only properties 1, 2 and 4 are valid for d we say that it is a pseudo-metric. If only properties 1, 3 and 4 apply or d we say it is a quasi-metric. If 1st, 2nd and 3rd mapping are valid d is called a semi-metric. For a semi-metric $d : X^2 \to [0, 1]$ in which one of the inequalities applies instead of an inequality of a triangle (property 4.):

4'. $(\forall x, y, z \in X) \, d(x, z) \geqslant T(d(x, y), d(y, z))$,
4''. $(\forall x, y, z \in X) \, d(x, z) \leqslant S(d(x, y), d(y, z))$,

(T is the t-norm, and S is the t-conorm) we say that there is a similarity.

The fuzzy metrics have been used in a variety of applications and especially recently to filter the color images, improving some filters when replacing the classic metrics. Unfortunately, the use of the fuzzy metrics in the engineering methods is severely limited, as there are only a few applications.

The fuzzy theory has become an active field of research in the last fifty years and the concept of fuzzy metrics has evolved into two different perspectives.

In this research instead classic distance or metric we apply the class of mappings $M_p : X^2 \rightarrow [0, 1]$ (see [5]), where $X = \mathbb{R}_0^+$:

$$
M_p(x, y, t) = \begin{cases} \frac{\sqrt[p]{\frac{x^p + y^p}{2}} + t}{\max\{x, y\} + t}, & t > 0 \\ 0, & t \leqslant 0 \end{cases} .
\tag{1}
$$

Each of these mappings contains a parameter t whose meaning is the distance M_p from point x to point y is one fuzzy number defined the domain of parameter t. Especially if t is a constant then the value is distance a crisp number. If this mapping satisfies certain characteristics (see [5]), then we say that it is a fuzzy metric. The mapping (1), for $p = 1$, is fuzzy metric. For other question about M_p is open.

2.2 Metaheuristics

The basic variable neighbourhood search (BVNS) is the most widespread variant of the environment variable method because it provides more preconditions for obtaining the better quality final solutions. For the basic VNS method, the basic steps are contained in a loop in which we change the index of the environment k, determine the random solution from that environment (function Shaking()), perform the procedure of local search (function LocalSearch()) and check the quality of the local minimum obtained. We repeat these steps until one of the stopping criteria is satisfied. Each time we select an environment, k the initial solutions are randomly generated to ensure that different regions are searched the next time they are deployed. We distinguish the environments by the number of transformations (distances) or by the type of transformations (metrics). The pseudocode BVNS can be represented as follows:

```
Generate initial solution x
do {
  k=1;
  while(k <= k_max) {
    x'=Shaking (x, k);
    x''=LocalSearch (x');
    if(f(x'') < f(x)) {
      x=x'';
      k=1;
    }
    else
      k=k+1;
```

```
  }
} while(!STOP);
```

The reduced variable neighbourhood search method does not contain a local search phase. It is method based solely on staging, which consists in systematically changing the environment and selecting one random solution in each environment. The decision-making steps are based on that one random solution. Its advantage is the speed of the execution as it avoids a detailed search, often of extremely large environments. This method is useful the for large-scale examples or for obtaining the quality initial solutions for another variant of the VNS method. Written in the pseudocode RVNS takes the following form:

```
Generate initial solution x
x_optimal = x;
f_optimal = f(x);
do{
  k=1;
  do{
    x'=Shaking (x, k);
    if(f(x')<f(x_optimal)){
      x_optimal=x';
      f(x_optimal)=f(x');
      k=1;
    }
    else
      k=k+1;
  }while(k!=kmax);
}while(!STOP);
```

The usual stop criterion (STOP) is the maximum number of iterations between the two enhancements [6].

3   Results

We used metaheuristics, namely the VNS (BVNS and RVNS) method, which is based on the fuzzy metric proposed in the section above. The size of the blocks, as far as the algorithm is concerned, can be resized, and we worked at 8×8 and 16×16 pixels. We clustered into two groups and investigate the validity and the quality of the clustering using false positive FP (False Positive) and false negative FN (False Negative) detected blocks. When we load an image in the RGB format then we divide it into non-overlapping blocks, squared shapes, fixed dimensions, $n \times n$, where n is 8 or 16. The block division is usually done to reduce the computing time and the complexity required for the pixel matching process that is, blocks. For testing purposes, all the images are 256×256 resolution, which does not affect the generality of the method, but allows for greater transparency of the method. The method can be applied to the images of the arbitrary resolution, say $2n \times 2n$. Given the resolution of 256×256, each image is divided into 128 dimensions 16×16 or 256 dimensions 8×8. The fuzzy distance from (1) is then calculated and for $p = 1$ and $p = 2$. The distances are sorted and VNS is applied. The RVNS and BVNS applied to the image block clustering problem are implemented as follows [6]. First, the distances between the blocks were calculated using the formula in (1). In the preprocessing stage, the types of the distance matrices are sorted in descending order. Those data were used to deploy the deployment operators more efficiently. Namely, as each solution is characterized by a set of centroids, the deployment operator consists in replacing the corresponding number of centroids. More specifically, a k-environment in k means that the centroids (block) are replaced by the randomly selected non-centroid objects. They are mostly k places a change. Each step considers the replacement of all the centroids, with no replacement occurring if the randomly selected object is closest to that centroid (in fact, it is the centroid itself). The local search consists of systematically replacing one centroid with a non-centroid object. It starts from the solution obtained by rolling and is executed on the principle of best improvement as long as there are improvements [3].

The performance of the proposed methods is most often measured in terms of precision and recall. The precision indicates the probability that the blocks which have been changed, have really been detected. The revocation indicates the probability (possibility) of detecting altered blocks in an image. The true positive (TP) is the number of blocks that have been modified, which have been classified as modified. The false positive (FP) represents the number of original (authentic) blocks that have been classified as modified, while the false negative (FN) represents the number of blocks that have been modified but classified as original (authentic):

$$\text{Precision} = \text{TP}/(\text{TP} + \text{FP}) \tag{2}$$

$$\text{Revocation} = \text{TP}/(\text{TP} + \text{FN}) \tag{3}$$

Specifically, the precision is calculated as the quotient of the number of modified blocks that are classified as modified, and sums the number of blocks of modified blocks that are classified as the modified and the number of original blocks that are classified as modified. On the other hand, the recall is calculated as a quotient of the number of the modified blocks that are classified as modified, and sums the number of the modified blocks that are classified as modified and the number of modified blocks that are classified as original.

The publicly available Image Manipulation Dataset database was used. The Image Manipulation Dataset image database consists of 48 images[1]. In the paper, nine images are shown from that base. The reason of that is the comparison with the results of the other researchers. The resolution of the tested image is 256×256. The images were modified so that a portion of the image was copied and pasted to another part of the same image (copy move). The parts of the image being copied can be geometrically transformed before applying, applying rotation and scaling. The copied parts can also be of different sizes

---

[1] https://www5.cs.fau.de/research/data/image-manipulation/.

(small, medium or large). This chapter presents the results for a single image from a publicly available database that other authors have used. In each image from that database, one or more regions are copied. Also, the size of the copied regions varies from image to image. The original and modified image is shown below.

Fig. 1. Images tested I, II, III, IV, V, VI, VII, VIII, IX.

The success of the proposed method is shown in Table 1 in comparison with the results obtained in [4] for the blocks of dimensions 8×8. Table 2 shows the results obtained for the same set of images, only for the blocks of dimensions 16×16.

Table 1. The results obtained when the image is divided into blocks of dimensions 8×8.

Picture	Results from [4]		Results achieved by the proposed BVNS method for $p = 1$ in fuzzy metric		Results achieved by the proposed RVNS method for $p = 1$ in fuzzy metric		Results achieved by the proposed BVNS method for $p = 2$ in fuzzy metric		Results achieved by the proposed RVNS method for $p = 2$ in fuzzy metric	
	Precision (%)	Revocation (%)	Precision (%)	Revocation (%)	Precision (%)	Revocation (%)	Precision (%)	Revocation (%)	Precision (%)	Revocation (%)
I	87.62	99.48	90.12	92.56	98.73	99.48	91.20	99.48	90.12	99.48
II	100	97.53	100	98.11	100	100	100	100	100	100
III	26.58	94.85	96.32	97.12	96.32	97.12	100	100	100	100
IV	63.32	88.47	77.56	89.77	77.56	89.77	80.06	92.91	84.76	92.91
V	59.25	98.68	66.22	96.34	64.92	98.68	70.62	98.68	70.62	98.68
VI	49.80	100	87.31	100	100	100	100	100	100	100
VII	95.19	96.25	98.99	97.40	96.01	96.25	98.99	97.40	98.99	97.40
VIII	62.53	97.30	59.31	100	66.09	100	79.02	100	77.99	100
IX	41.49	93.37	56.34	94.11	54.67	95.37	71.88	95.01	71.88	95.01

Table 2. The results obtained when the image is divided into blocks of dimensions 16 × 16 [7].

Picture	Results achieved by the proposed BVNS method for $p = 1$ in fuzzy metric		Results achieved by the proposed RVNS method for $p = 1$ in fuzzy metric		Results achieved by the proposed BVNS method for $p = 2$ in fuzzy metric		Results achieved by the proposed RVNS method for $p = 2$ in fuzzy metric	
	Precision (%)	Revocation (%)	Precision (%)	Revocation (%)	Precision (%)	Revocation (%)	Precision (%)	Revocation (%)
I	90.12	92.56	98.73	99.48	91.20	99.48	90.12	99.48
II	100	98.11	100	100	100	100	100	100
III	96.32	97.12	96.32	97.12	100	100	100	100
IV	77.56	89.77	77.56	89.77	80.06	92.91	84.76	92.91
V	66.22	96.34	64.92	98.68	70.62	98.68	70.62	98.68
VI	87.31	100	100	100	100	100	100	100
VII	98.99	97.40	96.01	96.25	98.99	97.40	98.99	97.40
VIII	59.31	100	66.09	100	79.02	100	77.99	100
IX	56.34	94.11	54.67	95.37	71.88	95.01	71.88	95.01

The tables above show the success of the proposed algorithm with respect to the results achieved by others [4]. The superiority of our method, when it comes to precision and revocation, is shown in almost all images. One can see the movement of precision and revocation when changing p over the proposed distance as well as dividing the image into blocks of different dimensions.

4   Conclusion

This paper describes the possibility of clustering images into different areas based on the fuzzy metrics and metaheuristics. The results that are better than those of the other authors who have dealt with the same problem in their research, show the success of the proposed method.

The future research should consider using some image-related parameters to reduce the computing speed. Attention should also be paid to examining the success of the method under varying degrees of JPEG compression as well as in detecting the other types of change in images.

Acknowledgments. This work was supported by the Serbian Ministry of Education, Science and Technological Development through Mathematical Institute of the Serbian Academy of Sciences and Arts, and Faculty of Technical Sciences.

References

1. Warbhe, A.D., Dharaskar, R.V., Thakare, V.M.: A survey on keypoint based copy-paste forgery detection techniques. Procedia Comput. Sci. **78**, 61–67 (2016)
2. Warif, N.B.A., et al.: Copy-move forgery detection: survey, challenges and future directions. J. Netw. Comput. Appl. **75**, 259–278 (2016)
3. Glišović, N., Davidović, T., Rašković, M., Clustering when data are missing using the environment variable method. In: XLIV Symposium on Operations Research SYMOPIS, Zlatibor, pp. 158–165, September 25–28, 2017. ISBN: 978-86-7488-135-4
4. Alkawaz, M.H., Sulong, G., Saba, T., Rehman, A.: Detection of copymove image forgery based on discrete cosine transform. Neural Comput. Appl., 1–10 (2016). https://doi.org/10.1007/s00521-016-2663-3
5. Ralević, N., Karaklić, D., Pištinjat, N.: Fuzzy metric and its applications in removing the image noise. Soft Comput. **23**(22), 12049–12061 (2019)
6. Davidović, T., Glišović, N., Rašković, M.: Bee colony optimization for clustering incomplete data. In: The 7th International Conference on Optimization Problems and Their Applications, OPTA-2018, July 8–14, 2018
7. Pavlović, A., Glišović, N., Gavrovska, A., Reljin, I.: Copy-move forgery detection based on multifractals. Multimedia Tools Appl. **78**(15), 20655–20678 (2019). https://doi.org/10.1007/s11042-019-7277-1

Author Index

Printed in the United States
By Bookmasters